The Phthalocyanines

Volume I
Properties

Authors:

Frank H. Moser
Consulting Chemist
Holland, Michigan

Arthur L. Thomas
Editor
Hull & Co.
Greenwich, Connecticut

CRC Press, Inc.
Boca Raton, Florida 33431

Library of Congress Cataloging in Publication Data

Moser, Frank Hans, 1907—
The phthalocyanines.

Bibliography: p.
Contents: v. 1. Properties -- v. 2. Manufacture and applications.
Includes index.
1. Phthalocyanins. I. Thomas, Arthur Louis. II. Title.
QD441.M623 1983 547.8′6 82-20736
ISBN 0-8493-5677-6 (v. 1)
ISBN 0-8493-5678-4 (v. 2)

Direct all inquiries to CRC Press, Inc., 2000 Corporate Blvd., N.W., Boca Raton, Florida, 33431.

International Standard Book Number 0-8493-5677-6 (Volume I)
International Standard Book Number 0-8493-5678-4 (Volume II)

Library of Congress Card Number 82-20736
Printed in the United States

INTRODUCTION

We have brought together the wealth of new information in the phthalocyanine field which has been generated since the early 1960s to about 1975. The thrust has been to provide as much topical review and discussion as possible for the 3000 to 4000 references from the journal and patent literature in this time interval. This intense activity in a very exciting field of chemistry, is at a steady, high tempo, and this study should be of interest to both the practitioner and theoretician of phthalocyanine chemistry. At a glance he can determine what work has been accomplished and he can then determine his program quickly and effectively.

Phthalocyanine compounds have, since their final discovery and elucidation of structure in 1934, achieved a notable success as a model structure for theoretical and experimental organic and physical chemical study and as high achievers as pigments and dyes in the green and blue portions of the spectrum. Because of the ease of their manufacture and the ready availability of the raw materials for their manufacture, they have displayed success as coloring matters, displacing, to a large extent, their predecessors, ultramarine blue and iron blue. Today about 12 million lb of phthalocyanine blue and about 3 million lb of phthalocyanine green are produced annually in the U.S. alone at a unit value of $7- to $10/lb.

Phthalocyanine blues and greens are used as pigments because of their outstanding stability to light, heat, acids, and alkalies, and, of course, their insolubility in water and organic solvents. They are used extensively in printing inks, paints, coatings, and plastics. The G shade of blue has the proper shade and transparency to use in color printing. They have given us beautiful automobile colors, and can be used in the matching colors in plastics for car interiors.

The phthalocyanines are also used as catalysts in sulfur oxidations in the petroleum industry and find use in lasers, lubricants, medicines, photography, as photo- and semiconductors, in xerography, and as indicators.

The phthalocyanines are eulogized as a perfect molecule for chemical study. There has been a steady outpouring of research publications based on the phthalocyanines, including the topics of catalysts, crystals, magnetic properties, semiconductors, spectra, including luminescence, lasers, and chemical reactions. There has also been an interest to prepare phthalocyanine polymers with useful properties for practical applications.

The authors' wish is that this monograph, like its predocessor volume, published in 1963, will respond to the demands of its readers for a ready access to the world literature in the field of phthalocyanine chemistry and technology.

Material quoted in the text is used with the permission of the copyright owner. Smaller quotations are followed by a reference number to indicate the original source; larger quotations are followed by their respective credit lines, and permission to use is acknowledged.

Frank H. Moser
Holland, Michigan

Arthur L. Thomas
Greenwich, Connecticut

FOREWORD

It is an interesting fact that the phthalocyanines were first discovered accidentally during an industrial operation at Grangemouth in Scotland. This occurred during the production of phthalimide. When ammonia was passed through molten phthalic anhydride in iron vessels, traces of a dark blue substance of great stability were noticed. This compound was isolated and proved to be iron phthalocyanine. The preparation and chemical study of many other members of the series have been fully described by Linstead. These macrocyclic organic pigments are of extraordinary stability, the copper derivative, for example, subliming without decomposition at 580°C.

The phthalocyanines and their metal derivatives form a beautifully crystalline series of compounds which, with the exception of the platinum derivative, are closely isomorphous. They are also of great historical interest in connection with crystal and molecular structure determination because this series of compounds is ideal for the development of the isomorphous replacement and heavy atom methods of structure determination. In fact, the first complete development and application of these powerful methods to the determination of organic structures were carried out on these compounds.

It is a remarkable fact that the very complex phthalocyanine molecule was the first organic structure to yield to an absolutely direct X-ray analysis, which did not involve any assumptions, not even regarding the existence of discrete atoms in the molecule.

The isomorphous replacement method has since been of extraordinary value, especially in the determination of very complex biological structures, such as the proteins. The heavy atom method, first developed in the case of the platinum compound, has been of universal application in nearly all organic structure determinations.

J. Monteath Robertson
Inverness, Scotland
October 1982

ACKNOWLEDGMENT

The authors thank BASF Wyandotte Corporation, Pigments Division (formerly Chemetron Corporation, Pigments Division) for their aid in supplying many patents and for the use of their library and duplicating facilities. The gracious permission of Mr. J. E. Counihan, President and General Manager, and the help of Mr. W. R. Rhodes, Technical Service Manager, Graphic Arts, and of Mrs. Lynn L. Read, Librarian, is especially appreciated.

We acknowledge with many thanks the courtesy of the libraries of Columbia University, of Hope College, Holland, Michigan, of Monmouth College, West Long Branch, New Jersey, of The Greenwich Library, Greenwich, Connecticut, of The New York Public Library, and of The Chemists' Club Library, to make use of their services, and we express our appreciation to Hull & Company, Greenwich, Connecticut, for permission to use their duplicating facilities. We also thank Mr. Robert S. Tannehill, Jr. and Mr. James Ward, Librarians, Chemical Abstracts, for their generosity in providing us with substantial patent and journal literature. Thanks are extended also to Mr. J. R. Allerton, Information Specialist, Chemical Abstracts. For their skillful guidance, we are grateful to Mr. Robert F. Gould, former Head, Books Department, American Chemical Society, to Ms. Sandy Pearlman, Managing Editor, CRC Press, Inc., and Mr. Paul Gottehrer, Senior Editor, also of CRC Press, Inc.

I (ALT) also express my appreciation to my wife, Charlotte Thomas, and to my parents, and I (FHM) express my appreciation to my friend Clarissa E. Moser for their encouragement in the completion of this work.

Frank H. Moser
Arthur L. Thomas
December 1982

THE AUTHORS

Frank H. Moser, Ph.D., is a consulting chemist residing in Holland, Michigan.

Dr. Moser graduated with Bachelor of Arts degree from Hope College, 1928, Holland, Michigan, the Master of Science degree in 1929, and the Ph.D. degree in 1931, both from the University of Michigan, Ann Arbor, Michigan. While at the University he was a teaching and a lecture assistant from 1928 to 1931. He worked for the Research Department of the School of Engineering at the University from 1931 to 1932, and in 1932 joined the Research Department of National Aniline and Chemical Company, Buffalo, New York, then, as now, a part of Allied Chemical Company. While there he worked on anthraquinone and triphenylmethane dyes.

Dr. Moser joined the Standard Ultramarine Company in Huntington, West Virginia in 1938. He became head of the Intermediate Department in 1940 and began work on phthalocyanine in 1943. He became Supervisor of the Intermediate, Iron Blue and Phthalocyanine departments in 1948. He has several patents in the field of phthalocyanine chemistry and is co-author with Dr. Arthur L. Thomas of American Chemical Society Monograph No. 157 ''Phthalocyanine Compounds'' published by Reinhold Publishing Corporation, New York, in 1963.

He became Research Director at the Standard Ultramarine and Color Company in 1958 and Research Director for Holland Color and Chemical Company in 1966, after the Standard Ultramarine and Color Company was acquired by the Holland Color and Chemical Company. In 1968 he became Technical Director of the same company, then known as the Pigments Division of Chemetron Corporation. He retired in 1972 and has since been doing consulting work in several fields of chemistry; phthalocyanine chemistry and special glass coatings.

He has co-authored an article on phthalocyanine in the *Kirk-Othmer Encyclopedia of Chemistry,* 3rd edition, with Mr. W. H. Rhodes.

Dr. Moser is a 50-year member of the American Chemical Society and has in the past been first chairman (1947) of the Tristate Section in Huntington, West Virginia. More recently he has been chairman of the West Michigan Section of the Society. He has been a councillor for 12 years and a member of the national constitution and bye-laws committee for 9 years. He is also a Fellow of the American Institute of Chemists and a member of the New York Academy of Science and of Sigma Xi.

Arthur L. Thomas, Ph.D., is an Editor at Hull & Co., Greenwich, Connecticut, a market research firm for the chemical industry both in the U.S. and overseas.

Dr. Thomas graduated with a Bachelor of Arts degree in chemistry from Columbia University, New York City, in 1951 and the Ph.D. degree in chemical engineering in 1956 from Princeton University, Princeton, New Jersey, where he assisted Professor Richard H. Wilhelm in research for the U.S. Air Force at the Forrestal Research Center. He joined E. I. du Pont de Nemours & Company, Inc. in Parlin, New Jersey, in 1956, as a chemical engineer and later as a research supervisor. In 1960 he joined the Standard Ultramarine & Color Company, Huntington, West Virginia, as a chemical engineer. It was here that he became involved in the research and development of phthalocyanine pigments, a field of continuing interest.

In 1965 to 1968 Dr. Thomas was involved in arc chemistry development at MHD Research, Inc. and Plasmachem, Inc., both in Newport Beach, California. In 1969 he joined the chemistry faculty at California Polytechnic State University in San Luis Obispo, California as instructor and assistant professor of chemistry. In 1973 Dr. Thomas was visiting professor at Columbia University, studying immunochemistry under Dr. Konrad C. Hsu. From 1974 to 1977 Dr. Thomas was Science and Engineering Editor at The Ronald Press Company in New York City. Since 1978 he has been with Hull & Co.

Dr. Thomas is a member of various scientific societies, including the American Association for the Advancement of Science, as well as a member of The Chemists' Club of New York. He is an honorary member of Tau Beta Pi, the national engineering society, and a life member of Sigma Xi, the scientific research society. Publications include journal articles as author or co-author and a monograph in chemistry as co-author.

To

Sir Reginald P. Linstead

and

J. Monteath Robertson

TABLE OF CONTENTS

Volume I

Volume II

Chapter 1
Preparation 1

Chapter 1

INTRODUCTION

Phthalocyanine is derived from the Greek terms for naphtha (rock oil) and for cyanine (dark blue). Dioscorides stated naphtha to be a clear, combustible rock oil produced from Babylonian asphalt. The word cyanine was used by Homer among other ancient Greek writers.

Gordon found naphthalene in coal tar in 1819. It was examined by Kidd[1] of Oxford University who gave its name in 1821, in a paper entitled "Observations on Naphthaline", a peculiar substance resembling a concrete essential oil, which is apparently produced during the decomposition of coal tar, by exposure to a red heat. Kidd's final remarks were, "It remains for me to propose a name for the white concrete substance which has been described in this paper and, unless a more appropriate term should be suggested by others, I would propose to call it naphthalene."

Laurent obtained *o*-phthalic acid[2] in 1836 from the oxidation of naphthalene by nitric acid.

The term "phthalocyanine" was first used by Linstead[3] of the Imperial College of Science and Technology in 1933, to describe the class of organic compounds which are the subject of this monograph.

Phthalocyanine was first obtained in 1907 by Braun and Tcherniac[4] at the South Metropolitan Gas Company in London, as a by-product of the preparation of *o*-cyanobenzamide from phthalamide and acetic anhydride:

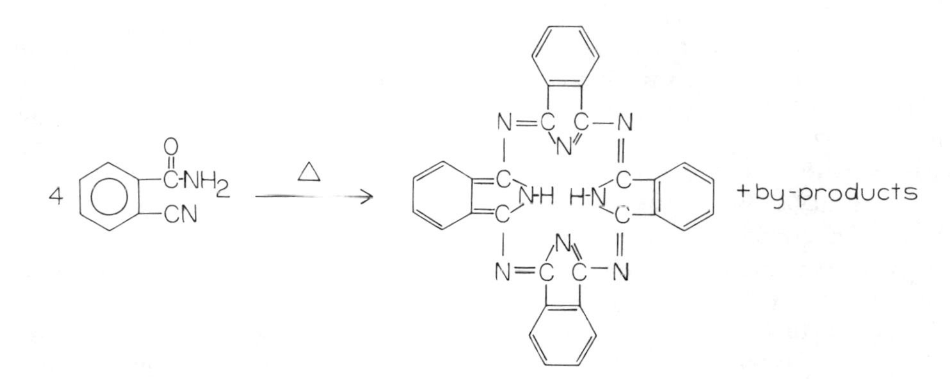

The next recorded preparation of a phthalocyanine (pc) was by de Diesbach and von der Weid[5] in 1927. They obtained copper pc in a 23% yield by the reaction of *o*-dibromobenzene and copper cyanide in pyridine:

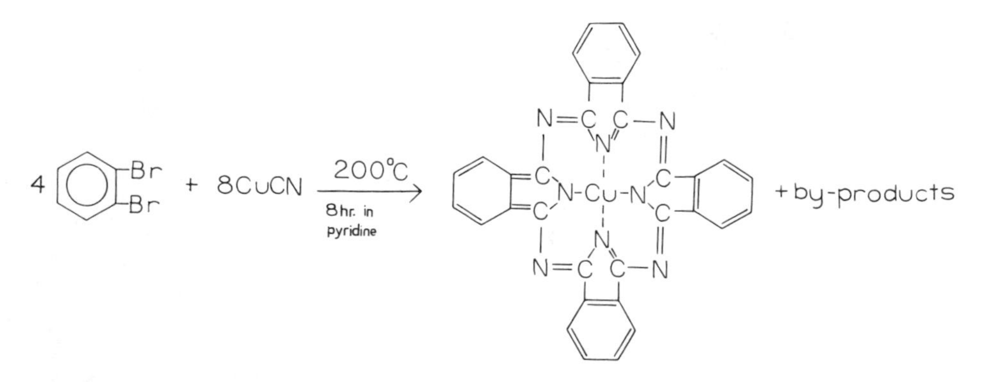

They also observed the exceptional stability of their product to alkalies, sulfuric acid, and heat.

The third observation of a pc was made during the preparation of phthalimide from phthalic anhydride and ammonia at Scottish Dyes, Ltd. in Grangemouth, in 1928. The glass lined reactor was chipped. A greenish blue impurity was observed in the product in the chipped area. Dunsworth and Drescher carried out the preliminary examination of the iron compound and found that it was very stable and that the iron it contained was not removed by sulfuric acid.

In 1929 the first patent with respect to compounds we now know as pcs was granted to Dandridge, Drescher, and Thomas of Scottish Dyes, Ltd.[6]

Linstead and his students at the University of London, supported by grants from Imperial Chemical Industries, starting in 1929, determined the structure of pc and several metal pcs, and announced their structure in 1933 to 1934.[3,7-12]

Copper pc was first manufactured by Imperial Chemical Industries in 1935, followed by Badische Anilin- & Soda- Fabrik in 1936, and E. I. duPont de Nemours & Company, Inc. in 1937.

The relationship of pcs to other porphins,[13] the determination of the structure of pcs,[9,10,12] and their planar nature[14-18] were also studied at this time. Robertson determined the dimensions and planarity of the pc molecule by X-ray analysis in a series of classic papers.[14-18]

In the 1930s, 1940s, and 1950s, X-ray spectra, polymorphism, absorption spectra, magnetic properties, catalytic properties, oxidation and reduction, photoconductivity, physical properties such as solubility, and photochemical, dielectric, and semiconductor properties were investigated. Also, in the same time frame, 45 metal pcs in all groups of the Periodic Table were described.

Commercial methods of manufacture of pc and copper pc as well as methods for eliminating crystallization and flocculation of the pigment form evolved. Also, partially chlorinated and fully chlorinated copper pcs and methods for their manufacture were developed.

The first pc dye was a pc polysulfonate; it was described in the first pc patent.[6] Since then, in the 1930s, 1940s, 1950s, and early 1960s, pc sulfonic acid dyes, pc sulfonic chlorides, other water-soluble pc dyes, quaternary and ternary ammonium salts of pyridyl derivatives of pc, solvent-soluble dyes, sulfur dyes, azo dyes, vat dyes, leuco dyes, chrome dyes, triazine dyes, and dye precursors were demonstrated and patented.

Also, at this time, the use of pcs as colorants for paint, plastics, printing inks, and textiles, as well as their use in writing and ball point pen inks, roofing granules, leather, gasoline, oil, photographic color prints, and even food coloring were announced.

In addition, it was found possible to polymerize the pcs, and about 20 references discussed the novel developments in this field.

From the time of their discovery to about 30 years later, in 1962, about 1200 literature references and patents described the progress of pc science and technology.

This monograph discusses the pcs from about 1963 to 1975. About 3500 journal and patent citations mark the advances in the subject of the pcs in this period.

REFERENCES

1. Kidd, J., *Phil. Trans. R. Soc. London,* 209, 1821.
2. Laurent, A., *Ann. Ber.,* 19, 47, 1836.
3. Linstead, R. P., *Br. Assoc. Adv. Sci. Rep.,* p. 465, 1933.
4. Braun, A. and Tcherniac, J., *Ann. Ber.,* 40, 2709, 1907.
5. de Diesbach, H. and Von der Weid, E., *Helv. Chim. Acta,* 10, 886, 1927.
6. Dandridge, A. G., Drescher, H. A., and Thomas, J., (to Scottish Dyes Ltd.), British Patent 322,169, November 18, 1929.
7. Byrne, G. T., Linstead, R. P., and Lowe, A. R., *J. Chem. Soc.,* p. 1017, 1934.
8. Dent, C. E. and Linstead, R. P., *J. Chem. Soc.,* p. 1027, 1934.
9. Dent, C. E., Linstead, R. P., and Lowe, A. R., *J. Chem. Soc.,* p. 1033, 1934.
10. Linstead, R. P., *J. Chem. Soc.,* p. 1016, 1934.
11. Linstead, R. P. and Lowe, A. R., *J. Chem. Soc.,* p. 1022, 1934.
12. Linstead, R. P. and Lowe, A. R., *J. Chem. Soc.,* p. 1031, 1934.
13. Chem. Eng. News, p. 20, 11 July 1960, and p. 35, 1 August 1960.
14. Robertson, J. M., *J. Chem. Soc.,* p. 615, 1935.
15. Robertson, J. M., *J. Chem. Soc.,* p. 1195, 1936.
16. Robertson, J. M., Linstead, R. P., and Dent, C. E., *Nature (London),* 135, 506, 1935.
17. Robertson, J. M. and Woodward, I., *J. Chem. Soc.,* p. 219, 1937.
18. Robertson, J. M. and Woodward, I., *J. Chem. Soc.,* p. 36, 1940.

Chapter 2

PREPARATION AND PROPERTIES OF METAL PCS

I. THE SCANDIUM FAMILY

A. Sc, Y, and the Lanthanides

$H(pc_2Sc)$ and $pcScCl \cdot H_2O$ are prepared and their electronic spectra are drawn.[1] The spectra exhibit sharp maxima at 614 and 676 nm, respectively.

The reaction of 4 mol of *o*-phthalonitrile with 1 mol of YCl_3 at 270 to 80°C for 1 hr is used to prepare the pc of yttrium.[2] Treatment of the product with a 10% solution of ammonium hydroxide yields a product in which the chlorine is replaced by an hydroxyl group.

Rare earth pcs are synthesized from their chlorides and *o*-phthalonitrile.[3-5] They are studied particularly with respect to their absorption spectra in the visible region in α-bromonaphthalene. The relation of the change of the absorption intensity maximum, λ, at its given wavelength vs. the atomic number of the rare earths is derived.[4] The hypsochromic shift of the position of λ is a function of the increase of the atomic number and can be interpreted by the enhancement of the stability of the metal-ligand bond of rare earth pcs with increasing atomic number.

Rare earth pcs are also prepared from rare earth acetates and *o*-phthalonitrile.[6,7] The crude products are dissolved in $HCON(CH_3)_2$, adsorbed on an alumina column, eluted with methyl alcohol, followed by attempted isolation in the solid state by the evaporation of methyl alcohol and drying at 100°C. Visible absorption spectra in several solvents and IR spectra are drawn and studied.

The UV and visible absorption bands of rare earth pcs shift to a shorter wavelength with the increase in atomic number of the rare earth and the IR band at 870 to 890 cm^{-1} which is assigned to the metal-nitrogen rocking vibration shifts to a higher frequency.[8] Ferrocene sandwich structures are suggested for di pc rare earths. Pc_2Ln(lanthanide)H is converted to pcLnX in the presence of X = acetate or chloride ion, but it is difficult to prepare the acetates or chlorides in pure states.[8]

Following their synthesis, the thermal stability of di pc complexes of lanthanides, hafnium, thorium, and uranium are studied by DTA and thermogravimetric analysis.[9]

The preparation and magnetic and spectral properties of one ytterbium and two gadolinium pcs are described and comparisons made with previous studies.[10] The susceptibility of all three compounds follows the Curie-Weiss law. Structures proposed are $ClpcYbCl \cdot 2H_2O$ for ytterbium and pcGdpcH and an associated anionic form for gadolinium. A simple acid-base equilibrium is proposed for the gadolinium compounds. From their experience in attempting to prepare rare earth pcs the authors caution that it is difficult to synthesize them: "Preparation of rare earth pc compounds [have been reported]. However, no analysis figures were given and compounds which were claimed to be the pc derivative of the same rare earth ion possessed different properties."[10]

In a review of lanthanide and actinide pc complexes,[11] another author makes the same caveat: "It can be regarded as a general tendency in lanthanide and actinide chemistry that these two series of elements are no longer restricted to a narrow range of plain reactions and simple compounds."

B. Actinide Phthalocyanines

"Although many metal porphyrins have been prepared no record of a uranyl pc could be found when this work was started. The properties of such a compound might be expected to provide information about the structure of the uranyl group and, because the U—O bonds are very sensitive to their environment, a study of their spectra may provide information about the interaction of uranium with the porphyrin ring.''[12] The authors also caution that it is difficult to prepare actinide pcs: ''The preparation of uranyl pc from uranyl chloride and dilithium pc is described. The infrared and electronic spectra of the complex are quite different from that previously reported for uranyl pc and it is suggested that previous workers obtained mainly a mixture of pc and inorganic uranyl compounds.''[12] In a footnote the authors state that they have been informed ''that to prepare uranyl pc from uranyl nitrate dimethyl formamide complex and dilithium pc in dimethyl formamide, DMF, solvent extreme precautions must be taken to remove the last traces of water from the solvent by repeated fractional distillation. In our attempt to prepare pc UO_2 by this method we dried our DMF over Linde's 5Å molecular sieve and distilled until our gas chromatograph failed to detect water (less than 50 ppm).''*

Complexes of uranyl ion with DMF and semicarbazide are described.[13]

UO_2pc is obtained by reacting $UO_2(OAc)_2$ and phthalonitrile, followed by observing the kinetics of the destruction of UO_2pc in solutions of 14 to 17.6 *M* H_2SO_4. ''The reactions indicated that UO_2 pc is a labile pc. The solubility of the compound in H_2SO_4 was determined. The pK was not constant, which also indicated that the compound is labile.''[14]

ThI_4 or UI_4 in *o*-phthalonitrile are heated at 240 or 530°C, respectively, to yield bipyramidal Th di pc and U di pc. The complexes are characterized by IR, electronic, and mass spectra, and by magnetic measurements.[15]

The solution of pc_2U in 18 *M* H_2SO_4 is orange and its spectrum does not show the presence of the free pc^{2-} ligand even after standing 6 days.[16]

II. THE TITANIUM FAMILY: Ti, Zr, AND Hf

Ti pc Cl is formed by reaction of $TiCl_3$ with Li_2pc in boiling quinoline.[17] It is stable in air and slightly soluble in hot quinoline. By oxidation of the quinoline solution, Ti pc O is formed as violet diamagnetic crystals.

Ti pc is formed by reaction of $C_6H_4(CN)_2$ with $TiCl_4$.[18]

$ZrCl_4$ and $HfCl_4$ refluxed with *o*-phthalonitrile yield $C_{32}H_{15}N_8ClZr(OH)_2 \cdot 2H_2O$ and $C_{32}H_{15}N_8ClHf(OH)_2 \cdot 2H_2O$, respectively.[19] An attempt to prepare the compounds from $ZrOCl_2$ and *o*-phthalonitrile was not successful.

Refluxing $ZrCl_4$ and $HfCl_4$ with acetic acid *o*-phthalonitrile yield $[Zr(AcO)_2]$ pcCl and $[Hf(AcO)_2]$ pcCl, respectively.[20] They are soluble in α-chloronaphthalene, quinoline, and concentrated H_2SO_4. In aqueous quinoline $[Zr(OH)_2]$ pcCl is formed.

III. THE VANADIUM FAMILY: V, Nb, AND Ta

The preparation and partial characterization of oxovanadium(IV) ion·complexes, presumably including pc, are described.[21]

M pcs, where M = H_2, Na_2, Cu, Co, AlCl, GaCl, FeCl, VCl, CrOH, and VOCl, form with VCl_3 in nitrobenzene two forms of complexes. The first form, characterized by several new narrow bands within the 780- to 1100- and 490- to 550-nm regions, is transformed into the second form with one broad asymmetric band at 470 to 590 nm. In the EPR spectrum, the first form exhibits a strong signal in the region of radicals which disappears during the transformation into the second form. The transformation rate is determined by the M pc/VCl_3 mole ratio R and by the nature of the central

* From Bloor, J. E., Schlabitz, J., Walden, C. C., and Demerdache, A., *Can. J. Chem.*, 42(10), 2201, 1964. With permission.

ion. Thus, the first form appears to be a classical π-6 charge transfer complex M $pc^{\oplus}$. . . $VCl_3^{\ominus}$, whereas the second form is an ion pair M pc . . . $VCl_2^{\oplus}$. . . $VCl_4^{\ominus}$."[22]

"With the exception of vanadium, the pc salts of the Group IVB and VB metals cannot be prepared . . . from the elemental form or from the oxides, because of the nonreactivity of these materials with pc forming reagents such as phthalonitrile . . ." A process is developed, therefore, "for the preparation of pcs of vanadium, columbium (niobium), tantalum, titanium, zirconium, and hafnium."[23] Such pcs include pc Nb Cl_x $(OH)_{n-x}$ where x = 1 to n, n = integer denoting the electrovalence of the niobium atom, and include pc Ta Cl_x $(OH)_{n-x}$ where x = 1 to n, n = integer denoting the electrovalence of the tantalum atom.

The preparation procedure involves reacting, in an inert atmosphere, molten phthalonitrile with an anhydrous halide of vanadium, niobium, or tantalum[23] as well as of titanium, zirconium, or hafnium. All these pcs are partially hydrolyzed salts.

IV. IRON, CHROMIUM, AND MANGANESE PHTHALOCYANINES

Cr(II)pc is prepared from $Cr(OAc)_2$ and *o*-phthalonitrile and is characterized by analysis, X-ray, IR, and magnetic data.[24] "On exposure to air, the Cr(II)pc oxidizes to a blue-violet compound of uncertain structure. It is suggested that the Cr(II)pc is an isomer of the Cr(II) previously described."[25]

Square planar complexes of the first row transition metals are reviewed with 100 references.[26]

When β-Cr(II)pc is sublimed, a purple-red product is obtained which, unlike β-Cr(II)pc, does not react with air.[27] It is α-Cr(II)pc. "The different stability in air of the two forms of Cr(II) pc can be reasonably explained on the basis of their different crystal features."*

The following octahedral chromium pcs are synthesized:

pc Cr—O—Cr pc
H_2O H_2O

pc Cr—OH—Cr pc
NH_3 OH

and

$C_{32}H_{12}N_8Cr$—OH—$CrC_{32}H_{12}N_8$
$(SO_3NA)_4$ OH H_2O $(SO_3NA)_4$

The structures are suggested from magnetic susceptibility measurements, EPR studies, and from examination of the IR absorption spectra.[28]

The synthesis, composition, and properties of iron, chromium, and manganese pcs are studied.[29] The kinetics of dissolution and dissociation, and the spectra of solutions in 17.5 *M* H_2SO_4 are determined. "The data indicate that the pc complexes of Fe^{2+}, Mn^{2+}, and Fe^{3+} are labile, while those of Cr^{3+} and Mn^{4+} are highly stable."

Although Cr, Mn, and Fe may be treated together in a study of pcs, they are members of VIB, VIIB, and VIIIB transition metal families. Rhenium, a member of the manganese family, is the subject of a research entitled "Rhenium (IV) pcs complexes. Hydroxychlororhenium tetrasulfonated pc."[30] They are formed by reacting potassium oxochlorohenate, potassium chlororhenate, urea, and tetrasulfonated phthalic acid at 220 to 240°C. The pc product is soluble in water and ethanol solutions. Their absorption spectra in UV and visible light show two bands ascribed to monomer-dimer equilibria. The dimerization constant is about 1.07×10^4.

* From Ercolani, C., Neri, C., and Porta, P., *Inorg. Chim. Acta,* 1(3), 415, 1967. With permission.

Mn pc and a solution of dilithiobenzophenone in tetrahydrofuran, THF, under exclusion of oxygen and moisture give Li(Mnpc)·6 THF. *In vacuo,* they give off THF. They decompose and give off THF in air.[31] Their ionic character is shown by measurements of their magnetic moments in CH_3CN at 293 K.

V. THE IRON FAMILY: Fe, Co, AND Ni PHTHALOCYANINES

The kinetics of dissolution and dissociation, and the spectra of solutions of Fe pc in 17.5 *M* H_2SO_4 are determined.[29] The data indicate that the pc complexes of Fe^{2+} and Fe^{3+} are labile.

Experiments in DMS and dimethylacetamide show that "cholroferric pc" is probably a hydrochloride of ferrous pc.[32] The chloride ion may act as an axial ligand at the ferrous center while the proton is situated adjacent to the pc ring, attached to an azomethine bridge.

Iron complexes of octaphenyltetraazaporphine, pc, and mesotetrapyridylporphine are the subject of a thesis.[33]

"The behavior of transition metal derivatives of porphyrin calls for biological significance when it forms additive complexes with various n-donor molecules which have the ability of coordinating to the central metal ion."[34]

LFepcL are prepared where L = CH_3NH_2, $(CH_3)_2NH$, $C_2H_5NH_2$, $C_3H_8NH_2$, hexylamine, pyridine, and γ-picoline; also prepared are $Fepc(C_6H_5NH_2)_6$, and $FepcL_4$ where L = $C_2H_5NH_2$, $C_3H_8NH_2$, hexylamine, and γ-picoline. The syntheses are made by the dispersion of Fepc in the various amines. The compositions are determined by weight-loss measurements and from the thermal behavior of the additive complexes as observed by DTA and X-ray powder diffraction.

Other adducts of Fe^{2+} pc have been prepared.[35] They include mixed-ligand complexes LFepcL′, where L = a nitroso aromatic group, L′ = *n*-butylamine or *N*-methylimidazole; L = cyclohexyl isocyanide, L′ = *N*-methylimidazole; L = a phosphite, L′ = *n*-butylamine, and bis adducts L_2Fepc where L = isocyanide or phosphite. The complexes are characterized by elemental analysis, electronic spectra, IR spectra, and proton magnetic resonance spectra. Models for the nitroso-iron bonding in the nitroso adducts are shown in Figure 1.

Fe^{2+} pc reacts with tri-*n*-butyl phosphite and tri-*n*-butylphosphine to form low spin bis adducts, Fe pc$(PR_3)_2$.[36] Replacement of $P(OBu)_3$ by PBu_3 occurs rapidly and follows a strictly dissociative mechanism.

The binding energies of Fe 1s electrons are measured for a series of Fe compounds including Fe pc and correlated with the $2p_3/_2$ and 3p binding energies.[37] The 1s energies are also related to the atomic charge of the Fe atoms.

Four Co pc dye intermediates are prepared by halogenating a Co pc with Cl, Br, I, $SOCl_2$, SO_2Cl_2, PCl_5, or POCl at 0 to 150°C in an inert organic solvent free of OH and NH_2 groups.[38] For example, 571 parts Co pc in 2500 parts *o*-$C_6H_4Cl_2$ is treated with Cl_2 at 40 to 50°C for 1 hr to give the dichloro derivative, where both chlorine atoms are attached to the Co atom, one on each side of the plane of the pc molecule.

Sixteen Co(III) complex pcs of the formula [pc Co-halogen$_2$)]$^{\oplus}$(halogen)$^{\ominus}$ where Y = NH_3, $H_2NCH_2CH_2NH_2$, $H_2N(CH_2)_3N(CH_3)_2$, $H_2N(CH_2)_3NH(CH_3)$, $[H_2N(CH_2)_3]_2NH$, pyridine are prepared by treating Co pcs with an oxidizing agent in the presence of an anion and an amine.[39] For example, a suspension of Co pc in *o*-$C_6H_4Cl_2$ is treated 1.5 hr with Cl_2 at ≤30°C to give dichloro-Co pc which is suspended in *o*-$C_6H_4Cl_2$ and treated with NH_3 to give the amine complex [$(NH_3)_2$-Co pc]$^{\oplus}$(halogen)$^{\ominus}$, where the NH_3 ligands are on opposite sides of the pc ring.

Chlorine and bromine complexes of Co pc are prepared by the direct halogenation of Co pc powder with Cl_2 or Br_2 gas in the absence of solvent or catalyst.[40]

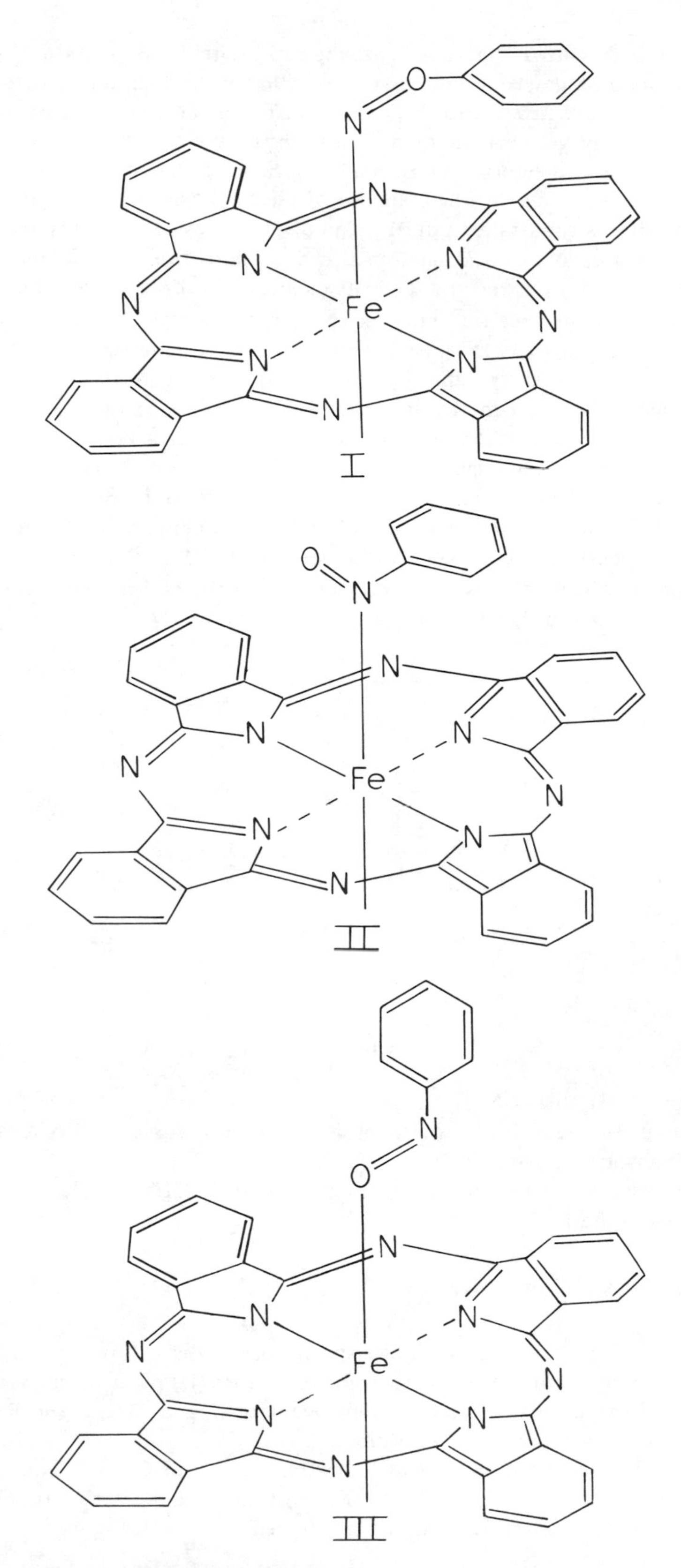

FIGURE 1. Models for the nitroso-iron bonding.

"It has long been known that metals are the active sites in many naturally occurring substances containing macrocycles, such as metalloporphyrins, and that their biological functions often are determined by the nature of, or the exchange of, labile extraplanar ligands, or by the variable oxidation states of the metal. To show the validity of the assumptions and interpretation made concerning the complicated natural systems, there is a need to clarify the behavior of relatively simple materials that do not share the multifold structural complexity and diverse sites of reactivity with such substances as the hemoproteins and vitamin B_{12}. Studies of this kind should help define those relationships that produce the delicate balance of properties required for a complex to perform some specific function in a dynamic system. The complexes of 4,4′,4″,4‴-tetrasulfo pc offer the possibility of this type of investigation."*

The influence of such extraplanar ligands as pyridine, nitrite, thiocyanate, cyanide, imidazole, benzimidazole, α-methyl imidazole, and histidine on the magnetic and spectral properties and on the relative stabilities of the +2 and +3 oxidation states of Co 4,4′,4″,4‴-tetrasulfo pc has been investigated. "While none of the ligands alters the ground state of the cobalt (II) complex from the doublet observed in the presence of water alone, the strong ligands, cyanide and imidazole, facilitate oxidation of the compound by atmospheric oxygen to cobalt (III) derivatives."[41]

Five binuclear cobalt (IV) pcs with oxygen or NH_2 bridges, and binuclear sulfonated Co(IV) pc with an OH bridge are obtained.[42] They are

[pc Co—O—Co pc] SO_4
H_2O H_2O

pc Co—O—Co pc
OH OH

[pc Co—NH_2—Co pc] Cl_3
NH_3 H_2O

[pc Co—NH_2—Co pc] SO_4

[pc Co—O—Co pc] $ClNH_3OH$

and

[PTS Co—OH—Co PTS] SO_4

where PTS = $C_{32}H_{12}N_8(SO_3Na)_4$.

Their structure is "established on the basis of magnetic susceptibility measurements, ESR, and IR absorption spectra."

The effects of electrolytes on the optical activity of Co(III) complexes including pcs is the subject of a thesis.[43]

VI. PLATINUM FAMILY PCS: Pt, Pd, Rh, Os, Ru, AND Ir

In a study of Ru pc, isolation of Ru species other than those of Ru(II) was not accomplished,[44] such that the relative stability of the Ru(II) pc must be associated with the features of the aromatic, planar, quadridentate ligand. Thus, the Ru pc amine adducts are probably octahedrally coordinated.

Refluxing *o*-phthalonitrile with $RuCl_3$ or $IrCl_3$ gives pc ClRuCl and pc ClIrCl.[45] Reprecipitation from 96% H_2SO_4 gives (HSO_4)RupcCl and (HSO_4)IrpcCl; their absorption spectra and acid-base properties are studied.

* From Weber, J. H. and Busch, D. H., *Inorg. Chem.*, 4(4), 472, 1965. With permission.

$Ru(OH)_2$ pc $(SO_3Na)_4 \cdot 4H_2O$ is prepared starting with fusion of potassium chloro-ruthenate, ammonium sulfophthalate, and urea.[46] The magnetic susceptibility of the ruthenium pc compound in the solid state is 0.883×10^{-6}. "Deviations from Beer's law and changes in the intensity ratio at 630 and 682 mμ in the absorption spectrum indicate the equilibrium between the two forms of this ruthenium pc: monomer (630 mμ) and dimer (682 and 575 mμ). The dimerization constant is 5.75×10^5."

Rhodium(III) pc is prepared by refluxing $RhCl_3$ with an excess of *o*-phthalonitrile,[47] washing with acetone and reprecipitating from 96% sulfuric acid, to yield $RhpcHSO_4$. The pK for the reaction

$$\mathrm{Rhpc(HSO_4) + H_2SO_4 \rightleftharpoons Rh(pcH)\,(HSO_4)^{\oplus} + HSO_4^{\ominus}}$$

is calculated to be 1.01 in 14 to 16.5 $M\,H_2SO_4$.

Heating *o*-phthalonitrile with OsO_4 and sulfuric acid gives $Os(SO_4)pc$[48] "in which a polymeric structure probably exists, indicating a reduction of Os to the tetravalent state."* The product probably contains repeating units of Os pc bound to each other through SO_4 links. The magnetic susceptibility and the rate constants of dissociation in sulfuric acid are determined. The solubility of the Os pc is much greater than that of Pt or Pd, "attributed to the peculiar lattice of this complex and the possibility of partial protonation of individual links in the polymer chain."*

The equilibrium constant of PdpcCl and PtpcCl in 13.5 to 17.7 *M* sulfuric acid is measured in terms of the reaction[49]

$$\mathrm{M\,pc + H_2SO_4 \rightleftharpoons M\,pc\,H^{\oplus} + HSO_4^{\ominus}}$$

"The fact that the spectra of H_2SO_4 solutions did not change with time indicated that Pd pc Cl and Pt pc Cl were in the category of stable pcs."

The chlorides of the platinum metals Pt, Pd, Rh, Os, Ru, and Ir are reacted with *o*-phthalonitrile to give Ptpc, PdpcCl, RhpcCl, Ir pc $Cl \cdot C_6H_4(CN)_2$, $RupcCl \cdot C_6H_4(CN)_2$, and $OspcCl_2 \cdot C_6H_4(CN)_2$.[50] The complex character of the Ir, Ru, and Os pcs is revealed "by the fact that the dicyanobenzene could only be removed by strong heating below about 500° and that it could be replaced with aniline by boiling in aniline."

VII. THE COPPER FAMILY: Cu, Ag, AND Au

Cu hexadecafluoropc is prepared from tetrafluorophthalonitrile and CuCl.[51] It is a dark blue powder and is more stable than Cu pc to such oxidizing agents as HNO_3, $KMnO_4$ and $Ce(SO_4)_2$.

The symmetry and UV spectra of complexes of azosubstituted pc is the subject of a study.[52]

Phthalonitrile is treated with copper salt and ammonia saturated methyl alcohol at 80 to 140°C to give a brilliant reddish blue metastable Cu pc having a coloring power comparable with that of acid pasted or salt milled Cu pc.[53]

Lipophilic Cu pcs are prepared by reacting chlorosulfonated Cu pc with propylene diamine derivatives.[54]

Cu pc complexes of bis(dicarboxyphenyl) ketone are prepared. They are stable to heat and begin to lose weight at about 450°C. Similar pcs are obtained with Co and Ni.[55]

* From Berezin, B. D. and Sosnikova, N. I., *Dokl. Akad. Nauk. S.S.S.R.*, 146, 604, 1962. With permission.

Cu tetra-4-*tert*-butyl pc, soluble in organic solvents, is synthesized and characterized by electron absorption spectra.[56] The Zn, AlOH, Ni, and Pd complexes are similarly prepared.

"Gold pc has been prepared by the action of gold monobromide on molten 1,3 diiminoisoindoline, in the absence of solvent. . . An approximately 0.001 *M* solution of gold pc in 1-chloronaphthalene was examined by electron spin resonance at 77 K. . . Gold was detected in the divalent state."[57]

VIII. THE ALUMINUM FAMILY: Al, Ga, In, AND Tl

A series of Al pc compounds are prepared in which the apical organic and organic siloxy groups are bonded to the central Al atom.[58,59]

Ga pc Cl is prepared; it is practically insoluble in organic solvents whereas ClGapcCl is soluble in organic solvents "to a perceptible extent."[60]

Chloroaquoindium pc, $InCl \cdot H_2O$ pc is synthesized.[61] "This substance contains one molecule of coordination water attached to the central metal atom. It is interesting to note that a coordination water molecule is stable even in vacuo and at fairly high temperature (probably above 250°C.). However, this is not so strange if one considers that indium makes a stable complex compound of coordination number of six. Although coordination water in pc complexes of aluminum is known, this finding is the first case in In pc complexes."

IX. SILICON PHTHALOCYANINES AND THE GERMANIUM FAMILY: Ge, Sn, AND Pb

"The pc silicon compounds as a group are of interest because of the apparent hexacoordination of the silicon and the stability of the Si−N bonds."[62] Pc silicon compounds are made that contain chemically and thermally stable Si−N bonds.

A series of dialkoxy pc silicon derivatives are described. They, "together with the silicon pcs previously described, serve to illustrate the way in which the properties of ring unsubstituted trivalent and tetravalent metal pcs vary with the nature of the groups attached to the metal. Thus, for example, $(pcSiO)_x$ is thermally stable and insoluble in organic solvents, $pcSiF_2$ is chemically inert, $pcSi(OH)_2$ is chemically reactive, and $pcSi(OC_{18}H_{37})_2$ is benzene soluble. . . In all these cases the properties can be accounted for in terms of the nature of the side chain or group and the properties of the macrocyclic ring."*

"An extensive series of pcs is formed by tin. . . Methods of preparing and purifying the compounds $pcSnF_2$, $pcSnBr_2$, $pcSnI_2$, and $pcSn(OH)_2$ are given together with improved procedures for the synthesis and purification of $pcSnCl_2$, pc_2Sn, pcSn, and pcPb. Pc_2Sn is shown to have two polymorphic modifications and to form a 1:1 solvate with 1-chloronaphthalene. Both polymorphs are found to undergo thermal decomposition at elevated temperatures with the formation of Snpc and H_2pc."**

Syntheses and reactions of X_2Sipc where X = Cl, OR, OH, are discussed together with their thermal, electrical, and NMR and IR spectral properties.[65]

Cl_2Sipc is prepared from $SiCl_4$ and diiminoisoindoline or *o*-cyanobenzamide.[66] Cl_2Sipc is sparingly soluble in most common solvents. When heated in H_2SO_4, it polymerizes to form −OSipcOSipcO− chains.

"In the originally reported syntheses of dichloropcsilicon, $pcSiCl_2$, *o*-phthalonitrile was allowed to react with either silicon tetrachloride or hexachlorodisiloxane in quinoline. These syntheses proved to be experimentally quite inconvenient, and their defi-

* From Krueger, P. C. and Kenney, M. E., *J. Org. Chem.*, 28(12), 3379, 1963. With permission.

** From Kroenke, W. J. and Kenney, M. E., *Inorg. Chem.*, 3, 251, 1964. With permission.

ciencies have impeded work on the silicon pcs, since for this series the chloro complex is the key compound."*

"For the pseudohalo compounds of silicon and germanium there is always at least a possibility that the groups are nitrogen bonded to the metal. This possibility has led, in the course of a search for silicon and germanium pcs having the central metal octahedrally coordinated by nitrogen to the study of the pseudohalo pc derivatives of these two metals."**

Pseudohalo pcs studied are[68,69]

Bisisocyanato pc silicon —	pc Si $(NCO)_2$
Bisisothiocyanato pc silicon —	pc Si $(NCS)_2$
Bisisoselenocyanato pc silicon —	pc Si $(NCSe)_2$
Bisisocyanato pc germanium —	pc Ge $(NCO)_2$
Bisthiocyanato pc germanium —	pc Ge $(NCS)_2$
Bisselenocyanato pc germanium —	pc Ge $(NCSe)_2$

"Tetravalent metal pcs are soluble in organic solvents if the *trans* groups attached to the metal are of a suitable kind. . . among the *trans* groups which impart solubility are methylsiloxy groups having two or more silicon atoms. . . use has been made of this knowledge in preparing a series of symmetrical siloxysilicon pcs of sufficient solubility to be useful for physical studies requiring solutions. These compounds have then been used for a nuclear magnetic resonance investigation of Si pcs. In addition, work has been carried out on the synthesis of a series of Si pcs where one, but not both, of the *trans* groups is a methyl or a phenyl group. These compounds have been used to obtain additional information on the nature of the octahedral silicon-carbon bond."*** A perspective view of such a compound is shown in Figure 2.

Additional silicon, germanium, and tin pcs are prepared.[71] "Of particular interest among these new compounds is pc Si Br_2, since the silicon atoms in it can reasonably be assumed to be coordinated in the trans position by bromine atoms, atoms which are neither small nor very electronegative. The relative thermal stability of the compound suggests that eventually a variety of octahedral silicon compounds containing heavy metal ligand atoms will become known. In fact, as the number of known Si pcs grows, the rather broad conclusion is emerging that a reasonably close parallel exists between the chemistry of tetrahedral and octahedral silicon compounds."†

During the course of NMR studies on porphyrins and porphyrin-like compounds, "the shielding region above the ring surface has figured prominently. . . The nuclear magnetic resonance spectra of two unsymmetrically substituted Si pcs have been studied. The results indicate that the shielding furnished by the ring drops off without major reversals in the region above its surface and near its fourfold axis. In addition, the spectra of two germanium porphins and two analogous germanium pcs have been compared. The results from this work show that the porphin ring offers more shielding than the pc ring."‡

Ge pc is synthesized and characterized.[73] "It exists in α and β polymorphs and it is very light sensitive but not easily oxidized. Ge(II) is stabilized by the ring system and

* From Lowery, M. K., Starshak, A. J., Esposito, J. N., Krueger, P. C., and Kenney, M. E., *Inorg. Chem.*, 4(1), 128, 1965. With permission.

** From Starshak, A. J., Joyner, R. D., and Kenney, M. E., *Inorg. Chem.*, 5(2), 330, 1966. With permission.

*** From Espositio, J. N., Lloyd, J. E., and Kenney, M. E., *Inorg. Chem.*, 5(11), 1979, 1966. With permission.

† From Rafaeloff, R., Kohl, F. J., Krueger, P. C., and Kenney, M. E., *J. Inorg. Nucl. Chem.*, 28(3), 899, 1966. With permission.

‡ From Kane, A. R., Yalman, R. G., and Kenney, M. E., *Inorg. Chem.*, 7, 2588, 1968. With permission.

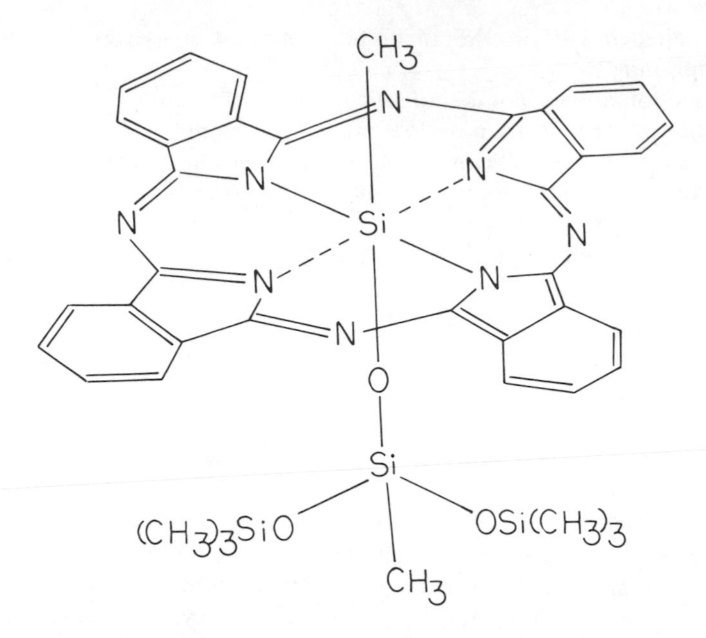

FIGURE 2. A perspective view of the structure of $PcSi(CH_3)-(OSi(CH_3)(OSi(CH_3)_2)_2)$.

oxidation to Ge(IV) occurs only after the ring is destroyed.'' Ge pc F_2, Ge pc Br_2, and Ge pc Cl_2 and $(pc\ Ge\ O)_x$ are also prepared.[74]

Si, Ge and Sn pcs have been studied in several theses.[75-84]

Diphenyl tin pc and dichloro tin pc are prepared[84] as the basis for ''a new approach to polymer formation [which] may be a way of attaching a group or series of groupings to the central metal atom. If a stable grouping could be attached to the metal and polymerized to produce a flexible polymer with a large number of repeating units, many of the shortcomings [of pc polymers] could be eliminated.''[84]

The synthesis of Sn pc and its electrical properties are studied.[85]

X. THE HYDROGEN FAMILY

H_2pc can be prepared by heating 1,3-diminoisoindoline in the presence of such reducing agents as Na_2S and hydroxyquinone.[86]

The spontaneous formation of porphyrins under primitive earth conditions, on the Jovian planets, and in interstellar space is conjectured and discussed.[87]

XI. FLUORINATED PHTHALOCYANINES

Pcs prepared from 3- and 4-fluorophthalic acids exhibit a shift in shade toward the green, which is to be expected from experience with other halogenated pcs.[88]

In a research entitled ''Synthesis of Intercalation Compounds and New Inorganic Fluorine Compounds'', various organic, including pc, and inorganic fluorine containing compounds are prepared and characterized.[89]

REFERENCES

1. Kirin, I. S. and Moskalev, P. N., *Zh. Neorg. Khim.*, 16(11), 3179, 1971.
2. Plyushchev, V. E. and Shklover, L. P., *Zh. Neorg. Khim.*, 9(8), 2015, 1964.
3. Plyushchev, V. E. and Shklover, L. P., *Zh. Neorg. Khim.*, 9(2), 335, 1964.
4. Shklover, L. P. and Plyushchev, V. E., *Zh. Fiz. Khim.*, 39(12), 2924, 1965.
5. Shklover, L. P. and Plyushchev, V. E., *Tr., Vses. Nauch. Issled. Inst. Khim. Reaktivov Osobo Chist. Khim. Veshchestv*, 30, 89, 1967.
6. Kirin, I. S., Moskalev, P. N., and Makashev, Yu. A., *Zh. Neorg. Khim.*, 12(3), 707, 1967.
7. Kirin, I. S., Moskalev, P. N., and Ivannikova, N. V., *Zh. Neorg. Khim.*, 12(4), 944, 1967.
8. Misumi, S. and Kasuga, K., *Nippon Kagaku Zasshi*, 92(4), 335, 1971.
9. Moskalev, P. N., Mishin, V. Ya., Rubtsov, E. M., and Kirin, I. S., *Zh. Neorg. Khim.*, 21(8), 2259, 1976.
10. MacKay, A. G., Boas, J. F., and Troup, G. J., *Aust. J. Chem.*, 27(5), 955, 1974.
11. Lux, F., *Proc. Rare Earth Res. Conf., 10th* 2 (CONF-730402-P2), 871, 1973.
12. Bloor, J. E., Schlabitz, J., Walden, C. C., and Demerdache, A., *Can. J. Chem.*, 42(10), 2201, 1964.
13. Frigerio, N. A. and Coley, R. F., *J. Inorg. Nucl. Chem.*, 25(9), 1111, 1963.
14. Berezin, B. D., Zhukov, Yu. A., and Koifman, O. I., *Dokl. Akad. Nauk S.S.S.R.*, 167(6), 1318, 1966.
15. Lux, F., Dempf, D., and Graw, D., *Angew. Chem. Int. Ed. Engl.*, 7(10), 819, 1968.
16. Kirin, I. S., Kolyadin, A. V., and Moskalev, P. N., *Zh. Neorg. Khim.*, 16(10), 2731, 1971.
17. Taube, R., *Z. Chem.*, 3(5), 194, 1963.
18. Shklover, L. P., Plyushchev, V. E., Rozdin, I. A., and Novikova, N. A., *Zh. Neorg. Khim.*, 9(2), 478, 1964.
19. Plyushchev, V. E., Shklover, L. P., and Rozdin, I. A., *Zh. Neorg. Khim.*, 9(1), 125, 1964.
20. Muehl, P., *Z. Chem.*, 7(9), 352, 1967.
21. Selbin, J. and Holmes, L. H., Jr., *J. Inorg. Nucl. Chem.*, 24, 1111, 1962.
22. Bialkowska, E., Graczyk, A., and Sobczynska, J., *Pr. Nauk Inst. Chem. Nieorg. Metal. Pierwiastkow Rzadkich Politech Wroclaw*, 28, 141, 1976.
23. Burbach, J. C., (to Union Carbide Corporation), U.S. Patent 3,137,703, June 16, 1964.
24. Ercolani, C., *Ric. Sci.*, 36(10), 975-9 (English) 1966.
25. Elvidge, J. A. and Lever, A. B. P., *J. Chem. Soc.*, p.1257, 1961.
26. Musker, W. K., *Adv. Chem. Ser.*, 62, 469, 1967.
27. Ercolani, C., Neri, C., and Porta, P., *Inorg. Chim. Acta*, 1(3), 415, 1967.
28. Przywarska-Boniecka, H. and Wojciechowski, W., *Mater. Sci.*, 1(2), 35, 1975.
29. Berezin, B. D. and Sennikova, G. V., *Izv. Vyssh. Ucheb. Zaved., Khim. Khim. Tekhnol.*, 10(5), 563, 1967.
30. Przywarska-Boniecka, H., *Rocz. Chem.*, 41(10), 1703, 1967.
31. Taube, R. and Munke, H., *Angew. Chem.*, 75(13), 639, 1963.
32. Jones, J. G. and Twigg, M. V., *J. Chem. Soc. A*, 9, 1546, 1970.
33. Dickens, L. L., Iron Complexes of Octaphenyltetraazaporphine, PC, and Meso-Tetrapyridylporphine, thesis, 1974, avail. Xerox Univ. Microfilms, Ann Arbor, Mich., Order No. 75-11,510; *Diss. Abstr. Int. B*, 35(11), 5304, 1975.
34. Kobayashi, T., Kurokawa, F., and Uyeda, N., *Bull. Inst. Chem. Res. Kyoto Univ.*, 53(2), 186, 1975.
35. Watkins, J. J. and Balch, A. L., *Inorg. Chem.*, 14(11), 2720, 1975.
36. Sweigart, D. A., *J. Chem. Soc. Dalton Trans.*, 15, 1476, 1976.
37. Johansson, L. Y. and Blomquist, J., *Chem. Phys. Lett.*, 34(1), 115, 1975.
38. Mertens, P. and Vollmann, H., (to Farbenfabriken Bayer AG), U.S. Patent 3,651,082, March 21, 1972.
39. Vollmann, H. and Mertens, P., (to Farbenfabriken Bayer AG), U.S. Patent 3,636,040, January 18, 1972.
40. Farbenfabriken Bayer AG, French Addition 2,114,073, August 4, 1972.
41. Weber, J. H. and Busch, D. H., *Inorg. Chem.*, 4(4), 472, 1965.
42. Przywarska-Boniecka, H. and Wojciechowski, W., *Mater. Sci.*, 1(1), 27, 1975.
43. Smith, H. L., Thesis, Univ. Microfilms, Ann Arbor, Mich., Order No. 66-10,088; *Diss. Abstr. B*, 27(7), 2261, 1967.
44. Krueger, P. C. and Kenney, M. E., *J. Inorg. Nucl. Chem.*, 25(3), 303, 1963.
45. Berezin, B. D. and Sennikova, G. V., *Dokl. Akad. Nauk S.S.S.R.*, 159(1), 117, 1964.
46. Przywarska-Boniecka, H., *Rocz. Chem.*, 42(10), 1577, 1968.
47. Berezin, B. D., *Dokl. Akad. Nauk S.S.S.R.*, 150(5), 1039, 1963.
48. Berezin, B. D. and Sosnikova, N. I., *Dokl. Akad. Nauk S.S.S.R.*, 146, 604, 1962.

49. Berezin, B. D., *Zh. Neorg. Khim.*, 7, 2501, 1962.
50. Keen, I. M., *Platinum Met. Rev.*, 8(4), 143, 1964.
51. Wotton, D. E. M., (to Imperial Smelting Corporation (N.S.C.) Ltd.), British Patent 1,037,657, August 3, 1966.
52. Cho, N.-S., Kim, K.-H., and Hahn, C.-S., *Daehan Hwahak Hwoejee*, 16(6), 378, 1972.
53. Tezuka, Y. and Akamatsu, T., (to Sumitomo Chemical Company, Ltd.), Japanese Kokai 73, 10, 127, February 8, 1973.
54. Kienzle, P., (to Pechiney Ugine Kuhlmann), French Demande 2, 168, 198, May 10, 1973.
55. Korshak, V. V., Rogozhin, S. V., and Vinogradov, M. G., *Izv. Akad. Nauk S.S.S.R. Otd. Khim. Nauk*, p. 1473, 1962.
56. Mikhalenko, S. A., Barkanova, S. V., Lebedev, O. L., and Luk'yanets, E. A., *Zh. Obshch. Khim.*, 41(12), 2735, 1971.
57. MacCragh, A. and Koski, W. S., *J. Am. Chem. Soc.*, 87(11), 2496, 1965.
58. Owen, J. E. and Kenney, M. E., *Inorg. Chem.*, 1, 331, 1962.
59. Owen, J. E., Aluminum pcs, thesis, Univ. Microfilms, Ann Arbor, Mich., Order No. 61-3309; *Diss. Abstr.*, 22, 731, 1961.
60. Berezin, B. D., *Izv. Vyssh. Ucheb. Zaved. Khim. Khim. Tekhnol.*, 7(6), 982, 1964.
61. Yoshihara, K., Shiokawa, T., Kishimoto, M., and Suzuki, S., *Shitsuryo Bunseki*, 22(3), 231, 1974.
62. Joyner, R. D. and Kenney, M. E., *Inorg. Chem.*, 1, 236, 1962.
63. Krueger, P. C. and Kenney, M. E., *J. Org. Chem.*, 28(12), 3379, 1963.
64. Kroenke, W. J. and Kenney, M. E., *Inorg. Chem.*, 3, 251, 1964.
65. Yamaguchi, M., Takahashi, H., and Shiihara, I., *Shinku Kagaku*, 16(2), 78, 1969.
66. Akopov, A. S., Berezin, B. D., Klyuev, V. N., and Solov'ev, A. A., *Zh. Neorg. Khim.*, 17(4), 981, 1972.
67. Lowery, M. K., Starshak, A. J., Esposito, J. N., Krueger, P. C., and Kenney, M. E., *Inorg. Chem.*, 4(1), 128, 1965.
68. Starshak, A. J., Joyner, R. D., and Kenney, M. E., *Inorg. Chem.*, 5(2), 330, 1966.
69. Starshak, A. J., Kenney, S. J., and Kenney, M. E., NASA Accession No. N65-22513, Rep. No. TR-3, 1965.
70. Esposito, J. N., Lloyd, J. E., and Kenney, M. E., *Inorg. Chem.*, 5(11), 1979, 1966.
71. Rafaeloff, R., Kohl, F. J., Krueger, P. C., and Kenney, M. E., *J. Inorg. Nucl. Chem.*, 28(3), 899, 1966.
72. Kane, A. R., Yalman, R. G., and Kenney, M. E., *Inorg. Chem.*, 7, 2588, 1968.
73. Stover, R. L., Thrall, C. L., and Joyner, R. D., *Inorg. Chem.*, 10(10), 2335, 1971.
74. Esposito, J. N., Sutton, L. E., and Kenney, M. E., U.S.N.T.I.S., AD Rept. No. 604117, 1964.
75. Joyner, R. D., Germanium and Silicon pcs, thesis, Univ. Microfilms, Ann Arbor, Mich., Order No. 61-3317; *Diss. Abstr.*, 22, 731, 1961.
76. Kroenke, W. J., The pc-Tin System and the pc Oxides of Tin, Germanium, and Silicon, thesis, Univ. Microfilms, Ann Arbor, Mich., Order No. 64-13; *Diss. Abstr.*, 24, 2682, 1964.
77. Krueger, P. C., Pc Silicon and pc Ruthenium Derivatives, thesis, Univ. Microfilms, Ann Arbor, Mich., Order No. 64-14; *Diss. Abstr.*, 24, 2682, 1964.
78. Esposito, J. N., The Unsymmetrically Substituted Silicon pcs and the Germanium Hemiporphyrazine System, thesis, Univ. Microfilms, Ann Arbor, Mich., Order No. 66-14,287; *Diss. Abstr. B*, 27(7), 2284, 1967.
79. Kane, A. R., Silicon pcs and Germanium Porphins, thesis, 1969, avail. Univ. Microfilms, Ann Arbor, Mich., Order No. 70-5101; *Diss. Abstr. Int. B*, 30(9), 4045, 1970.
80. Haynie, R. L., Free Radical Sites in Silicon pcs, thesis, 1972, avail. Univ. Microfilms, Ann Arbor, Mich., Order No. 72-26,163; *Diss. Abstr. Int. B*, 33(4), 1452, 1972.
81. Bernal-Castillo, J., Functional Silicon pcs, thesis, avail. Univ. Microfilms, Ann Arbor, Mich., Order No. 74-16, 475; *Diss. Abstr. Int. B*, 35(1), 123, 1974.
82. Douglass, S. L., Molecular Structure of and Motional Behavior in Some Silicon pcs, 1975, thesis, avail. Xerox Univ. Microfilms, Ann Arbor, Mich., Order No. 75-19,198; *Diss. Abstr. Int. B*, 36(3), 1207, 1975.
83. Sutton, L. E., Germanium and Tin pcs and Hemiporphyrazines, thesis, avail. Univ. Microfilms, Ann Arbor, Mich., Order No. 67-11, 571; *Diss. Abstr. B*, 28(3), 829, 1967.
84. Bonderman, D. P., Preparation and Characterization of Some Novel pc Compounds, thesis, Order No. 68-16, 783, avail. Univ. Microfilms, Ann Arbor, Mich., *Diss. Abstr. B*, 29(6), 1968, 1968.
85. Shorin, V. A., Borodkin, V. F., and Al'yanov, M. I., *Tr. Ivanov. Khim. Tekhnol. Inst.*, 15, 133, 1973.
86. Lokhande, H. T., *Colourage*, 17(3), 39, and 48, 1970.
87. George, P., *Ann. N.Y. Acad. Sci.*, 206, 84, 1973.

88. Hopff, H., *Chimia,* 15(1), 194, 1961.
89. Lagow, R. J. and Adcock, J. L., U.S.N.T.I.S., AD Rept. No. 780919/7GA, avail. NTIS, *Gov. Rep. Announce. (U.S.),* 74(17), 59, 1974.

Chapter 3

PURIFICATION, DETERMINATION OF PURITY, AND ANALYSIS

I. PURIFICATION AND DETERMINATION OF PURITY

Sublimation continues to be of interest to purify pc compounds. It is a simple, effective, and inexpensive procedure for small quantities of materials which are the subject of laboratory interest. The vapor pressure and temperature stability characteristics of the pcs lend themselves readily to purification by sublimation.

A device is described[1] for studying the kinetics of the vacuum sublimation of organic compounds of low volatility. Cast iron turnings may also be used to improve heat transfer in vacuum sublimation,[2,3] with sublimation at 10^{-5} to 10^{-6} torr, at 450 to 500°C for several hours. After three sublimations, mass spectrometry (MS) reveals the residual impurity to be 10^{-3} to 10^{-4}%. A device for determining temperature conditions during these sublimations is also discussed.[4] Purification of pcs is also made by successive purifications in organic solvents and sulfuric acid followed by vacuum sublimation.[5] Gas chromatography (GC) reveals the residual impurity to be 0.1% after chemical purification and to be less than 0.1% after sublimation.

Typical residual volatile impurities in the preparation of pcs are phthalonitrile, phthalimide, and phthalic anhydride. It is found by analysis by GC that pcs purified by sublimation contain less of these impurities than pcs prepared by washing in organic solvents or by precipitation from concentrated sulfuric acid.[6] Another method for purifying pcs involves treatment in boiling sodium chloride and hydrochloric acid solutions followed by sublimation at 350 to 450°C at 5×10^{-5} torr.[7]

Noncomplexed ionic copper can be removed from crude copper pc by heating copper pc containing 155 to 350 ppm copper in a mineral acid with calcium pc, complexing the copper ion, filtering and washing to neutrality, giving final ionic copper content of 20 to 37 ppm.[8]

By GC analysis, volatile impurities in pcs can be detected, such as phthalonitrile, phthalic anhydride, phthalic acid, and phthalimide. After treating crude copper pc with water and ethanol, and reprecipitating thrice from concentrated sulfuric acid, and subliming four times, these impurities can be reduced to undetectable amounts "with only 0.0009 percent of an unidentified volatile impurity being detected."[9]

Copper in copper pc dyes can be determined by atomic absorption spectrophotometry with good reliability because "previous analyses of copper chelated to the pc nucleus have required initial destruction of the organic matter."[10]

Factors involved in the quality or purity of Phthalocyanogen Blue are determined as follows. The Dean and Stark method is used to determine moisture content. Free alkali is determined by treating Phthalocyanogen Blue with neutral aqueous HCHO and titrating the free alkali with 0.1 *N* sulfuric acid.[11]

The possible use of reagents of low purity for determining the composition of complexes such as titanium pc tetrasulfonic acid is suggested.[12]

A method is described for the determination of the metal content of Cu and Ag pc complexes.[13] The Cu complex is decomposed by treatment with a hot sulfuric acid-nitric acid mixture and digested. The mixture is diluted, adjusted to pH 5, and titrated with EDTA to a PAN indicator end point. The silver complex is decomposed by treatment with a hot sulfuric acid-nitric acid mixture with perchloric acid. After dilution and adjustment of the solution to pH 9, $K_2Ni(CN)_4$ is added with an excess of EDTA. The EDTA is then backtitrated with a standard Pb^{2+} solution to a PAN end point. The method gives acceptable results on 50- to 100-mg samples.

II. USE IN ANALYSIS

The tetrasodium salt of cobalt (II) 4,4′,4″,4‴-tetrasulfo pc dihydrate is suggested as a new redox indicator in cerimetry.[14] The color change at the end point in the titration of 0.005 to 0.1 *N* Fe (II) or hydroquinone in 1 *N* sulfuric acid with $Ce(SO_4)_2$ is from turquoise blue to purple.

REFERENCES

1. Al'yanov, M. I., Khoinov, Yu. I., Borodkin, V. F., and Benderskii, V. A., *Izv. Vyssh. Ucheb. Zaved. Khim. Khim. Tekhnol.*, 14(12), 1911, 1971.
2. Al'yanov, M. I., Borodkin, V. F., Benderskii, V. A., and Khoinov, Yu. I., Russian Patent 311,937, August 19, 1971.
3. Al'yanov, M. I., Borodkin, V. F., Benderskii, V. A., and Khoinov, Yu. I., *Izv. Vyssh. Ucheb. Zaved. Khim. Khim. Tekhnol.*, 14(10), 1606, 1971.
4. Khoinov, Yu. I., Al'yanov, M. I., Borodkin, V. F., and Benderskii, V. A., *Izv. Vyssh. Ucheb. Zaved. Khim. Khim. Tekhnol.*, 15(10), 1595, 1972.
5. Khoinov, Yu. I., Fedorov, M. I., Al'yanov, M. I., Borodkin, V. F., Benderskii, V. A., and Larionov, V. R., *Izbran. Dokl. Nauch. Tekh. Konf. Ivanov. Khim. Tekh. Inst.*, p.12, 1973.
6. Al'yanov, M. I., Borodkin, V. F., Kalugin, Yu. G., Larionov, V. R., Khoinov, Yu. I., Yashin, Ya. I., and Smirnov, R. P., *Izv. Vyssh. Ucheb. Zaved. Khim. Khim. Tekhnol.*, 16(10), 1604, 1973.
7. Kiryukhin, I. A., Lobanova, K. N., Popov, Yu. A., Shaulov, Yu. Kh., and Benderskii, V. A., *Zh. Fiz. Khim.*, 50(3), 649, 1976.
8. Plankenhorn, E., German Offenlegungsschrift 2,401,088, July 24, 1975.
9. Krutoyarova, O. K., Al'yanov, M. I., Yashin, Ya. I., Borodkin, V. F., Benderskii, V. A., and Khoinov, Yu. I., *Izv. Vyssh. Ucheb. Zaved. Khim. Khim. Tekhnol.*, 14(9), 1379, 1971.
10. Blagrove, R. J., Grossman, V. B., and Gruen, L. C., *J. Soc. Dyers Colour*, 89(1), 25, 1973.
11. Egorov, N. V., Polyakov, V. G., Lifentsev, O. M., and Mel'nikov, B. N., *Tekst. Prom. (Moscow)*, 32(8), 76, 1972.
12. Lazarev, A. I. and Lazareva, V. I., *Zaved. Lab.*, 41(5), 534, 1975.
13. Cheng, K. L., *Microchem. J.*, 7(1), 29, 1963.
14. Gowda, H. S. and Achar, B. N., *Proc. Natl. Acad. Sci. India Sect. A*, 43(Parts 1 and 2), 125, 1973.

Chapter 4

CRYSTALS

I. R FORM CRYSTAL PHTHALOCYANINES

The R form of Cu pc was announced in U.S. Patent 3,051,721 on August 28, 1962,[1] in the following words: "This invention relates to (1) an unsubstituted, new, solvent stable, noncrystallizing, metal pc having a characteristic structure, as shown by X-ray diffraction and infra red absorption, herein defined as the 'R' form; and (2) one method for synthesizing said pc using high shear stresses; and (3) a continuous method of producing metal pcs not necessarily in said new form. The new 'R' form of Cu pc exists as a red shade blue pigment which has improved properties over the old red shade blue pigment regarding both strength or tinctorial power, resistance to crystallization, and flocculation resistance although the flocculation resistance is to a considerable extent a function of the solvent system." "One method of producing the new 'R' form of metal pcs involves the fusion of a mixture containing phthalonitrile, urea and a metalliferous reagent with intensive agitation under conditions of high shear at the time of reaction. A high shear appears to be essential to the production of the new 'R' form."

II. δ FORM CRYSTAL PHTHALOCYANINES

The δ form of Cu pc was announced in U.S. Patent 3,150,150[2] on September 22, 1964 and in U.S. Patent 3,160,635[3,4] on December 8, 1964, which were filed February 10, 1961 and August 9, 1960, respectively, prior to the issuance of U.S. Patent 3,051,721 "Pigmentary Cu Pc In the 'R' Form and Its Preparation". Thus, U.S. Patent 3,150,150 and U.S. Patent 3,160,635 recognized three existing forms of crystalline pcs, α, β, and γ, but not R.

In U.S. Patent 3,150,150 it is stated the α, β, and γ forms of pcs can be characterized by X-ray diffraction spectra. Also, "the novel δ-crystal form of Cu pc may be characterized by its pronounced redness of shade as compared with the previously known crystal forms, and by the fine needle-like appearance of its crystals." The δ form of Cu pc is made by treating the α form with a water-immiscible organic liquid in the presence of a surface active agent and optionally in the presence of water for at least 5 hr. Suitable solvents include benzene and cyclo-1:3-hexadiene and suitable surface active agents including sulfonated methyl oleate and sulfated caster oil. In U.S. Patent 3,160,635 the δ form is obtained in a manner similar to acid pasting if, prior to addition of the pc, a mild neutralizing agent such as urea, is added to the acid to partially neutralize it. The milling procedure requires 6 to 36 hr.

III. X FORM CRYSTAL PHTHALOCYANINES

A novel crystalline form of metal-free pc is described in U.S. Patent 3,357,989[5,6] which was published on December 12, 1967. This patent recognizes the α, β, and γ forms of metal pcs and metal-free pc, as well as the R, and two δ forms of metal pcs. In U.S. Patent 3,357,989 a new metal-free pc crystal form, the X form, is described. It is characterized as such by X-ray diffraction and IR spectra.

The X form is prepared by extended milling of the α or β forms or a mixture of α, β forms of metal-free pc. For example, "commercial Monolite Fast Blue GS [a mixture of α, β forms of metal-free pc] was solvent extracted with dichlorobenzene, then washed with acetone and dried. The pigment was then dissolved in sulfuric acid and precipitated in ice water. The precipitate was washed with methanol and dried. The pigment was then neat milled in a ball mill for 7 days, washed with dimethylformamide and then with methanol."

The X form is "especially useful as photoconductive materials in electrophotography;" it may also be used to prepare polymers and is also useful as a pigment in inks, paints, and varnishes.

A preparation[7,8] for the X form of metal-free pc, useful as a photoconductor for electrophotography, is described.

A spectroscopic characterization of the X form metal-free pc in the IR, visible, and near IR regions shows distinct differences among the α, β, and X polymorphs, and indicates that the X H_2 pc is a dimer.[9]

IV. π FORM CRYSTAL PHTHALOCYANINES

The π form of metal pcs was announced in U.S. Patent 3,708,292 on January 2, 1973.[10] It is characterized by X-ray diffraction and IR spectra. Its chief virtues are described "as a brighter appearance, greater stability to recrystallization in the presence of heat and strong solvents, and increased electrical photosensitivities."

The preparation of the π form of metal pcs involves "mixing, at a suitable reaction temperature, phthalonitrile, a metal salt and ammonia in an alkylalkanolamine solvent or 1,3-diimino-iso-indoline and a metal salt in a nonammonia saturated alkyl-alkanolamine solvent, and heating the mixture to about reflux temperature." 2-Dimethylaminoethanol is a preferred solvent; 135 to 150°C is a preferred temperature range, and total reaction time is 10 to 70 min.

The π form metal pc "may be used to prepare electrophotographic plates and to be used in electrophotographic processes."

π form Co pc and Cu pc for electrostatic copying plates are prepared.[11] π form metal-free pc is used in mixtures with binding agents as heat stable photosensitive coatings on an aluminum layer in electrocopying processes.[12]

V. α FORM CRYSTAL PHTHALOCYANINES

The IR absorption and reflection spectra in polarized light of thin films (0.5 to 0.6 μm) of Cu pc, deposited *in vacuo* on a KBr surface are characteristic of α Cu pc and indicate planar orientation of the molecules in the α Cu pc layer.[13] The spectra of the β Cu pc in KBr pellets indicate that the $\beta \rightarrow \alpha$ transformation takes place during grinding.

An attempt is made to define a numerical index of the α form of crystallinity[14] of Cu pc pigments by using X-ray diffraction analysis, because "the applicability of α Cu pc blue is strictly associated with the spectral configuration of the pigment."

α Cu pc fine particles of particle diameter 0.1 to 0.5 μm are made by seeding pc precursor reaction mixture with an aliquot of the fine α Cu pc powder.[15] For example, 23 parts 3-amino-1-imino-1H-isoindole, 7 parts $Ca(OAc)_2 \cdot H_2O$, 1 part α form powder (0.1 to 0.5 μm diameter), and 200 parts

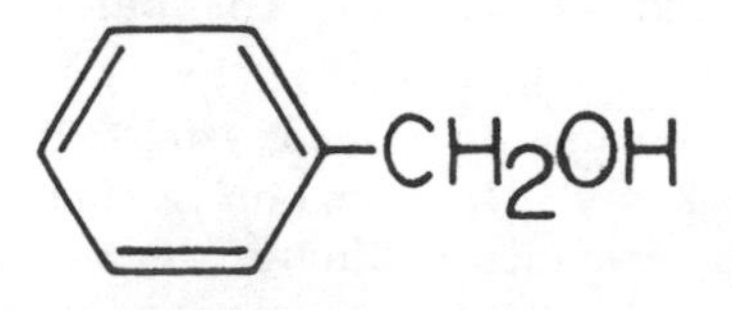

are stirred 2 hr at 120°C to give 17.6 parts α form Cu pc with particle diameter 0.1 to 0.5 μm compared with 10 to 50 μm for the similar process without seeding.

Fine α Cu pc particles are made by another method.[16] Large β Cu pc crystals are stirred with a C_{11-13}-alkyl-benzenesulfonic acid and a straight chain alkylbenzene at 100 to 110°C for 3 hr and slurried in ethanol to give fine α crystals. Similar treatment of large α crystals gives fine α crystals.

VI. β FORM CRYSTAL PHTHALOCYANINES

Single crystals of β Cu pc are grown in an inert gas after purification by sublimation at 10^{-2} torr pressure and evaporation at 500°C.[17]

β Cu pc crystals are prepared by heating phthalonitrile with a metal salt and alkali in a hydrophilic solvent and treating with HCl, H_3PO_4, or HOAc. For example, a mixture of 32 parts phthalonitrile, CuCl 6.2, and NaOH 5.0 in methanol 120 parts is stirred 25 hr at 30°C, treated with 40 parts HCl, and refluxed 1 hr. The solid is washed in hot 0.5% aqueous NaOH to give 31.5 parts β Cu pc.[18]

Measurements on oriented β Cu pc crystals in films show that the transmission color is reddish blue in the direction of the acicular axis and greenish blue at right angles to it.[19] As the asymmetry increases, the transmission color along the needle axis exerts a greater effect on color, even if the particles are statistically unoriented.

Density determinations are made on single crystal sections of β form pcs prepared by vacuum sublimation[20] in a density gradient column consisting of mixtures of 1,2-dibromoethane and carbon tetrachloride with calibrated glass markers. The method, which can detect differences in densities on the order of 1 in 10^5, shows that the various β metal pc crystals have densities as much as ±5% lower or higher than the maximum and many crystals contain minute voids, but others free of voids have densities 1 to 3% less than the maximum.

The crystal structure of β Cu pc is determined from three-dimensional X-ray diffraction data, with standard deviation of bond lengths of 0.005 Å.[21] The structure is isomorphous with those of the metal-free and Ni pcs.

Purification of Cu pc and growth of β Cu pc crystals in a quartz tube under reduced pressure are described.[22] Growing crystals in an inert gas from the vapor phase appears to be more successful than growing crystals from a solvent such as quinoline, α-chloronaphthalene, or sulfuric acid. "Crystals grown from organic solvents are far from being perfect. Evidently it is very difficult to control the nucleation. Our investigations showed that traces of the solvents are built in as part of the complexes, which is also evident in the [large variations in electrical properties]."

VII. ε FORM OF CRYSTAL PHTHALOCYANINES

The ε form of metal pcs was announced in German Patent 1,181,248 on November 12, 1964. The R form, announced on August 28, 1962 in U.S. Patent 3,051,721, is considered to be the same as the ε form. According to British Patent 1,411,880 of October 29, 1975, "the 'R' form is identical with the ε modification." Preparation of ε form Cu pc pigments are described.[23,24]

A process is developed to produce Cu pc pigment of the pure ε or virtually pure ε form[25,26] that are more reddish than the α, γ, or δ forms and are fairly resistant to conversion into β Cu pc.

VIII. γ FORM CRYSTAL PHTHALOCYANINES

X-ray powder diffraction patterns, IR absorption spectra, and ESR spectra show the existence of at least two pc crystal forms, α and β.[27] Experiment indicates that, with respect to these data, the α and γ forms differ only in particle size. Thus, the

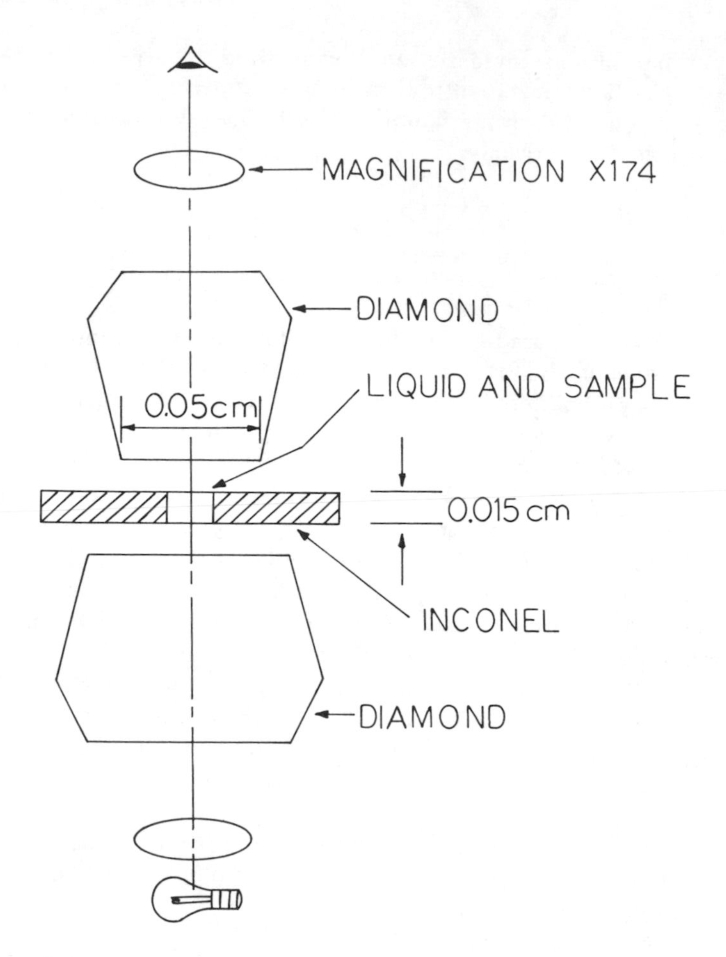

FIGURE 1. Schematic diagram for the generation of pressure and observation of samples within the diamond high-pressure cell.

debate continues whether, with respect to α, β, and γ pcs, there exists only α and β pcs.

γ Cu pc is prepared by heating α or β Cu pc with 30 to 44% sulfuric acid in butyl or methyl cellosolve or tetrahydrofuran (THF). For example, 10 g α or β Cu pc in 100 g 98% sulfuric acid is diluted with 227 g tetrahydrofuran, heated 2 hr at 60°C and poured into water to give γ Cu pc.[28]

γ Cu pc or γ Ni pc are prepared by heating phthalonitrile with a Cu or Ni salt and alkali in a hydrophilic solvent, optionally with a chelating agent, and treated with sulfuric acid. Sulfuric acid concentrations of 0.5 to 42% or 58 to 75%, based on solvent, gives the β phase whereas 53 to 57% concentration gives the γ phase.[29]

The excess electron and excess hole energy bands of the γ polymorph of Pt pc are calculated.[30] The γ Pt pc molecule is treated as square planar with D_4 symmetry. The energy bands obtained are highly anisotropic.

IX. HIGH PRESSURE PHTHALOCYANINE CRYSTAL FORMS

Under high pressures macroscopic single crystals of metal-free pc can be transformed to new crystal structures.[31] A diagram of the diamond high pressure cell is shown in Figure 1.

Crystals were placed individually into the pressure transmitting liquid, which was heptane or water. "The most dramatic effect of high pressure on the pc crystals is the transformation at 7.1 ± 0.7 kbar. The 001 plane appeared to shorten by twenty percent and to thicken by twenty percent in a transformation that occurred so quickly that the whole crystal appeared to change at once.'' The transformation is reversible.

X. EFFECT OF SOLVENT ON PHTHALOCYANINE CRYSTAL STRUCTURE

The properties of the phenomenon of crystal transformation of α pcs to β pcs in organic solvents is a subject of recurring interest, impelled by the fact that the most desirable pigment in terms of color properties in a given vehicle may not be the most crystallization resistant.

The process of the $\alpha \rightarrow \beta$ transformation of Cu pc in various organic media is studied by electron microscopy and X-ray diffractometry.[32] The preliminary growth of metastable α crystals precedes the actual transition into stable β crystals.

The solubility parameter of any given solvent in a paint vehicle for Cu pc is the determining factor in the extent of the transition from the α to β form.[33] A zone of transformation is defined experimentally at 80 and 120°C for 18 solvents including glycols, alcohols, amines, water, dioxane, dimethylsulfoxide, and benzene derivatives.

The $\alpha \rightarrow \beta$ transition of Cu pc in organic media is studied by visible spectroscopy, X-ray diffractometry, electron microscopy, and BET determination of the specific surface areas.[34] The transition can be described in terms of the classical theories of nucleation and crystal growth. "Stage I, during which Ostwald ripening of the α phase and nucleation of β crystals takes place as competing reactions, is followed by Stage II which is characterized by the disappearance of the α phase and the production of β crystals on a large scale. Final crystallization according to a second order equation and ripening of the β crystal define Stage III of the transition process.''

From an experimental study of the $\alpha \rightarrow \beta$ transition of metal pcs in organic solvents, by X-ray diffraction and electron diffraction, it is indicated that the behavior of the metal pcs differs even in the same solvent and depends on the central ion.[35] "Selected area electron diffraction also indicates that the crystal structure of the β form, so far unknown because of the impossibility of obtaining single crystals for X-ray analysis, is actually isomorphic with that of Pt pc, whose structure had been determined by Robertson and Woodward and refined by Brown.''*

Treatment of β Cu pc with boiling solvents causes the β crystals to grow; treatment of α Cu pc with solvents causes the α crystals to grow before the phase transition occurs to the β form and the β crystals grow.[36]

X-ray diffraction and electron micrograph studies also reveal that when α Cu pc is mixed with such metal oxides as alumina, magnesia, and zinc oxide, and suspended in an organic solvent, the $\alpha \rightarrow \beta$ transition is accelerated. "Particularly, alumina has a remarkable effect on the acceleration of β Cu pc formation.''[37]

The influence of organic solvent media on the crystal growth and phase change of pcs including procedures to guide the growth of particle size is studied by X-ray analysis and reflection spectra.[38]

Solvent stable, floculation resistant α, β, and γ modifications of Cu and Ni pc compositions are prepared by combining phenyl-, chloro-, and sulfo-substituted tetraazaporphins with the pc.[39] Thus, a mixture of 50 parts phthalic anhydride and 8 parts α, α'-diphenylmaleic anhydride is treated with NH_2CONH_2, NH_4Cl, $CuSO_4$, and ammonium molybdate in nitrobenzene to give a crude Cu pc (*I*) composition which, when milled with Na_2SO_4 in the presence of toluene, gives a solvent stable β modification.

* From Uyeda, N., *Progr. Org. Coatings,* 2(2), 131, 1973. With permission.

Treatment of (*I*) with sulfuric acid gives an α modification which shows no crystallization or color change when refluxed with toluene.

XI. EFFECT OF TEMPERATURE ON PHTHALOCYANINE CRYSTAL STRUCTURE

The kinetics of the thermal α → β polymorphic transition in metal-free pc is studied.[40] First-order rate constants are determined at 260 to 343°C for the α → β thermal conversion by observing the growth in the IR absorption bands at 724 and 782 cm^{-1}. The X form is not an intermediate in the conversion.

The α → β transformation of vapor deposited Cu pc films is studied by IR spectroscopy, X-ray diffraction, and electron microscopy. The absorbance at 783 cm^{-1} is used to measure the degree of transformation which begins at 185 to 216°C.[41]

Also, a high temperature optical cell which allows simultaneous monitoring of thermally induced transformations by using differential thermal analysis is designed, constructed, and used to study the α → β transformation in evaporated thin films and in bulk samples. The rate constants are about three times larger for the bulk sample than for the evaporated films.[42]

Transmission spectra of sublimed thin films of the α forms of pc, Cu pc, Ni pc, Co pc, and Zn pc in the 0.5 to 0.9 μm region changes abruptly as a result of α → β transformation induced by heat treatment above 300°C.[43]

Cu pc forms thin films of the α form with epitaxial orientation when vacuum deposited onto a cleaved face of muscovite.[44] The thermal transformation of these films into the β form is observed by electron microscopy and electron diffraction.

Thin films of pc crystals formed on muscovite surfaces by vacuum condensation are oriented in one direction if the substrate is preheated at 300°C for 1 hr and maintained at 150°C, but consist of crystals of different orientations if the substrate is preheated at 400°C for 1 hr and then maintained at 150°C. The monoclinic unit cell constants of Pt pc, vacuum condensed on muscovite are a = 23.18 Å, b = 3.818 Å, c = 23.84 Å, and β = 91.9°C. Films of Cu, Co, Fe, Ni, and H_2 pcs have an orientation similar to that of the Pt derivative. They are all isomorphous and occur in the metastable form of their dimorphs. In both the Cu and Pt pc films the crystals appear to transform from the monoclinic to the triclinic structure as they reorient on the muscovite surface.[45]

Thermal treatment or neutron irradiation can be used to convert α indium pc to β indium pc.[46]

Thermally induced phase transitions in VO pc are observed by differential scanning calorimetry and the phases are characterized by X-ray, IR, visible, and UV spectroscopy.[47] Three phases are prepared by nonequilibrium vapor quenching, equilibrium crystal growth, and melt quenching. Dimeric structures are suggested for all three phases, as shown in Figure 2.

Lower temperature phase changes are observed also.[48] "We observed the temperature dependence of the dielectric constant, the nmr spin lattice relaxation, the lattice constants, and the ac conductivity in the range from −140°C to +150°C. Fe pc and Co pc showed no anomaly in this range, whereas all the other samples undergo a phase transition. The transition temperature of Cu pc (10°C) and of the solid solution 9 H_2 pc −1 Fe pc (18°C) is higher than that of D_2 pc (−12°C) but lower than that in H_2 pc (5°C)."*

* From Dudreva, B. and Grande, A. S., *Ferroelectrics*, 8(1—2), 407, 1974. With permission.

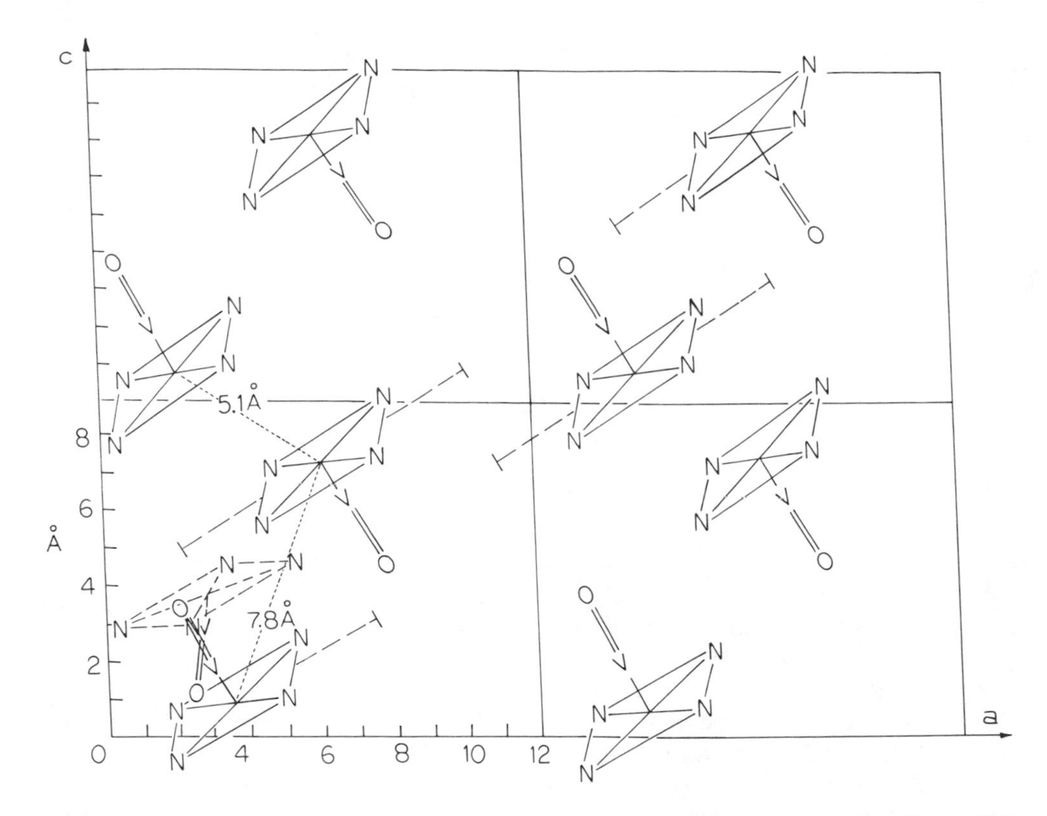

FIGURE 2. Structural model for Phase II VOPc showing also the ac plane disorder produced by the 180° rotation about the [101] twin axis (dotted lines). Dashed lines show the actual diameter of the phthalocyanine molecule. Central nitrogen (N) square drawn almost twice oversize.

XII. POLYMORPHISM OF VARIOUS PHTHALOCYANINES

The crystal structure and polymorphism of pcs are reviewed.[49-52]

IR absorption spectra are used to indicate the existence and distinction between α and β chloroindium pc.[46]

Polymorphism in vanadyl pc is observed for three phases. Phase transitions among them are studied by preparing them by vapor quenching, crystal growth, and melt quenching, and by observing them by differential scanning calorimetry, X-ray scattering, and optical absorption spectroscopy.[47]

XIII. POLYMORPHISM OF Cu Pc AND CRYSTALLIZATION RESISTANCE

Structural relationships among five forms of Cu pc — α, β, γ, δ, ε — are discussed in terms of measurements of dichroism of electronic spectra.[53] Relationships exist between α and γ and between δ and ε Cu pc. "Correlating with those relationships is the relative gradation of lattice energies for the individual modifications, which is obtained from the solubility in benzene, and the heat of solution. According to this, the stability increases in the sequence $\alpha \approx \gamma < \delta < \varepsilon < \beta$."

The concern to prepare crystallization resistant pcs is expressed by practical methods to provide mixtures among compounds of the formula pc $(-CH_2XA)_n$ where X = oxygen or sulfur atom, A = aryl moiety, and n = 1 to 8, with Cu pc in the δ, ε, γ form.[54] Other crystallization resisting adjuvants are

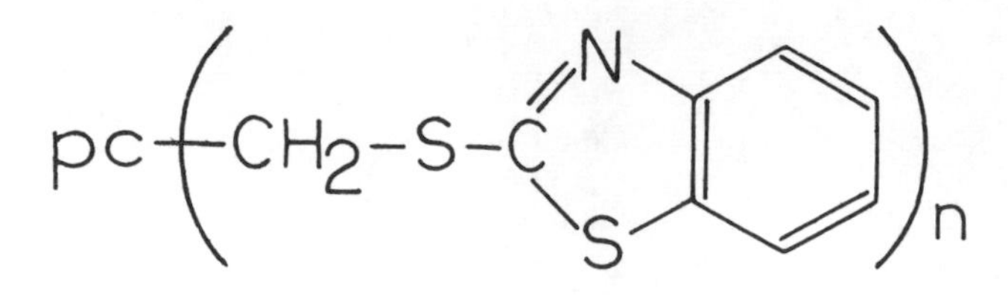

where n = 1 to 8[55] and

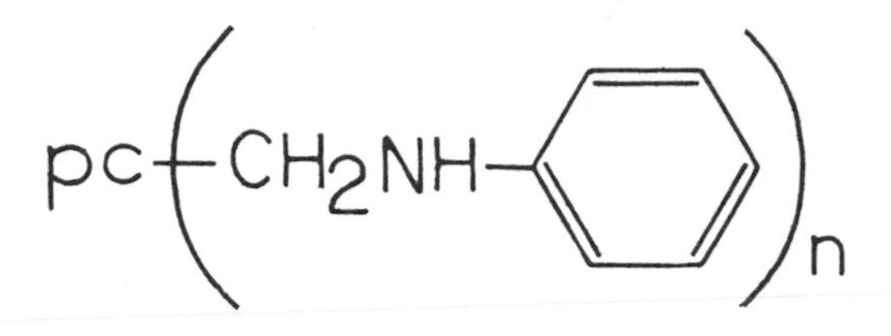

where n = 1 to 4.[56]

Another method to obtain crystallization resistant α Cu pc[57] is to prepare the Cu pc as a mixture starting with 1.5 to 10 mol % of a tert-butyl, isopropyl, or cyclohexyl substituted phthalic acid derivative to 98.5 to 90 mol % of a Cl, SO_3H, Br, cyclohexyl, or methyl substituted phthalic derivative.

Crystallization of α Cu pc in aromatic hydrocarbons is also inhibited by incorporating 1 to 20% Cu tris(arylaminomethyl) pcs.[58]

From the reflection and absorption IR spectra scanned in polarized light, the nature of the spectral characteristics of the sublimed α pc layer (0.5 to 0.6 μm diameter particle) on polished KBr plates is studied at 700 to 800 cm^{-1}.[59]

The influence of the crystal structure on absorption spectra, photo-EMF, and photoconductivity in metal and metal-free pcs is studied.[60] The α and β forms provide different absorption and photoeffect curves in the visible region of the spectrum.

The structures of the modifications of Pt pcs and Cu pcs and some of their chloro derivatives reveal that the unstable modifications of Cu pc are isomorphous with the stable modification of 4-Cl Cu pc and Pt pc.[61]

The structure and polymorphous transformation of Cu pc are reviewed with 34 references.[62]

The configuration and dimensions of well-crystallized organic pigments including Cu pc are studied.[63] The equivalent diameter of aggregates of well-crystallized, approximately isometric to parallelepipedal Cu pc and polychlorinated pc, determined from adsorption measurements ranges from 0.03 to 0.3 μm. The dimensions of crystals are shown by electron microscopy or line broadening in X-ray diffraction to be smaller by a factor of ½ and ¼ than those of crystallites and agglomerates, respectively.

XIV. CRYSTAL STRUCTURE OF VARIOUS PHTHALOCYANINES AND THEIR DERIVATIVES

Crystals of the 2:3 complex of Zn pc and normal hexylamine are found by X-ray analysis using Weissenberg photographs with Ni filtered Cu Kα radiation to be monoclinic, space group $P2_1/c$, with cell dimensions a = 12.40Å, b = 15.76Å, c = 20.05Å, β = 93.2°C, with four pc and six hexylamine molecules per cell. The pc molecule is not planar; the central zinc ion is displaced from the plane by 0.48 Å toward the nitrogen atoms of the amine. The zinc ion is of a square pyrimidal five coordination. There are two bonding states for the amine in the complex crystal. One is weakly bonded and can easily be released from the lattice and the other is strongly bonded and coordinated directly to the zinc ion.[64]

The structure of a 4:1 amine complex between Fe^{2+} pc and 4-methylpyridine is determined[65] by X-ray crystallography. The crystals are orthorhombic, with space group Pbca, a = 25.19 Å, b = 17.86 Å, c = 10.30 Å, and d = 1.339 g/cm³. It is indicated that two of the four 4-methylpyridine molecules are coordinated to an octahedral Fe^{2+} and the other two 4-methylpyridine molecules are occluded.

The structure of bis (4-methylpyridine) Co pc is also determined by X-ray analysis.[66] The crystals belong to the space group Pbca with a = 10.40 Å, b = 25.07 Å, c = 17.99 Å, and d (calculated) = 1.073 g/cm³ for Z = 4. The four molecules in the unit cell are centrosymmetric.

Crystal data are obtained for Pb pc prepared by the reaction of PbO and phthalonitrile in refluxing α-chloronaphthalene. "Dark purple needle shaped crystals were grown by sublimation at about 250°C in flowing nitrogen at 5 to 7 torr. The crystal data were obtained from oscillation, Weissenberg, and precession photographs about the long axis of the crystal, and powder diagrams calibrated with silver powder." The crystals are monoclinic, $P2_1/b$, a = 25.48 Å, b = 25.48 Å, c = 3.73 Å, γ = 90°C, with density = 1.98 g/cm³. "The non-planar molecule of lead pc stacks linearly, to form a molecular column parallel to the c axis."*

Thin films of pcs on muscovite surfaces placed there by vacuum condensation are single-directionally oriented if the substrate is preheated at 300°C for 1 hr and kept at 150°C, but consist of crystals of different orientation if the substrate is preheated at 400°C for 1 hr and then kept at 150°C. Films of Cu, Co, Fe, Ni, Pt, and H_2 pcs have similar orientations.[68]

The Si_{pc} − O − Si_{CH3} bond angle common to the shift reagent compounds $(CH_3)_3SiO(pcSiO)_xSi(CH_3)_3$ (x = 1 to 5) is determined by an induced shift technique and the structure of $pcSi[OSi(CH_3)_3]_2$ is determined by X-ray crystallography.[69]

Figure 3 shows the packing of molecules in the unit cell of $pcSi[OSi(CH_3)_3]_2$.

"The data on the nearest neighbor arrangement around the silicon in $pcSi[OSi(CH_3)_3]_2$ and in other silicon pcs provide direct proof of the approximately octahedral coordination of the silicon in these compounds. It is accordingly now incontestably clear that the largest known group of octahedral silicon compounds is the pc group. The number of well characterized silicon pcs in the literature is over 40."** The essential planarity of the rings in Figure 3 indicate the silicon atom is small enough to fit into the pc ring without distorting it. Also, "similar to that found earlier for other pcs is the nonequivalence of the six interior angles of the benzo rings. Probably the nonequivalence arises as a result of strain on the rings by their fusion with the pyrrole rings."**

Mixed pc crystals are also studied. Cu pc/VO pc mixed crystal films contain mixed crystal grains that are not large enough to be detected by X-ray methods.[70] "By infrared spectroscopy and scanning electron microscopy or after recrystallization by texture goniometer investigations the orientations of microcrystals on the substrate was obtained."

A detailed discussion of the crystal structure and absorption spectra of pcs, especially H_2 pc and Cu pc is given.[71]

The crystal structure of stannic pc: an eight coordinated tin complex, $Sn(pc)_2$ is determined by three-dimensional X-ray analysis.[72,73] The orthorhombic cell dimensions and unit cell content are a = 10.547 Å, b = 50.743 Å, and c = 8.9046 Å, and Z =

* From Ukei, K., *Acta Crystallogr. Sect. B,* 29(10), 2290, 1973. With permission.

** From Mooney, J. R., Choy, C. K., Knox, K., and Kenney, M. E., *J. Am. Chem. Soc.,* 97(11), 3033, 1975. With permission.

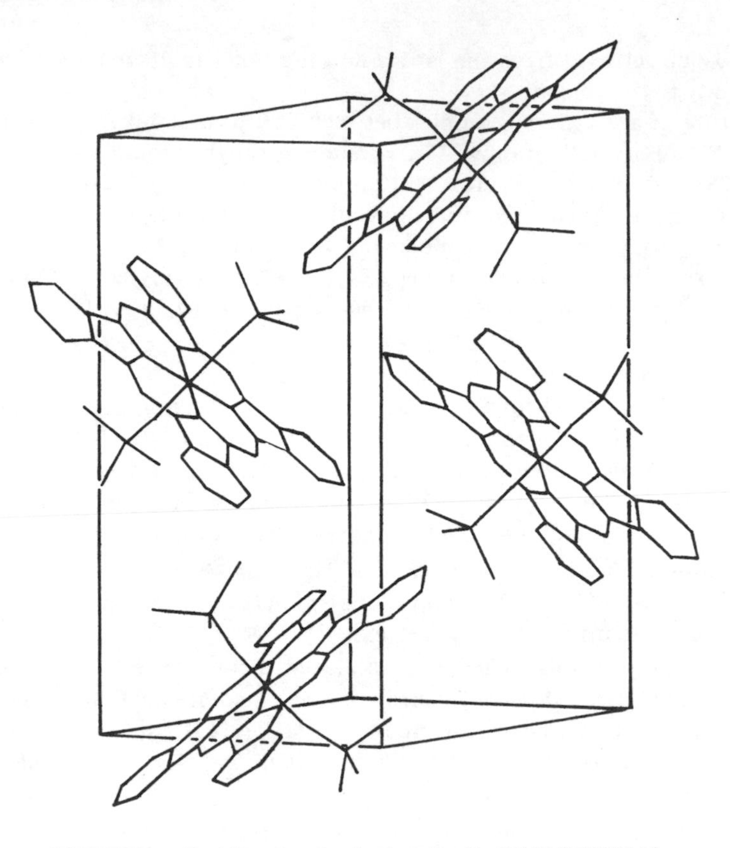

FIGURE 3. Packing of molecules in unit cell of $PcSi[OSi(CH_3)_3]_2$.

4. The density is 1.60 g/cm³. The cores of the pc moieties are separated from each other by 2.70 Å and are rotated 42°C with respect to each other.

Crystals of dichloro pc tin (IV) are monoclinic space group A2/a, with a = 21.104 Å, b = 11.060 Å, c = 11.392 Å, β = 96.04°C, and Z = 4. "The phthalocyanine ring of the dichlorophthalocyaniniatotin (IV) molecule is shown to be substantially 'crumpled' — a stepped deformation — apparently owing to an oversize tin atom."[74]

The three-dimensional X-ray diffraction study of Pt pc reveals there are two closely similar crystalline structures, the only difference between the structures being the manner of packing the molecules in space.[75]

The relations between the structures of the modifications of Pt and Cu pcs and of some chlorine derivatives are studied.[76]

The configuration of atomic planes bordering cracks is studied experimentally using a Cu pc platelet containing a wedge-shaped crack. A formula is derived to show the interactions involved.[77]

An X-ray crystal structure determination of sublimed U pc_2 yields crystals with space group C2/c and with a = 18.74 Å, b = 18.73 Å, c = 15.61 Å, β = 113.6°, and d = 1.66 g/cm³.[78]

Crystals of U pc_2 and Th pc_2 are found to belong to space group B2/b with Z = 4. U pc_2 has a = 18.73 Å, b = 15.60 Å, c = 18.70 Å, γ = 113.4°, and d = 1.70 g/cm³. Th pc_2 has a = 18.74 Å, b = 15.79 Å, c = 18.73 Å, γ = 113° 20′, and d = 1.67 g/cm³. "These data show that the replacement of U^{4+} by the larger Th^{4+} ion causes an increase of only the b unit cell parameter."[79]

Transfer integrals, including exchange, are calculated for a H_2 pc single crystal. "Only one transfer integral, between the reference molecule and its nearest neighbor along the b axis is important, all of the others being smaller by two or more orders of magnitude.''[80]

A neutron diffraction study of pc shows that the two inner hydrogen atoms lie in the plane of the great ring.[81] These two hydrogen atoms are disordered so that on the average 0.5 of a hydrogen atom is associated with each pyrrole nitrogen atom. Crystal data indicate a molecular weight of 514, monoclinic structure, a = 19.98 Å, b = 4.73 Å, c = 14.80 Å, β = 122.0°, Z = 2, and space group = $P2_1/a$ - C^5_{2h}. The neutron wavelength is 1.237 Å. All atoms of the molecule are coplanar.

In a research entitled ''Mechanism of Energy Degradation in the Region of Plasma Excitations of Molecular Crystals'',[82] the spectra are studied for the characteristic losses in the reflection of electrons having an initial energy E_p = 100—450 eV from the surface of polycrystalline tetracene and metal free pc films. The plasmon destruction mechanism is used to interpret the mechanism of energy transfer to the material.

A discussion of pigments in terms of their crystalline properties considers the shape, size, and lattice structure of pigment particles.[83] The crystal properties are determined by electron micrographs, X-ray diffraction, and adsorption measurements. Modifications of the crystal structure and particle size produced by crystallization, grinding, and dispersion and the resulting changes in tinctorial properties are considered. Results in terms of Cu pc and other organic pigments are discussed.

The crystal structure of Pb pc is determined to be monoclinic, $P2_1/b$, a = 25.48(8), b = 25.48(8), c = 3.73(1) Å, γ = 90°C, $PbC_{32}N_8H_{16}$, Z = 4, density = 1.98 g/cm^3.[84] The nonplanar molecule of Pb pc stacks linearly to form a molecular column parallel to the c axis.

The crystal structure study of pc by X-ray analysis is the subject of a thesis.[85]

In a research entitled ''Electron Diffraction Problem in pc and its Metal Derivatives'', values of Fourier potentials of the indexes (00n) and (2n, 0, $\bar{n}$) (n = 1,2,3,4) are calculated for metal-free, Ni, Pt, and Cu pc crystals.[86] An electron diffraction theory is discussed for (00n) reflections of Pt pc by assuming a harmonic potential.

REFERENCES

1. Pfeiffer, F. L. (to American Cyanamid Company), U.S. Patent 3,051,721, August 28, 1962.
2. Brand, B. P. (to Imperial Chemical Industries Ltd.), U.S. Patent 3,150,150, September 22, 1964.
3. Knudsen, B. I. and Rolskov, H. S. (to Kemisk Vaerk Koge A/S), U.S. Patent 3,160,635, December 8, 1964.
4. Kemisk Vaerk Koge A/S, British Patent 981,364, January 27, 1965.
5. Byrne, J. F. and Kurz, P. F. (to Xerox Corporation), U.S. Patent 3,357,989, December 12, 1967.
6. Byrne, J. F. and Kurz, P. F. (to Xerox Corporation), U.S. Reissue 27,117 of U.S. Patent 3,357,989, April 20, 1971.
7. Brach, P. J. and Six, H. A. (to Xerox Corporation), U.S. Patent 3,657,272, April 18, 1972.
8. Xerox Corporation, British Patent 1,277,924, June 14, 1972.
9. Sharp, J. H. and Lardon, M. A., *J. Phys. Chem.*, 72(9), 3230, 1968.
10. Brach, P. J. and Six, H. A. (to Xerox Corporation), U.S. Patent 3,708,292, January 2, 1973.
11. Brach, P. J. and Six, H. A. (to Xerox Corporation), German Offenlegungsschrift 2,218,767, December 7, 1972.
12. Brach, P. J. and Lardon, M. A. (to Xerox Corporation), German Offenlegungsschrift 2,218,788, December 7, 1972.
13. Ogorodnik, K. Z., *Opt. Spektrosk.*, 37(3), 600, 1974.
14. Venchiarutti, D., *Pitture Vernici*, 48(5), 185, 1972.
15. Ninomiya, M., Komai, A., and Shirane, N., Japanese Kokai 74,53,624, May 24, 1974.
16. Suzuki, S. and Arai, Y., Japanese Kokai 72,22,429, October 7, 1972.
17. Hamann, C., *Krist. Tech.*, 6(4), 491, 1971.

18. Abe, Y. and Hosoda, T., Japanese Kokai, 71,43,636, December 24, 1971.
19. Hauser, P., Horn, D., and Sappok, R., *FATIPEC Congr.*, 12, 191, 1974.
20. Fielding, P. E. and Stephenson, N. C., *Aust. J. Chem.*, 18(10), 1691, 1965.
21. Brown, C. J., *J. Chem. Soc. A*, 10, 2488, 1968.
22. Hamann, C., *Krist. Tech.*, 6(4), 491, 1971.
23. Miyatake, M., Tamura, S., and Ishizuka, S. (to Toyo Ink Manufacturing Company, Ltd.), Japanese Kokai 73,76,925, October 16, 1973.
24. Kumano, I., Miyatake, M., Tamura, S., and Ishizuka, S. (to Toyo Ink Manufacturing Company, Ltd.), Japanese Kokai 74,59,136, June 8, 1974.
25. Badische Anilin- & Soda-Fabrik AG, British Patent 1,411,880, October 29, 1975.
26. Honigmann, B. and Kranz, J. (to Badische Anilin- & Soda-Fabrik AG), German Offenlegungsschrift 2,210,072, September 6, 1973.
27. Assour, J. M., *J. Phys. Chem.*, 69(7), 2295, 1965.
28. Arai, Y. and Suzuki, S., Japanese Kokai 72,01,026, January 12, 1972.
29. Abe, Y. and Hosoda, T., Japanese Kokai 71,43,637, December 24, 1971.
30. Mathur, S. C. and Singh, J., *Chem. Phys.*, 1(5), 476, 1973.
31. Kirk, R. S., No. 301, National Bureau of Standards, Washington, D.C., 1969, 389.
32. Suito, E. and Uyeda, N., *Kolloid-Z. Z. Polymere*, 193(2), 97, 1963.
33. Condorelli, G., *Pitture Vernici*, 49(7), 259, 1973.
34. Honigmann, B. and Horn, D., *Sci. Monogr.*, 38, 283, 1973.
35. Uyeda, N., *Prog. Org. Coatings*, 2(2), 131, 1973.
36. Aristov, B. G., Davydov, V. Ya., Ezhkova, Z. I., Konysheva, L. I., Silina, T. V., and Feizulova, R. G., *FATIPEC Congr.*, 13, 112, 1976.
37. Kawashima, N., Suzuki, T., and Meguro, K., *Bull. Chem. Soc. Jpn.*, 49(8), 2029, 1976.
38. Aristov, B. G., Davydov, V. Ya., Ezhkova, Z. I., Konysheva, L. I., Silina, T. V., and Feizulova, R. G., *FATIPEC Congr.*, 13, 112, 1976.
39. Leister, H. and Wolf, K. (to Farbenfabriken Bayer AG), German Offenlegungsschrift 2,104,200, August 17, 1972.
40. Sharp, J. H. and Miller, R. L., *J. Phys. Chem.*, 72(9), 3335, 1968.
41. Hamann, C. and Wagner, H., *Z. Anorg. Allg. Chem.*, 373(1), 18, 1970.
42. Griffiths, C. H. and Walker, M. S., *Rev. Sci. Instrum.*, 41(9), 1313, 1970.
43. Lucia, E. A. and Verderame, F. D., *J. Chem. Phys.*, 48(6), 2674, 1968.
44. Ashida, M., Uyeda, N., and Suito, E., *J. Cryst. Growth*, 8(1), 45, 1971.
45. Kirk, R. S., *Mol. Cryst.*, 5(2), 211, 1968.
46. Muehl, P., *Krist. Tech.*, 2(3), 431, 1967.
47. Griffiths, C. H., Walker, M. S., and Goldstein, P., *Mol. Cryst. Liq. Cryst.*, 33(1-2), 149, 1976.
48. Dudreva, B. and Grande, S., *Ferroelectrics*, 8(1-2), 407, 1974.
49. Suito, E., Uyeda, N., and Ashida, M., *Senryo Yakuhin*, 12(2), 41, 1967.
50. Tomita, M., *Senshoku Kogyo*, 21(12), 725, 1973.
51. Suito, E., *Kotai Butsuri, B.*, 1(3), 151, 1976.
52. Suito, E., *J. Electronmicrosc.*, 18(4), 341, 1969.
53. Horn, D. and Honigmann, B., *FATIPEC Congr.*, 12, 181, 1974.
54. Kienzle, J. A. P., Huille, M. E. A., and Cabut, L. A., German Offenlegungsschrift 2,424,531, December 12, 1974.
55. Kienzle, J. A. P., Huille, M. E. A., and Cabut, L. A., German Offenlegungsschrift 2,424,676, December 5, 1974.
56. Kienzle, J. A. P., Huille, M. E. A., and Cabut, L. A., German Offenlegungsschrift 2,424,675, December 12, 1974.
57. Leister, H., German Offenlegungsschrift 2,256,485, May 22, 1974.
58. Kienzle, J. A. P., Huille, M. E. A., and Cabut, L. A., German Offenlegungsschrift 2,157,555, May 25, 1972.
59. Ogorodnik, K. Z., *Opt. Spektrosk.*, 39(2), 396, 1975.
60. Myl'nikov, V. S. and Putseiko, E. K., *Fiz. Tverd. Tela*, 4, 772, 1962.
61. Honigmann, B., Lenne, H. U., and Schroedel, R., *100 (Hundert) Jahre BASF, Aus Forsch.*, 223, 1965.
62. Silina, T. V. and Aristov, B. G., *Zh. Vses. Khim. Obshchest.*, 19(1), 77, 1974.
63. Honigmann, B., *Farbe Lack*, 82(9), 815, 1976.
64. Kobayashi, T., Ashida, T., Uyeda, N., Suito, E., and Kakudo, M., *Bull. Chem. Soc. Jpn.*, 44(8), 2095, 1971.
65. Kobayashi, T., Kurokawa, F., Ashida, T., Uyeda, N., and Suito, E., *J. Chem. Soc. D*, 24, 1631, 1971.
66. Cariati, F., Morazzoni, F., and Zocchi, M., *Inorg. Chim. Acta*, 14(2), L31, 1975.

67. Ukei, K., *Acta Crystallogr. Sect. B,* 29(10), 2290, 1973.
68. Achida, M., Uyeda, N., and Suito, E., *Bull. Chem. Soc. Jpn.,* 39(12), 2616, 1966.
69. Mooney, J. R., Choy, C. K., Knox, K., and Kenney, M. E., *J. Am. Chem. Soc.,* 97(11), 3033, 1975.
70. Starke, M., Wagner, H., and Hamann, C., *Krist. Tech.,* 7(12), 1319, 1972.
71. Witkiewicz, Z. and Dabrowski, R., *Wiad. Chem.,* 27(3), 141, 1973.
72. Broberg, D. E., The Crystal Structure of Stannic Pc: An Eight Coordinated Tin Complex, thesis, Order No. 68-906, avail. Univ. Microfilms, Ann Arbor, Mich., *Diss. Abstr. B,* 28(8), 3204, 1968.
73. Bennett, W. E., Broberg, D. E., and Baenziger, N. C., *Inorg. Chem.,* 12(4), 930, 1973.
74. Rogers, D. and Osborn, R. S., *J. Chem. Soc. D,* 15, 840, 1971.
75. Brown, C. J., *J. Chem. Soc. A,* 10, 2494, 1968.
76. Honigmann, B., Lenne, H. U., and Schroedel, R., *Z. Krist.,* 122, 185, 1965.
77. Blekherman, M. Kh. and Natsvlishvili, G. I., *Kristallografiya,* 14(2), 351, 1969.
78. Gieren, A. and Hoppe, W., *J. Chem. Soc. D,* 8, 413, 1971.
79. Kirin, I. S., Kolyadin, A. B., and Lychev, A. A., *Zh. Neorg. Khim.,* 18(8), 2295, 1973; *Chem. Abstr.,* 79, 150455x, 1973.
80. Devaux, P. and Delacote, G., *Chem. Phys. Lett.,* 2(5), 337, 1968.
81. Hoskins, B. F., Mason, S. A., and White, J. C. B., *J. Chem. Soc. D,* 10, 554, 1969.
82. Bubnov, L. Ya. and Frankevich, E. L., *Khim. Vys. Energ.,* 10(4), 377, 1976.
83. Honigmann, B., *J. Paint Technol.,* 38(493), 77, 1966.
84. Ukei, K., *Acta Crystallogr. Sect. B,* 29(Part 10), 2290, 1973.
85. Bissell, E. C., X-ray Structural Studies of Hemiporphyrazine and pc, thesis, avail. Univ. Microfilms, Ann Arbor, Mich., Order No. 70-25,849; *Diss. Abstr. Int. B,* 31(6), 3295, 1970.
86. Miyake, S., Fujiwara, K., and Fujimoto, F., *Cryst. Crystal Perfect., Proc. Symp. Madras,* 1963, 259.

Chapter 5

REACTIONS

I. INTRODUCTION

The reaction chemistry of pcs is virtually solution chemistry, in aqueous or nonaqueous media. Regardless of the medium and despite the deeply penetrating opacifying quality of the pc solutes, one author has observed, "Today pcs have been prepared from nearly all metallic and semimetallic elements, but owing to their extreme low solubility, the development of their chemistry is not easy."[1]

II. ACID-BASE REACTIONS

Pcs are amphoteric. For example, pc may lose the two inner imino protons to form the anion pc^{-2}, "while possible protonation sites are the inner basic nitrogen atoms and the four bridging nitrogen atoms.

"Although metal pcs are formally salts of the parent tetradentate ligand, nothing is known about the acidity of pc itself. The possible protonation of the inner nitrogen atoms is of particular interest, since protonation of these atoms in some porphyrins is known to produce dramatic stereochemical changes, presumably as a result of steric crowding or electrostatic repulsion. We have investigated reactions of pc with some strong acids and bases using conductometric and spectrophotometric techniques. . . In the presence of strong bases, solutions of pc display the previously reported anomalous spectrum similar to that of a metal pc. Addition of acid to this solution produces the normal spectrum, and these changes are attributed to reversible deprotonation of the central imino groups. Electrical conductivity of solutions of pc and its copper derivative in chlorosulfuric acid show that in both cases only four chlorosulfate ions are produced per pc molecule, suggesting that in this acidic solvent only the bridging aza nitrogen atoms are protonated.

"Although pc only dissolves in most common organic solvents to the extent of 10^{-5}—10^{-4} *M*, its high extinction coefficients in the visible region of the spectrum enable its reactions to be monitored. The most intense band ($\pi \rightarrow \pi^*$) in the visible spectrum of metal pcs is normally between 600 and 700 nm, whereas the parent base, pc, has two approximately equally intense bands in this region. . . Pc and its metal derivatives are therefore easily distinguished by their characteristic spectra in the 600—700 nm region."* A caution in pc solution chemistry is that at concentrations above 10^{-4}—10^{-6} *M* pcs are known to dimerize.[2]

"Pc and its unsubstituted metal derivatives are generally insoluble in common organic solvents. However, they dissolve in highly acidic media such as concentrated sulfuric acid, presumably because of protonation of the bridging nitrogen atoms, which should resemble the basic nitrogen atom of pyridine. There is evidence that pc films react with gaseous hydrogen chloride and hydrogen bromide, and with concentrated hydrochloric and hydrobromic acids ferrous pc forms 1:1 adducts thought to involve protonation of one of these nitrogen atoms. Information about the species present in solutions of these compounds in highly acid solvents is lacking. . . In order to obtain reliable values for the degree of protonation of several pcs in highly acidic media, the authors have made conductometric measurements on HSO_3Cl solutions under nitrogen at 25.0°C. Since the SO_3Cl^- ion has an apparent high mobility in this

* From Ledson, D. L. and Twigg, M. V., *Inorg. Chim. Acta*, 13(1), 43, 1975. With permission.

medium, comparing the electrical conductivity of KSO_3Cl solutions with that of solutions of base provides a good estimate of the number of SO_3Cl^- ions produced in the reaction with the base, and hence the number of protons involved. The results now reported show that four SO_3Cl^- ions are produced for each molecule of dissolved pc or its copper derivative. This suggests that all of the bridging nitrogen atoms are protonated in this medium. It is noteworthy that HSO_3Cl is more acidic than H_2SO_4."[3]

The kinetics of the acid hydrolysis of 4,4′,4″,4‴-tetracarboxylated Cu pc are studied in 43.50 to 66.56% sulfuric acid (g H_2SO_4 in 100 g solution) at 25°C at a Cu pc concentration of 0.66×10^{-5} mol/ℓ. The study is done spectrophotometrically in the Beer's law region.[4]

Ligand basicity is studied for copper and other transition metal pc complexes by reacting these complexes with Lewis bases such as $AlCl_3$.[5]

The acid-base properties of metal di pcs are studied spectrophotometrically.[6]

III. REACTIONS OF FE(II) PHTHALOCYANINE

"Iron(II) pc has attracted considerable attention in recent years due to its similarity to iron porphyrin systems."[7]

The replacement of dimethyl sulfoxide by imidazole and N-substituted imidazole at the axial position of Fe pc is studied.[8-10] "Fe pc reacts with imidazole to form a bis-imidazole complex at conveniently measurable rates. When dissolved in DMSO, square planar Fe pc most probably exists as Fe pc $(DMSO)_2$, with two solvent molecules bound to the center. Displacement of the first solvent molecule is rate determining:

$$\text{Fe pc (DMSO)}_2 \xrightarrow{\text{slow}} \text{Fe pc (DMSO) (imid)} \xrightarrow{\text{fast}} \text{Fe pc (imid)}_2$$

and at a given temperature, observed pseudo first order rate constants are given by the equation[10]

$$k_{obs} = k[\text{imid}]^*$$

"An iterative technique employing computer processing of spectrophotometric data has been used to obtain the two formation constants for complexing of nitrogenous bases with Fe(II) pc in DMSO. Good correlations of K_f with pK_{BH^+} are obtained for the first formation step with a closely related set of ligands. π-bonding does not appear as important as in similar reactions with ferrous porphyrins."** Bases studied include imidazole, 4-aminopyridine, 4-methylpyridine, pyridine, 4-acetylpyridine, 4-cyanopyridine, and 2-methylpyridine.

Complexes of Fe pc with aromatic nitroso compounds, isocyanides, and phosphites are isolated.[12] They include mixed ligand complexes, L Fe pc L′, where L = a nitroso aromatic group, L′ = *n*-butylamine or *N*-methylimidazole; L = cyclohexyl isocyanide, L′ = *N*-methylimidazole; L = a phosphite, L′ = *n* -butylamine and bis adducts L_2 Fe pc where L = an isocyanide or a phosphite. The complexes are characterized by elemental analysis, electronic spectra, IR spectra, and proton magnetic resonance spectra.

Fe(II) pc, considered as a six coordinate low spin pc, complexes with pyridine, piperidine, and methylimidazole ligands which reversibly bind benzyl isocyanide in toluene solution by a dissociative mechanism.[13] A factor of 10^4 difference in the rate of CO and isocyanide dissociation from the iron pc complexes is attributed to the large

* From Jones, J. G. and Twigg, M. V., *Inorg. Chim. Acta,* 12(2), L15, 1975. With permission.

** From Jones, J. G. and Twigg, M. V., *Inorg. Chim. Acta,* 10(2), 103, 1974. With permission.

differences in the stability of the complexes. "Data on systems studied to date indicate that only an iron porphyrin appears to combine the biologically important features of large equilibrium constants coupled with remarkable lability for O_2, CO, and benzylisocyanide binding."[13]

"While the dissociation of benzyl isocyanide from Fe(II) pc is slow in the dark, the reaction is rapid in the presence of normal fluorescent room lighting. Over a thousandfold increase in rate is observed between dark and full illumination. The observed increase in the rate of dissociation in the presence of light results in a shift of the equilibrium

$$L_2\ \text{Fe pc} + \text{RNC} = \text{L Fe pc (RNC)} + \text{L}$$

where RNC = benzylisocyanide and L = piperidine.

"The effect is quite dramatic. Solutions containing concentrations of L and RNC to shift the equilibrium to the right appear blue. In the presence of light the color rapidly changes to the green color of L_2 Fe pc. On placing the solution in the dark, the color changes back to blue. The change in color from blue to green and back again has been repeated several hundred times without any loss in reversibility. The relative rate of 'true equilibrium' in the dark may be varied by changing the ligand L. . . We know of no previous suggestion that solar energy storage in biological systems may occur by a process similar to that observed in the iron pc model system. The remarkable versatility of such a process and the similarity between the iron pc and biological metal complexes make such a proposal very attractive. We are continuing our investigations of the kinetics and equilibria of substitution reactions of iron pc and porphyrin complexes with the goal of optimizing energy storage and determining how this stored energy may be used to drive unfavorable reactions in much the same way ATP is used to drive biological reactions."*

Benzyl isocyanide exerts trans effects on methyl imidazole dissociating from Fe(II) complexes of pc.[15]

Six coordinate low spin Fe(II) pc complexes (L_2 Fe pc where L = imidazole, pyridine, piperidine, and 2-methylimidazole) reversibly bind carbon monoxide in toluene solution by a dissociative mechanism.[16] Equilibrium constants for the reaction

$$L_2\ \text{Fe pc} + \text{CO} \rightleftharpoons \text{L Fe pc (CO)} + \text{L}$$

decreases in the order 2-methylimidazole > piperidine > pyridine > imidazole. "Among the pc systems, a most striking difference in the kinetic and equilibrium data occurs for the imidazole ligand systems. The considerably diminished lability of the amine ligand and the CO ligand for the L = imidazole systems is of particular interest since imidazole ligands are trans to the reactive ligand in heme proteins and vitamin B_{12}."**

"Fe(II) pc reacts with tri-*n*-butyl phosphite, $P(OBu)_3$, and tri-*n*-butylphosphine, PBu_3, to yield low spin bis adducts, Fe pc $(PR_3)_2$. The kinetics of the replacement of $P(OBu)_3$ by PBu_3,

$$\text{Fe pc}\ [P(OBu)_3]_2 + 2PBu_3 \rightarrow \text{Fe pc}\ (PBu_3)_2 + 2P\ (OBu)_3$$

are reported and the mechanism shown to be strictly dissociative. The rate of dissociation of $P(OBu)_3$ from Fe pc$[P(OBu)_3]_2$ is shown to be about 5000 times greater than dissociation of imidazole from Fe pc $(\text{imidazole})_2$."***

* From Stynes, D. V., *J. Am. Chem. Soc.*, 96(18), 5942, 1974. With permission.
** From Stynes, D. V. and James, B. R., *J. Am. Chem. Soc.*, 96(9), 2733, 1974.
***From Sweigart, D. A., *J. Chem. Soc. Dalton Trans.*, 15, 1476, 1976. With permission.

Fe(II) pc and concentrated HBr react to yield a 1:1 Fe Pc-HBr adduct.[17] In aprotic solvents adduct formation is reversible and is a reaction of bromide ion with protonated pc.

Fe(II) pc in DMSO or dimethylacetamide also reacts with a solution of HCl gas in the solvent to give a spectrum identical with that of a compound formed by reaction of aqueous HCl with solid Fe pc, indicating that the compound is a HCl adduct of Fe pc, the addition of HCl being a reversible process.[18] Solid Fe(II) pc also reacts with dry HCl gas at −78°C to give a product which after exposure to dry nitrogen and heating to room temperature gives the adduct Fe(II) pc · 4 HCl. This adduct can also be prepared by the reaction of HCl gas with solid Fe(II) pc at room temperature.[19]

Interactions of H_2[(4-*tert*-butyl pc Fe)$_2$O] with pyridine, morpholine, piperidine, imidazole, and cyclohexylamine, give complexes of the type (amine H)$_2$[(4-*tert*-butyl pc Fe)$_2$O].[20]

IV. REACTIONS OF COBALT PHTHALOCYANINE

Co(II) pc forms 1:1 and 1:2 adducts with pyridine and 4-methylpyridine.[21] The structure of the adducts is studied by magnetic susceptibility, thermal, and IR spectral data. The adducts react with oxygen at low temperatures to form 1:1 Co:O_2 adducts.[21]

The tendency to six coordination and the strength of the intermolecular interaction in Co(II) pc are discussed with reference to the properties of its adducts with 3-methylpyridine and pyridazine.[22]

Li_2[Fe pc] in tetrahydrofuran(THF) reacts with CH_3I in the absence of air and moisture to give Li[CH_3 Fe pc] · 6THF. Thermal decomposition of the complex yields Li[Fe pc], THF, and C_2H_6.[23]

The complex Li[Co pc] · 4.5 THF is stable in alcoholic (methanol, ethanol) solutions.[24] The relative nucleophilicity of the $[pcCo(I)]^{\ominus}$ anionic complex, determined from the kinetics of methylation with CH_3I and C_4H_9Br is more than two times greater than Cl and is about the same as that of cobaloxime, which also contains Co(I). The ion $[pcCo(I)]^{\ominus}$ is stable in neutral protonic solutions, in contrast to cobaloximes derivatives and vitamin B_{12}s which are stable only in alkaline media.

Salts of the metal pc anion $[Mpc]^{\ominus}$ with a central metal atom in a lower oxidation state, such as Li_2[Fe pc] · 5.5 THF and Li[Co pc] · 4.5 THF, react with alkyl halides in THF by a substitution reaction to form stable organo metal pcs:[25]

$$\text{Co (II) pc} \xrightarrow{+e^{\ominus}} [\text{Co (I) pc}]^{\ominus} \xrightarrow{+RX} \text{Co (III) pc} - R + X^{\ominus}$$

The methylation and phenylation of styrene with derivatives of di- and tervalent cobalt and iron pcs is studied in the presence of such palladium(II) salts as lithium chloropalladate.[26] All compounds tested are shown to transfer an organic ligand to the olefin:

$$\text{RM pc}^{(n-)} + C_6H_5CH{=}CH_2 \longrightarrow C_6H_5CH{=}CHR$$

where R = CH_3 or C_6H_5, M = Fe or Co, and n = 0,1,2.

The reduction, methylation, and photolysis of Co pc solutions are studied by visible spectrophotometry, ESR, NMR, and mass spectrometry.[27] Analogies are drawn with the reactions of vitamin B_{12}.

V. CHARGE-TRANSFER COMPLEXES

The formation of charge-transfer complexes can be followed by conductometric titration.[28] A theory is presented requiring the conductivity to pass through a maximum when donor and acceptor are present in the solution in the stoichiometry required for the formation of a charge-transfer complex. Conductometric titrations with conductivity peaks in agreement with theory are reported for pc-iodine in DMSO. The value of the conductivity peak is expressed quantitatively in terms of a molar conductivity coefficient, σ_M.[29]

The reaction

$$C_2H_2 + C_2D_2 \rightarrow 2\,C_2HD$$

is studied at 20 to 200°C in glass vessels coated with Cu, Na, Mg, and Co pcs.[30] No exchange is detected even with the addition of electron acceptors such as iodine and *p*-chloranil at higher temperatures and under UV illumination. When sodium vapor is brought into contact with the pc, a fast reaction is observed with activation energies of 2.7 and 3.1 kcal/mol for

$$C_2D_2 + \text{HpcNa} \rightarrow C_2HD + \text{D pc Na}$$

and

$$C_2H_2 + \text{DpcNa} \rightarrow C_2HD + \text{HpcNa}$$

The formation of a charge-transfer complex is proposed and it is suggested that the H in the C_6H_6 ring in the pc takes part in the reaction.[30]

VI. REACTIONS OF PHTHALOCYANINE

The kinetics of pc formation is the subject of a thesis.[31]

The reaction of pc with $FeCl_3$, $FeCl_2$, $SnCl_2$, and dry HCl is studied by determining the absorption curves of the solution in α-bromonaphthalene or α-chloronaphthalene.[32] Complex formation takes place only in air, indicating that oxygen has a unique role in the complex formation process.

The metalation of solid pc by aqueous metal salts of Zn, Mn^{2+}, Fe^{2+}, Ga, Al, VO^{2+}, Pt^{2+} and Pt^{4+} depends strongly on the nature of the metal salt and on the crystalline modification of the pc.[32] Al, Ga, Pt^{2+}, and VO^{2+} do not react with pc.[33]

The reaction of pc in the solid phase with $CuCl_2$ in solution proceeds by a rapid reversible formation of a solid intermediate which is slowly converted thermally in the presence of 96% H_2SO_4 to Cu pc.[34] The maximum yield of Cu pc occurs when the first step is carried out at 20°C for 10 min. at the stoichiometric ratio of reactants. β pc reacts faster and more completely than α pc.

V(III) reacts with pc tetrasulfonic acid to form a complex with maximum absorbance at 520 nm, at pH 2.5. Beer's law is obeyed for 5×10^{-6} to 5×10^{-5} *M* V(III).[35]

The reaction of pc with amines reveals a degree of conversion decreasing in the order $(CH_3)_2NH$ > piperidine > Et_2NH > NH_3 > morpholine > imidazole.[36] A strong solvent effect is also noted.

In a research entitled "Possible Mechanisms of Formation of Radical States in Biological Systems", the mechanism of radical formation from co-enzymes in fully oxidized states is studied in model systems.[37] The following scheme is suggested for reaction of OH^- with pc:

$$\text{pc} + OH^{\ominus} \rightarrow \text{pc}^{\ominus} + OH\cdot$$

The reaction is first order. The kinetics studied by determining the loss of $OH^{\ominus}$ potentiometrically and free radicals by EPR are in good agreement.

Sensitizers for carrying out liquid phase photochemical reactions presumably include pc tetracarboxylic acid and a Cu pc complex.[38] They are positioned preferably in a thin layer on a transparent support such as glass, polyethylene, or polypropylene.

VII. REACTIONS OF METAL PHTHALOCYANINES

The β form of Cr pc, Mn pc, Co pc, and Fe pc react with nitric oxide to form the corresponding mononitrosyl derivatives.[39] No reaction of nitric oxide is observed with the β form of Ni, Cu, or VO pc.

Pcs are aminated with NH_2OH salts in a mixed solvent of H_2SO_4 and lower aliphatic alcohols in the presence of a catalyst, such as ammonium molybdate or $FeSO_4$.[40] "Thus, a cooled mixture of Cu pc 5.7 in 2% fuming H_2SO_4 98 and isopropyl alcohol 20.6 was treated with ammonium molybdate 0.15 and 2 $NH_2OH \cdot H_2SO_4$ 6.6, the mixture stirred 5 hours at 30—40°, poured into ice water 500 parts, the precipitate filtered, and treated with aqueous alkali to give 6.8 parts Cu nonaamino pc." Similarly, other aminated pcs are prepared.

Disulfonyl imide salts are made by treating pc SO_2NH_2 successively with $Cl(CH_2)_3SO_2Cl$ and NaOH to give pc $SO_2N(Na)SO_2(CH_2)_3Cl$. A mixture of m-$O_2NC_6H_4SO_2N(Na)SO_2(CH_2)_3Cl$ and aqueous CH_3NH_2 is heated at 50°C for 10 hr to give m-$O_2NC_6H_4SO_2N^{\ominus}SO_2(CH_2)_3N^{\oplus}H_2CH_3$.[41]

The following reaction is run in DMF and the equilibrium constant K determined at reagent concentrations of 10^{-4} *M*:

$$MoO_2\ pc + Cd(NO_3)_2 \rightleftharpoons Cd\ pc + MoO_2(NO_3)_2$$

$$K = 2.8 \pm 0.4 \times 10^{-4}$$

Values of $K \times 10^4$ for the same reactions in which Zn^{2+}, Co^{2+}, and Ni^{2+} replace Cd^{2+} are 25 ± 5, 15 ± 2, and 23 ± 4.[42] "The relative magnitude of these K values does not correspond with the relative stability of the pcs formed, and it is suggested that data on the kinetics of dissociation of MoO_2 pc in proton donors and salt media are needed for an understanding."

Phenylthiomethyl chloride reacts with Fe pc to form the complex $Na[C_6H_5SCH_2\ Fe\ pc] \cdot 4$ THF which tends to eliminate CH_2, giving the corresponding thiophenolates.[43]

Dissociation of metal pcs which occurs only in concentrated sulfuric acid and at relatively high temperatures is expressed by the third order equation

$$-d\, c_{MpcH^{\oplus}} / dt = k_v c_{MpcH^{\oplus}} \cdot c^2 H_3O^+$$

probably with a trimolecular S_N2 mechanism.[44]

Rate constants are determined for the basic and acidic hydrolyses of dichlorosilicon pc.[45] The activation energy, E, and activation entropy, ΔS, for the basic hydrolysis are 15.6 kcal/mol and 19 entropy units, respectively, and for the acid hydrolysis in 16.23 *M* H_2SO_4 are 28.2 kcal/mol and −15 entropy units, respectively.[45]

A kinetic study is made of nitrogen donor complex formation with Zn pc by means of a quartz crystal microbalance and electron microscopy.[46] The nitrogen donors are pyridine, piperidine, and *n*-hexylamine.

The phototropic reactions of pc reactive dyes are studied.[47]

VIII. SULFONATION

For a reaction time of 4.5 hr, optimum temperatures and oleum concentrations are given for the formation of Cu pc mono- and disulfonic acids from oleum containing 6 to 58% SO_3 at 30 to 140°C.[48]

Sulfonated Cu pcs containing an average of 2.4 sulfo groups in the 3- and 4-positions, 1 to 1.8 of which are in the 4-position, are prepared.[49] For example, a 75% 4-sulfophthalic anhydride and 25% 3-sulfophthalic anhydride mixture 32.7, phthalic anhydride 39.7, urea 250, CuCl 12, and $TiCl_4$ 12 parts in nitrobenzene heated at 185°C give sulfonated Cu pc containing 1.3 sulfo groups (1.0 in 4-position) which is sulfonated in oleum to give a sulfo group content of 2.4.

IX. STABILITY

The stability of pcs to acid treatment is reviewed.[50] In particular, the removal of Cu and Co from their solid pcs by 25 to 65% H_2SO_4 at 25 to 70°C. Metal concentrations in solution range from 0.6 to 92×10^{-5} *M* after 5 to 100 hr.

The rate of decomposition of pc increases with increasing H_2SO_4 concentration to a maximum at about 80% H_2SO_4 (half-life of pc at 0°C in 79.1% H_2SO_4 is 1.1 hr; at 50°C, 0.0042 hr). With further increase in acid concentration, decomposition velocity decreases sharply and the half-life of pc in 98.5% H_2SO_4 is 140 hr at 50°C.[51] Dissociation constants of metal complexes of pc are also given for metal pcs. Stability increases in the following order: Zn pc < Cu pc < Co pc < Ni pc < Cu pc Cl_{15} < $Al(HSO_4)$ pc.

The stability of dyes including Cu pc disulfonic acid during combined bleaching and dyeing of cotton is determined.[52] The solutions contain NaOH 1.0, and surfactants (such as OP-10) 0.4, sodium silicate 6.0, and H_2O_2 3.0. The dyes are heated in the solutions at about 90°C for 45 min. Their stability is determined by ascending paper chromatography. The copper complex of pc disulfonic acid gives an unchanged chromatogram after testing.

The stability of dyes to light and water is the subject of a study conducted for the EPA (Environmental Protection Agency).[53] "The overall results show that most commercial colors are resistant to sunlight and water degradation and that it would take many weeks before appreciable dye degradation occurred in a natural aquatic environment."[53] The rate of photodegradation of Cu pc disodiumsulfonate is shown in Figure 1. The importance of this study is stressed from the observation that "The total dye consumption of the textile industry is over 100,000,000 lb/year. Since it has been estimated that a maximum of 90% of these dyes end up on fabrics and the remaining 10% goes to the waste stream, approximately 10,000,000 lb of dye per year are presently discharged to waste streams by the textile industry."

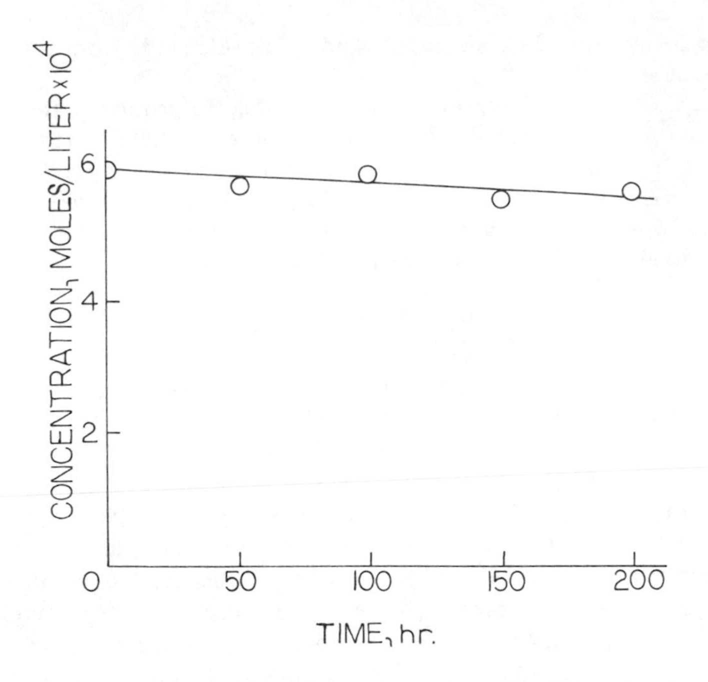

FIGURE 1. Rate of photodegradation of Direct Blue 86 in water at 50°C.

REFERENCES

1. Taube, R., *Pure and Applied Chem.*, 38(3), 427, 1974.
2. Ledson, D. L. and Twigg, M. V., *Inorg. Chim. Acta*, 13(1), 43, 1975.
3. Ledson, D. L. and Twigg, M. V., *Chem. Ind. (London)*, 3, 129, 1975.
4. Gaspard, S., Verdaguer, M., and Viovy, R., *C. R. Acad. Sci., Ser. C*, 275(11), 573, 1972.
5. Gaspard, S., Verdaguer, M., and Viovy, R., Proc. Int. Conf. Coord. Chem., 16th, Univ. Coll. Dublin, R75, 3 pp., 1974.
6. Moskalev, P. N. and Alimova, N. I., *Zh. Neorg. Khim.*, 20(10), 2664, 1975.
7. Sweigart, D. A., *J. Chem. Soc. Dalton Trans.*, 15, 1476, 1976.
8. Jones, J. G. and Twigg, M. V., *Inorg. Chem.*, 8(10), 2120, 1969.
9. Jones, J. G. and Twigg, M. V., *Inorg. Nucl. Chem. Lett.*, 5(4), 333, 1969.
10. Jones, J. G. and Twigg, M. V., *Inorg. Chim. Acta*, 12(2), L15, 1975.
11. Jones, J. G. and Twigg, M. V., *Inorg. Chim. Acta*, 10(2), 103, 1974.
12. Watkins, J. J. and Balch, A. L., *Inorg. Chem.*, 14(11), 2720, 1975.
13. Stynes, D. V., Proc. Int. Conf. Coord. Chem., 16th 1974, 1.15, 2 pp (English).
14. Stynes, D. V., *J. Am. Chem. Soc.*, 96(18), 5942, 1974.
15. Pang, I. W., Singh, K., and Stynes, D. V., *J. Chem. Soc., Chem. Commun.*, 4, 132, 1976.
16. Stynes, D. V. and James, B. R., *J. Am. Chem. Soc.*, 96(9), 2733, 1974.
17. Jones, J. G. and Twigg, M. V., *Inorg. Chim. Acta*, 4(4), 602, 1970.
18. Jones, J. G. and Twigg, M. V., *Inorg. Nucl. Chem. Lett.*, 6(2), 245, 1970.
19. Dickens, L. L. and Fanning, J. C., *Inorg. Nucl. Chem. Lett.*, 12(1), 1, 1976.
20. Bundina, N. I., Kaliya, O. L., Lebedev, O. L., Luk'yanets, E. A., Rodionova, G. N., and Ivanova, T. M., *Koord. Khim.*, 2(7), 940, 1976.
21. Cariati, F., Galizzioli, D., Morazzoni, F., and Busetto, C., *J. Chem. Soc. Dalton Trans.*, 7, 556, 1975.
22. Cariati, F., Morazzoni, F., and Busetto, C., *J. Chem. Soc. Dalton Trans.*, 7, 556, 1975.
23. Taube, R., Drevs, H., Duc-Hiep, T., *Z. Chem.*, 9(3), 115, 1969.
24. Eckert, H. and Ugi, I., *Angew. Chem.*, 87(23), 847, 1975.
25. Eckert, H. and Ugi, I., *Angew. Chem.*, 87(23), 847, 1975.
26. Vol'pin, M. E., Taube, R., Drevs, H., Volkova, L. G., Levitin, I. Ya., and Ushakova, T. M., *J. Organomet. Chem.*, 39(2), C79, 1972.
27. Day, P., Hill, H. A. O., and Price, M. G., *J. Chem. Soc. A*, 1, 90, 1968.
28. Gutmann, F. and Keyzer, H., *Electrochim. Acta*, 11(6), 555, 1966.

29. Gutmann, F. and Keyzer, H., *Electrochim. Acta,* 11(8), 1163, 1966.
30. Ichikawa, M., Soma, M., Onishi, T., and Tamaru, K., *J. Phys. Chem.*, 70(6), 2069, 1966.
31. Knauer, A. J., Kinetics of Pc Formation, thesis, avail. Univ. Microfilms, Ann Arbor, Mich., Order No. 69-15,743; *Diss. Abstr. Int. B,* 30(4), 1538, 1969.
32. Glikman, T. S. and Barvinskaya, Z. L., *Opt. Spektrosk.*, 8, 425, 1960.
33. Berezin, B. D., Shormanova, L. P., Klyuev, V. N., and Korzhenevskii, A. B., *Zh. Neorg. Khim.*, 19(10), 2745, 1974.
34. Berezin, B. D., Shormanova, L. P., and Fel'dman, R. I., *Zh. Neorg. Khim.*, 19(7), 1833, 1974.
35. Tserkovnitskaya, I. A. and Perevoshchikova, V. V., *Zh. Anal. Khim.*, 26(8), 1527, 1971.
36. Vul'fson, S. V., Lebedev, O. L., and Luk'yanets, E. A., *Zh. Prikl. Spektrosk.*, 17(5), 903, 1972.
37. Fomin, G. V., Davydov, R. M., Blyumenfel'd, L. A., and Sukhorukov, B. I., *Mekh. Dykhaniya. Fotosin. Fikatsii Azota,* 134, 1967.
38. Braun, A. and Lohse, F., (to Ciba-Geigy AG), German Offenlegungsschrift 2,210,108, October 5, 1972.
39. Ercolani, C. and Neri, C., *J. Chem. Soc. A,* 11, 1715, 1967.
40. Somiya, T. and Makita, M. (to Sumitomo Chemical Company Ltd.), Japanese Kokai 20,078, September 16, 1964.
41. Jaeger, H. and Dehmel, G., German Offenlegungsschrift 2,021,257, November 11, 1971.
42. Zhukov, Yu. A., Berezin, B. D., and Sokolova, I. N., *Izv. Vyssh. Ucheb. Zaved. Khim. Khim. Tekhnol.*, 13(11), 1566, 1970.
43. Taube, R. and Steinborn, D., *J. Organomet. Chem.*, 65(1), C9, 1974.
44. Berezin, B. D., *Zh. Fiz. Khim.*, 38(12), 2957, 1964.
45. Akopov, A. S., Berezin, B. D., Klyuev, V. N., and Solov'ev, A. A., *Zh. Neorg. Khim.*, 18(4), 965, 1973.
46. Saito, Y., Kobayashi, T., Uyeda, N., and Suito, E., *Bull. Inst. Chem. Res. Kyoto Univ.*, 49(4), 256, 1971.
47. Baranova, G. S., Romanova, M. G., and Chekalin, M. A., *Anilinokrasoch. prom — st,* 1, 3, 1973.
48. Aleshonkov, A. P., Borodkin, V. F., and Gusev, V. A., *Izv. Vyssh. Ucheb. Zaved. Khim. Khim. Tekhnol.*, 14(1), 151, 1971.
49. Von Tobel, H. (to Sandoz Ltd.), Swiss Patent 525,937, September 15, 1972.
50. Berezin, B. D., *Izv. Vyssh. Ucheb. Zaved. Khim. Khim. Tekhnol.*, 6(5), 841, 1963.
51. Berezin, B. D., *Izv. Vyssh. Ucheb. Zaved. Khim. Khim. Tekhnol.*, 7(1), 111, 1964.
52. Gromova, O. V. and Romanova, M. G., *Tekst. Prom. (Moscow),* 4, 57, 1973.
53. Porter, J. J., *Text. Res. J.*, 43(12), 735, 1973.

Chapter 6

OXIDATION-REDUCTION REACTIONS

I. COBALT PHTHALOCYANINE

A procedure to oxidize Co pc is as follows:[1] 57 g Co pc is dispersed in 1900 g 5% aqueous $KMnO_4$ at 20°C, 450 mℓ 6 $N H_2SO_4$ added at 20°C, and the precipitate filtered and washed with 2% aqueous Na_2CO_3 and water to give 63 g oxidized Co pc. Other oxidation reagents used are CrO_3, $Ce(SO_4)_2$, and $K_2Cr_2O_7$.

In the oxidation of Co pc with $KMnO_4$, $Ce(SO_4)_2$, or Cr_2O_3, a dark brown compound is obtained whose structure is proposed to be[2]

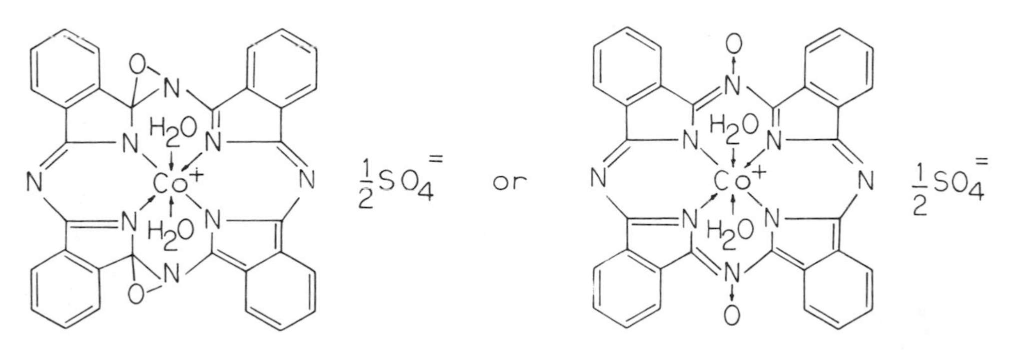

The reduction of Co pc with $Na_2S \cdot 9\ H_2O$ in DMS and with sodium metal in ammonia gives the monosodium salt (I) and a mixture of (I) and the disodium salt (II)[3]

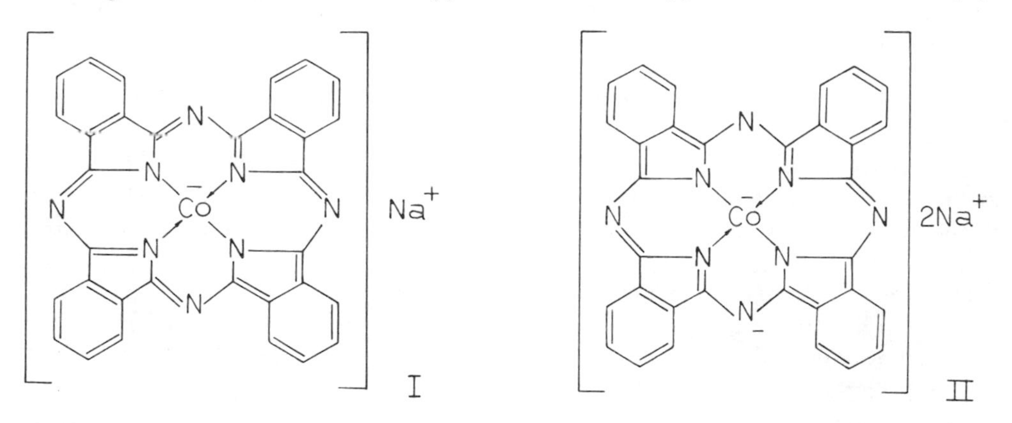

(I) gives a green solution in organic solvents whereas (II) gives a reddish violet solution.

The reversible reaction of Co(II) tetrasulfo pc with molecular oxygen to form a stable oxygen adduct is studied.[4]

The formation of a mononuclear 1:1 adduct of oxygen with Co (II) tetrasulfo pc at low temperatures in the presence or absence of ammonia is detected by ESR spectra. "This adduct is best described as a paramagnetic Co(III) complex with the superoxide anion $O_2^{\ominus}$ as a ligand."[5]

Other Co(II) pc 1:1 oxygen adducts with Co(II) pc are characterized by ESR spectroscopy at low temperatures.[6]

"Co(II) tetrasulfo pc forms in aqueous solutions an adduct with molecular oxygen, of stoichiometry Co:O = 2:1. The existence of the stable adduct has been proved by measuring absorption spectra under aerobic as well as anerobic conditions and by man-

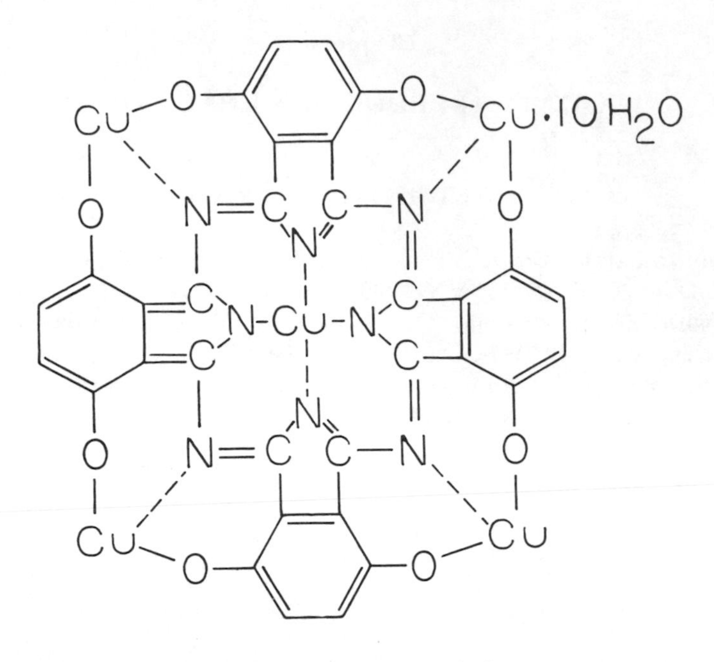

FIGURE 1. Cu pc oxy copper hydrate.

ometric measurements of oxygen absorption. The reaction leading to the formation of an adduct is of the first order with respect to Co(II) tetrasulfo pc, the formation of the intermediate product Co(II) tetrasulfo pc · O_2 being the rate determining step."*

Different reaction temperatures allow the isolation of 1:1 and 1:2 adducts of Co(II) pc with pyridine and 4-methylpyridine. All the adducts react with oxygen and form well-characterized 1:1 adducts with oxygen at low temperature.[8]

II. COPPER PHTHALOCYANINE

Controlled oxidation of Cu pc to phthalimide is obtained with 0.1 *N* potassium dichromate in sulfuric acid at 0°C, providing a quantitative method to estimate the pigment.[9] Further oxidation gives phthalic acid which is completely oxidized to carbon dioxide.

Oxidation of Cu pc is the subject of another research.[10]

Unusual highly metallized oxy derivatives of Cu pc such as shown in Figure 1 are prepared by adding a solution of $CuSO_4$, 150 g/ℓ, to a solution of 1.45 g of the sodium salt of 3,3′,3″,3‴-tetrahydroxy Cu pc in 200 mℓ water and heating to boiling until a colorless sample is noted on a filter paper test. The precipitate is filtered, washed with water, ethanol, and ether, and dried at 100 to 200°C. Such salts do not melt at <500°C and are insoluble in water and organic solvents. They ignite in air at 450 to 600°C and burn without flame.[11]

The reactivity of Cu pc toward ozone is studied in a double corona charging unit typical of direct electrostatic printing apparatus. Possible reactions of ozone with the Cu pc toner are suggested in Figure 2.

"The toner could react with ozone via the carbon-nitrogen double bond. The stability of this pigment indicates that the oxidant would have little or no effect on the structure."[12]

* From Wagnerova, D. M., Schwertnerova, E., and Veprek-Siska, J., *Collect. Czech. Chem. Commun.*, 39(8), 1980, 1974. With permission.

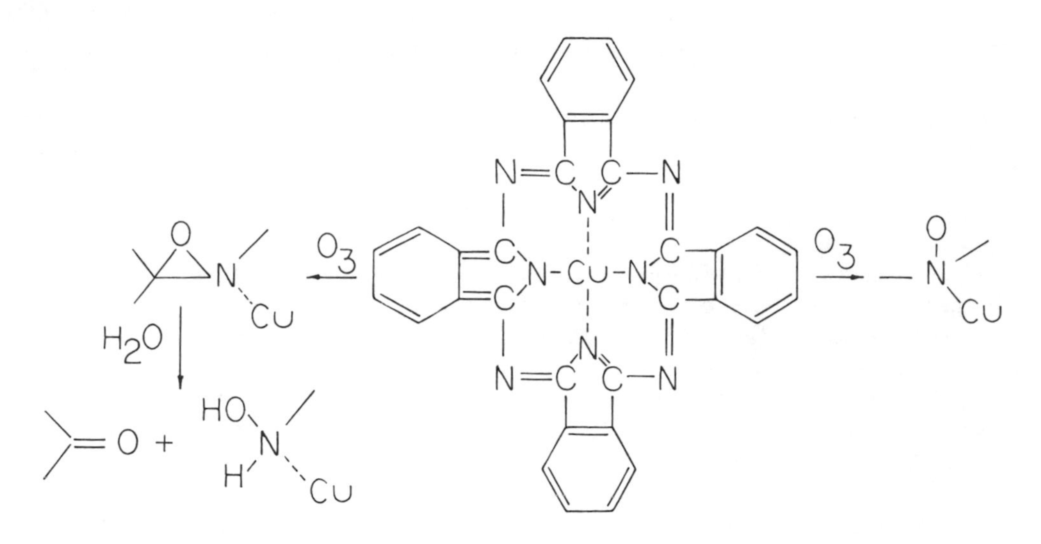

FIGURE 2. Possible reactions of ozone with toner.

Cu pc is reduced to Li [Cu pc] · 4.5 THF and to Li_2[Cu pc] · 6 THF in tetrahydrofuran with lithium as dilithiumbenzophenone as the reducing agent in the presence of stilbene, according to the reactions.[13]

$$\text{Cu pc} + \tfrac{1}{2}\text{Li}_2\text{Bzph} \xrightarrow{\text{THF}} \tfrac{1}{2}\text{Bzph} + \text{Li[Cu pc]} \cdot 4.5\ \text{THF}$$

and

$$\text{Cu pc} + \text{Li}_2\text{Bzph} \xrightarrow{\text{THF}} \text{Bzph} + \text{Li}_2\text{[Cu pc]} \cdot 6\ \text{THF} \cdot$$

Cu pc 3,3′,3″,3‴ pc tetrasulfonyl chloride is reduced at 70 to 95°C for 2 hr using Fe and HCl; thiourea; Fe, HCl, and thiourea; or ferrous salts and thiourea, giving such water-insoluble green pc dyes with polysulfide bridges and isothiourea groups as[14]

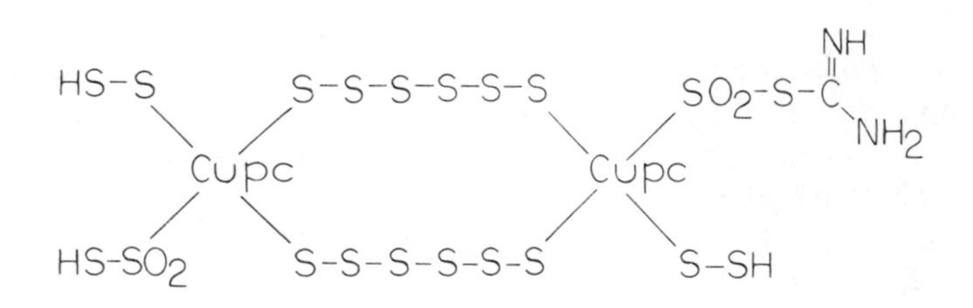

The kinetics of the photoreduction of Cu pc derivatives with amines in aqueous solution are measured at various reactant concentrations. "The photoreduction proceeds by an oxidation-reduction reaction between the electronically excited dye molecule as an electron acceptor and the amine as the donor. In addition to the action as electron donors, certain amines can also act as a reaction quencher with their activities increasing with increasing ionization energy. It is postulated that the effect as a quencher is the result of a reversible electron transfer between donor and excited dye molecule which makes it possible for the excited state of the dye to be deactivated without a chemical reaction."*

* From Eigenmann, G., *Helv. Chim. Acta,* 46, 855, 1963. With permission.

III. MANGANESE PHTHALOCYANINE

A crystalline solid formed by the oxidation of Mn(II) pc in pyridine is identified as pc pyridine Mn(III) -μ-oxo pc pyridine Mn(III) dipyridinate, $C_{74}H_{42}Mn_2N_{18}O(py)_2$ by X-ray diffraction. The molecule consists of two Mn pc complexes joined by an essentially linear Mn-O-Mn bridge. The structure may have some relation to oxidation processes in biological systems.[16]

The properties of a pc Mn oxide are reexamined from an X-ray structural analysis, and part of previously reported oxidation-reduction mechanism of Mn pc in pyridine is revised.[17] Also coordination of pyridine to Mn pc complexes is studied.

To study the influence of porphyrin-like ligands on the chemistry of transition metal ions, the behavior of Mn in chelates with pc, methyl pheophorbide-a, and etioporphyrin I, members of three distinct classes of porphyrins is investigated.[18] Mn(II) pc adds two equivalents of pyridine, in a reaction whose rate depends on the preparation of the sample.

The kinetics of reduction of Mn(III)-tetrasulfo pc with $N_2H_4 \cdot HCl$ and with $N_2D_4 \cdot DCl$ are determined.[19] The reduction rate is much slower with $N_2D_4 \cdot DCl$ than with $N_2H_4 \cdot HCl$.

A spectrophotometric study of the kinetics of reversible oxygen binding by Mn(II) pc in pyridine solution "shows that negative autocatalysis by the product [believed to be $C_5H_5N(Mn)O_2$] takes place."[20] At constant oxygen concentration the reaction is first order. Mn(II) pc also undergoes oxygen binding in aniline, in quinoline containing imidazoles, and in alcoholic sodium hydroxide solution.

"Recent interest in oxygen bridged porphyrin derivatives with special emphasis on the importance of the axial substituent, uncertainty concerning the magnetic properties of such a dimeric oxygen bridged Mn pc derivative . . ." prompts the authors "to report some preliminary results in a continuing study of the Mn pcs.

"The involvement of manganese in the photosynthetic liberation of oxygen from water is well established; although the detailed mechanism is still obscure, it is clear that an oxidation reduction reaction is involved. The ability of Mn pc to undergo oxidation reduction reactions in which water and oxygen are involved, and which are photochemically sensitive provides therefore an additional impetus to elucidate this system.

"Conflicting reports concerning the properties of (pyridine Mn pc)$_2$O have been elucidated, and a series of new compounds of the form (amine Mn pc)$_2$O with *N*-methylimidazole, 3-picoline, and piperidine are reported. Infrared spectroscopy is usually used for the assignment of mono-oxo bridged structures, however, these compounds provide an exception to the established criteria."*

A study is made of the effect of light and reduction reactions involving pc and etioporphyrin I Mn complexes. According to the authors, "Recent investigations have shown that Mn plays an important role in photosynthesis. In particular, Mn seems to be essential in the oxygen evolving systems. Recently it was reported that a Mn chelate related to the porphyrins, namely, Mn pc, apparently formed a peroxide reversibly which, in turn, seemed to be capable of dissociating the oxygen-oxygen bond in a reversible fashion. This observation was so unusual that it attracted our attention immediately. It seemed possible to incorporate such stages in the oxygen evolution scheme of photosynthesis with great ease."[22]

"It has been shown that the stable oxidation level of Mn may be shifted among the II, III, and IV oxidation states, depending on the nature of the fifth and sixth coordi-

* From Canham, G. W. R. and Lever, A. B. P., *Inorg. Nucl. Chem. Lett.*, 9(5), 513, 1973. With permission.

nating groups. Furthermore, photochemical oxidation as well as photochemical reduction of the Mn(III) pc has been observed, and photochemical reduction of the Mn(IV) compound demonstrated."* The interrelationship of Mn pc complexes is shown in Figure 3.

Mn(III) tetrasulfo pc is reduced in the dark and photochemically.[23] The reduction in the dark is readily obtained especially at high pH levels with such reducing agents as atomic hydrogen, Na_2S, $Na_2S_2O_3$, ascorbic acid and $N_2H_4 \cdot HCl$. Irradiation of air-free alkaline solutions of Mn(III) pc in the absence of added reducing agents leads to the same spectral changes as described for the dark reaction and reoxidation regenerates Mn(III) pc. Here the electron donor is OH^- ion and the photons accelerate the reduction. The reaction is relatively slow, however, even in strong light.[23]

Mn(II) pc can be reduced in THF with lithium under addition of stilbene according to the reaction

$$\text{Mn pc} + n^{\ominus} \rightarrow [\text{Mn pc}]^{(n-)}$$

in five steps. The preparation of the anionic Mn pc complexes as Li_2 Mn pc · xTHF is described, and their electronic structure is discussed on the basis of LCAO-MO theory.[24]

For example, as shown in Figure 4,[24] Li_4[Mn pc] · 10 THF is prepared as follows. About 200 mg stilbene and 80 mℓ THF in a stream of nitrogen are introduced to 3 g pure Mn pc in flask S_1 through the sidearm A. The sidearm is rinsed with alcohol and THF and then a piece of clean lithium (about 1 g) is introduced. Upon mechanical agitation, the first three reduction steps of Mn pc take place. As soon as the red brown crystals of Li_4[Mn pc] appear in a uniform condition (microscope lamp!), the vessel S_1 is connected to the frit F. Upon opening the stopcocks h_3, h_4, and h_5 the apparatus is made secure by stopcock h_6. The stopcock h_2 is opened and the precipitate spills onto frit F, and by careful evacuation through stopcock h_5 the mother liquor is sucked into vessel S_2. The lithium is retained in vessel S_1 by stopcock h_2. When the required wash THF is obtained, the mother liquor is led into flask S_3 upon closing stopcocks h_4 and h_6, then upon cooling flask S_2 with dry ice several milliliters THF are distilled and flow back to flask S_2. About 30 mℓ THF are distilled back to flask S_2. Stopcock h_5 is closed, and by turning the apparatus and warming flask S_2 the solvent flows back across the frit. The product is washed with agitation, then the wash solution, by cooling flask S_2 is again sucked back through the stopcock h_5, and recombined with the mother liquor in flask S_3.

IV. VANADYL PHTHALOCYANINE

Manometric measurements show that vanadyl tetrasulfo pc is slowly and irreversibly oxidized by oxygen at pH > 12.6 with possible intermediate formation of an adduct with oxygen.[25]

V. KINETICS OF OXIDATION WITH PEROXIDES

Rates of oxidation of pcs by H_2O_2 in 17 M H_2SO_4 are studied photometrically. The pcs are divided into three groups — labile complexes, stable complexes whose central atom can not be further oxidized, and stable complexes where further oxidation of the

* From Engelsma, G., Yamamoto, A., Markham, E., and Calvin, M., *J. Phys. Chem.*, 66, 2517, 1962. With permission.

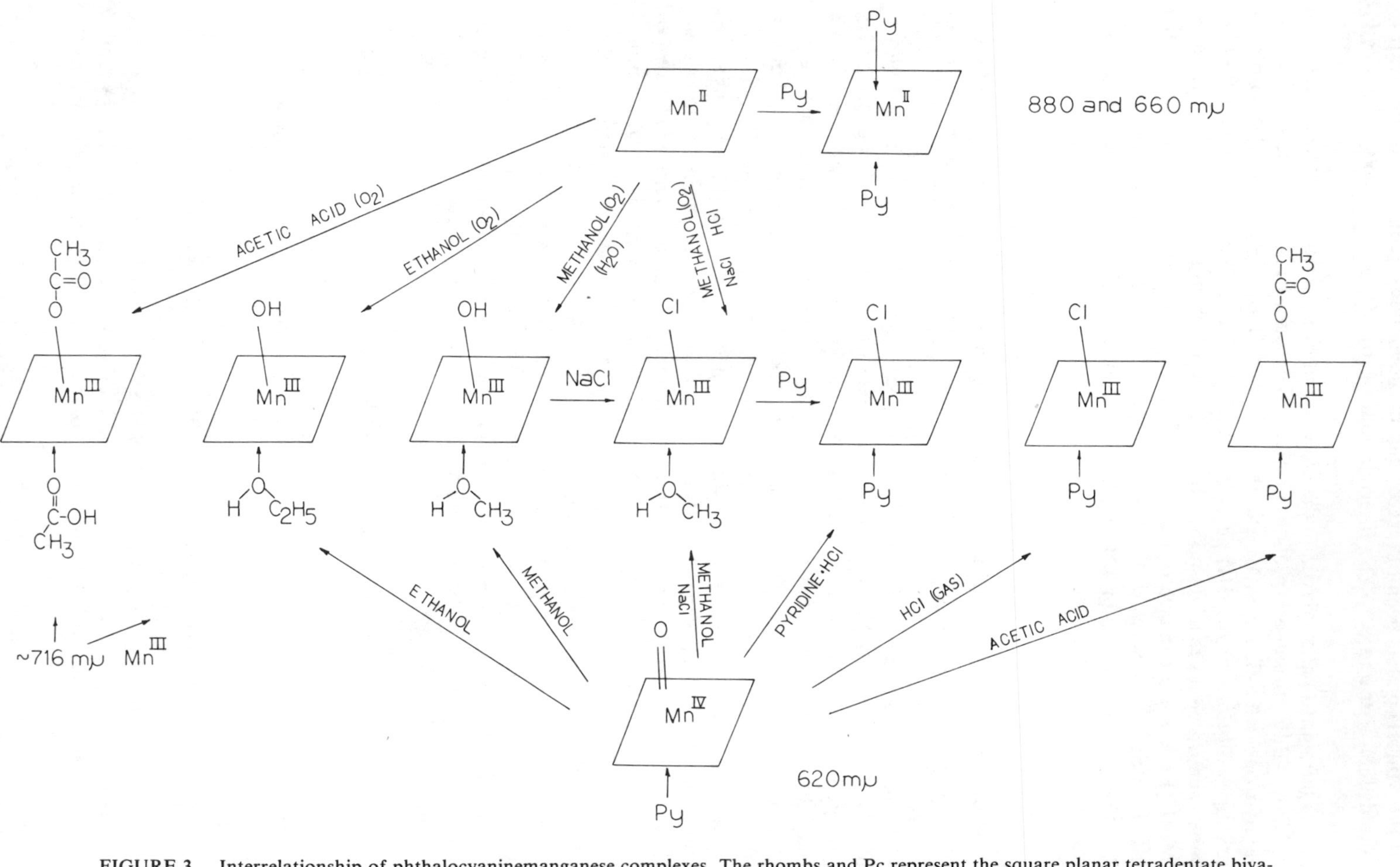

FIGURE 3. Interrelationship of phthalocyaninemanganese complexes. The rhombs and Pc represent the square planar tetradentate bivalent phthalocyanine ligand, $C_{32}H_{16}N_8$; Py for pyridine.

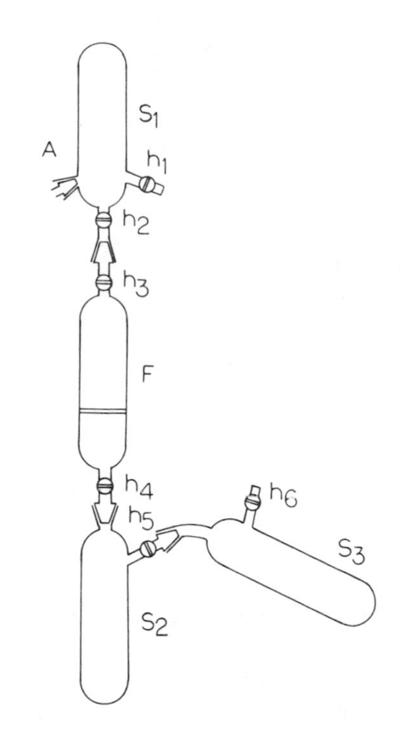

FIGURE 4. Apparatus for the preparation of Li_4 [Mn pc]·10 THF and Li_5 [Mn pc]·9 THF.

central atom is possible.[26] The order of the reactions is determined and a mechanism is proposed.

The kinetics of the oxidizing decomposition of pcs are measured photometrically. Pc or its metal complex at 2 to 6 × 10^{-4} *M* concentration is oxidized by 6 to 60 × 10^{-3} *M* $S_2O_8^{\ominus}$, or by 7 to 80 × 10^{-4} *M* NO_3^- or by 1.6 to 16.6 × 10^{-2} *M* H_2O_2 in 17 *M* H_2SO_4 at 25°C ($K_2Cr_2O_7$, $KBrO_3$, KIO_3, and $KMnO_4$ react too rapidly). The reaction is first order both in pc and oxidant.[27]

In a series of five publications, the redox reactions of pcs and related compounds are studied, with these titles:

I. Kinetics of the Reaction of Tetra-4-*Tert* Butyl Pc with Benzoyl Peroxide
II. Kinetics and Mechanism of the Reaction of Pcs with Benzoyl Peroxide
III. Nature of Intermediate Products of Pc Oxidation by Acylperoxides
IV. Reactivity of Porphine-Like Compounds in Reaction with Benzoyl Peroxide
V. Redox Potentials of Pcs and Mechanism of Their Oxidation. Some Properties of Intermediate Products[28-32]

Kinetic studies of the reaction of tetra-4-*tert*-butyl pc with benzoyl peroxide at 20 to 60°C in nitrobenzene with and without diphenylpicrylhydrazine show that tetra-4-*tert*-butyl pc and benzoyl peroxide initially form a complex, the formation rate constant of which is 0.022 ± 0.0031 mol sec at 25°C.[28] The reaction without diphenylpicrylhydrazine does not involve free radicals.

The kinetics of decomposition of the reaction intermediate is studied by EPR spectrometry for the reaction between tetra-4-*tert*-butyl pc, its Cu derivatives, and its Ni derivatives, at 0.1 to 25.0 × 10^{-5} *M* and $(p\text{-}RC_6H_4CO)_2O_2$ at 0.01 to 10.0 × 10^{-2} *M* in the presence of diphenylpicrylhydrazine at 0.83 × 10^{-4} *M* in carbon tetrachloride at 20°C or in nitrobenzene at 77 K, where R = CH_3CH_3O, Cl, or NO_2.[29] The dependence of the reaction constant k on R is correlated with good agreement by the Hammett equation.

The oxidation of metal complexes of tetra-4-*tert*-butyl pc by benzoyl peroxide is first order in Cu, Zn, or H_2 tetra-4-*tert*-butyl pc and second order in Ni, Pd, or Pt tetra-4-*tert*-butyl pc.[30] The suggested mechanism involves the reversible formation of a complex of benzoyl peroxide with the tetra-4-*tert*-butyl pc which either decomposes directly for the Cu, Zn, or H_2 complexes, or by reaction with another mole of tetra-4-*tert*-butyl pc for the Ni, Pd, or Pt complexes.

The kinetics of reaction of benzoyl peroxide with tetra-4-*tert*-butyl pc and tetra-6-*tert*-butyl-2,3-naphthalocyanine are determined.[31] Oxidation takes place by reversible complex formation between benzoyl peroxide and the pc.

The values of the first-order redox potentials E′ of a number of pc derivatives are found by potentiometric titration.[32] The reaction capabilities of the studied porphins in oxidation reactions are compared with values of E′. The electronic absorption spectra are obtained of intermediates in the oxidation of pcs-π-cation radicals, dications, and analogs of isoporphyrin. A mechanism of the oxidation of porphyrin-like compounds is proposed.

VI. FURTHER REDOX CONSIDERATIONS

"Redox processes play a critical role in living systems and frequently involve metalloporphyrins. Fairly subtle biological events can influence electron transfer chains by modifying the redox couples of their cytochrome components. There is a gross understanding of how variation of solvent (or environment) and of axial ligation can affect redox couples, but many fine details remain to be elucidated. The ability to bind molecular oxygen is also related to the redox couple but in an ill-defined way. The pcs represent a class of compounds of similar structure to the porphyrins, but whose redox characteristics, as will be demonstrated here, exhibit greater sensitivity towards the factors mentioned above. They may be conveniently used as a probe to gain a more detailed understanding of the electronic processes involved. . . the redox potentials of a range of metal pcs in non-aqueous media with analogous data for some porphyrin complexes provides evidence for extensive back donation in metallo pcs. The potentials for some couples are exceedingly sensitive to axial ligation. It is suggested that this sensitivity may be used to 'tune' a redox couple for a specific purpose such as designing a reversible oxygen breathing system. . . Whilst it seems that the redox potentials of iron porphyrins are less sensitive to the environment than those of iron pc, dramatic changes can still be observed. Thus the 'tuning' mechanism may have a biological role through, for example, protein bound groups moving closer to, and further away from, the metal ion, with conformational change."*

One-electron oxidation and reduction states of pcs are the subject of a thesis.[34] In a series of two publications the redox behavior of metallic complexes of various tetrapyrrolic rings are studied.[35,36] Their titles are "Halfwave Potentials and Nature of the Complex,"[35] and "Monoradical Identification and Stability".[36] In the first research, the redox reactions of metallic derivatives (Zn, Mg, and Cd) of tetraphenylporphine and pc are studied by voltammetry on a rotating Pt electrode in DMF. Chlorophylls a and b and protochlorophyll a are also studied.[35]

The monoradicals of Zn, Mg, and Cd tetraphenylporphines and pcs generated by constant voltage electrolysis are characterized by their ESR and optical absorption spectra.[36] Their electronic structure can explain some of their redox properties and reactivity.[36]

* From Lever, A. B. P. and Wilshire, J. P., *Can. J. Chem.*, 54(15), 2514, 1976. With permission.

The presence of methyl viologen accelerates photooxidation in ethanol of chlorophyll and its analogs, presumably including pc.[37]

VII. OXIDATION CONSIDERATIONS OF TRANSITION METAL PHTHALOCYANINES

A. Higher Oxidation States

"Higher oxidation states of iron and cobalt are of interest with regard to their role in the catalytic function of enzymes. Thus, iron in the oxidised form of horseradish peroxidase and radish peroxidase, once formulated as iron (V), has recently been reformulated, on the basis of Moessbauer studies, to contain iron (IV), a very rare oxidation state for iron. One oxidising equivalent is assumed to reside on the ligand, rather than on the metal ion. The existence of similar Co(IV) compounds has not been established, although a Co(IV) intermediate has been suggested in a model system for vitamin B_{12} to explain the oxidative cleavage of a cobalt-carbon bond. Recently there was a report of halide complexes derived from Co(II) pc formulated [P. Mertens and H. Vollman, U.S. Patent 3,651,082/1972] as Co pc X_2. Such complexes were obtained by reaction of Co(II) pc with thionyl chloride, bromine, or thionyl bromide. We have re-prepared these complexes and have observed that reaction of Fe(II) pc with thionyl chloride yields the complex Fe pc Cl_2."[38] The complexes Co pc Cl_2, Co pc Br_2, and Fe pc Cl_2. . . are characterised as six-coordinate derivatives involving halogen co-ordinated to the metal on either side of the macrocyclic ring."[38] "In summary, reaction of Co(II) pc with thionyl chloride and with bromine leads to the formation of Co pc Cl_2 and Co pc Br_2, respectively; magnetic, ir, oxidative titration, and esr data indicate these complexes to contain a pc radical species, characterised in the solid state for the first time."*

Also, "radical pc (1 -) complexes of Cr(III), Fe(III), Co(III), and Zn(II) are reported and characterized by electronic and vibrational spectroscopy, magnetism, esr, and Moessbauer spectroscopy and through oxidative titrations. . . It is suggested that most of the previously reported Mn(III) pc derivatives are five-coordinate. A survey of the first-row transition metal pcs reveals that only the above-mentioned ions will form radical pc species under the conditions used."[39,40]

B. Lower Oxidation States

In a research entitled "Pc Complexes of Transition Metals with Unusually Low Oxidation Numbers for the Central Atoms," the authors react 0.5 to 2 mol of Li_2 benzophenone with pc complexes of Ti to Ni in THF to give intensely colored air- and moisture-sensitive complexes. The following were isolated and analyzed: Li[M pc] · 5 THF (M = Fe, Co), Li[Ni pc] · 4 THF, Li_2[M pc] · 6 THF (M = Fe, Co, Ni).[41] "All are low spin complexes with 0 or 1 unpaired electron. In the mono lithium compounds, which behave as electrolytes in CH_3CN, the central atom is in oxidation state $+1$; in the dilithium compounds, which are sparingly soluble in THF, the oxidation state is zero."[41] In another study, Ni, Co, Fe, and Mn pcs are reduced in THF with Li and Na and their addition compounds with $C_{10}H_8$ and benzophenone.[42]

It is also demonstrated that pc can be reduced to Li_3pc and Li_4pc according to the reactions:[43]

$$Li_2pc + \frac{1}{2}Li_2BzPh \xrightarrow{THF} \frac{1}{2}BzPh + Li_3pc \cdot 6\ THF$$

* From Canham, G. W. R., Meyers, J. F., Lever, A. B. P., *J. Chem. Soc. Chem. Commun.*, 14, 483, 1973. With permission.

and

$$Li_3pc \cdot 6\ THF + Li_2BzPh \xrightarrow{THF} LiBzPh + Li_4pc \cdot 8\ THF$$

$Li_3pc \cdot 6$ THF and $Li_4pc \cdot 8$ THF are "both sensitive to oxygen and moisture, lose the THF quantitatively in vacuum at 150°C, and, like Li_2pc, are electrolytes."

VIII. REDUCTION

"In view of their natural importance in photosynthesis and as oxygen carriers, and their use as pigments and in catalytic oxidation, metal porphyrin and pc complexes, together with related 'model' macrocyclic systems, continue to be a field of current interest. Their redox behaviour is still not fully understood. Polarographic studies for a number of metal porphyrins have shown that it is possible to form anionic species where up to four electrons may be added to the porphyrin molecule. In general reduction occurs by electron addition to a vacant orbital of the ring system and the disproportionation energies for a given ion charge have been found to be almost constant. The relative magnitude of these energies have been explained in terms of successive electron addition to the $e_g - \pi^*$ ligand orbital. However, when the central metal ion is transitional, partially filled or vacant *d* orbitals are available for electron uptake and there is therefore the possibility that reduction of the metal may occur. Indeed, it is now well established that the first reduction step for the Co(II) pc complex involves addition to an orbital which is essentially metal in character. Reduction potential data for some tetrasulphonated pcs in DMSO have been reported, but although these complexes have the advantage of being considerably more soluble in aprotic solvents than the unsulphonated compounds, only two or three reduction steps could be observed due to the presence of Na^+ counter ions. On the other hand, higher reduction products of metal pcs have been isolated as crystalline solids by chemical reduction using sodium or radical anions. Assignments of the electronic structure of these ions were made on the basis of magnetic susceptibility measurements. In addition both electron spin resonance and electronic spectra have been used in an attempt to rationalise the behaviour of metal pcs toward reduction. Polarographic data for the unsubstituted metal pcs have not previously been reported presumably because of the very low solubility of the majority of these compounds in solvents such as dimethylformamide or dimethylsulphoxide. In this paper we present such results with a view to assigning the electronic structures of pc anions.[44]

In summary, "The reduction potentials of a number of metal pc complexes have been measured at a dropping mercury electrode in dimethylformamide.[44,45] For the divalent transition metal pcs Mn to Cu inclusive, and for free base pc, all of which were too insoluble to be examined by conventional polarographic methods, the reduction potentials have been measured by generating their soluble mono-negative ions by controlled potential electrolysis, and measuring the subsequent reductions and oxidation back to the neutral complex. The results of the investigation are discussed in the light of previous spectroscopic work with a view to correlating the electrode processes with either ligand or metal reductions."*

Pcs can be reduced in polar solvents with formaldehydesulfoxylate, $NH_2 = NHSO_2H$, $Na_2S_2O_4$, or $Na_2S_2O_5$ together with $NaBH_4$, Zn, or Fe.[46]

Blue Fe(II) pc solutions in ethanol and in isopropanol containing sodium hydroxide are reduced on refluxing to red solutions which are stable in the absence of oxygen, but rapidly convert to the original blue solutions in the presence of a trace of oxygen.[47]

* Clack, D. W., Hush, N. S., and Woolsey, I. S., *Inorg. Chim. Acta*, 129, 1976. With permission.

Also reduced are Co, Mn, Ni, and Cu pcs and sulfonated Co, Ni, and Cu pcs. Methanol and t-butanol containing sodium hydroxide do not change colors.[47]

IX. PHOTOREDUCTION CONSIDERATIONS

Mg, Zn, Cd, and Cu pcs are reduced by the light of an electric bulb in degassed pyridine solutions in the presence of hydrazine or H_2S, whereas H_2 pc and Be pc do not photoreduce under these conditions.[48] In the presence of oxygen the photoreduced products are regenerated to the original pc complexes.

In a pc dye image transfer process, leuco pcs are reduced in the print material to lightfast dye images by 3-pyrazolidine developing agents.[49] The print material paper is impregnated with a 6% acetic acid solution of 10 g/ℓ of Phthalogen Blue 1B (leuco pc Co ethylene diamine complex) to which 12 cc of a 30% SiO_2 hydrosol has been added.

Positives by thermal tanning development makes use of a leuco Co pc in the negative layer and agents that reduce it to a pc dye which is incorporated into the final Ag image in the transfer sheet.[50] The reducing agent is 1-phenyl-3-pyrazolidinone.

X. TETRASULFO PHTHALOCYANINES

The mechanism of oxygen reduction in terms of the binding ability of metals or metal chelates for oxygen is studied.[51] When the metal is Fe^{++} and the chelate is pc tetrasulfonic acid, a binuclear complex with oxygen is demonstrated in aqueous solution. This complex is not stable in the presence of H_2O_2 and its stability constant is 10^7/mol whereas the stability constant of oxyhemoglobin is 10^5/mol.

In the reversible fixation of oxygen to Fe(II) pc tetrasulfonic acid, pc 4,4′,4″,4‴ tetrasulfonic acid is treated under nitrogen with one equivalent $Fe(NH_2)(SO_4)_2 \cdot 6H_2O$ at 60°C and the solution adjusted to pH 6 to 7, causing the immediate formation of Fe(II) pc tetrasulfonate as a green complex. On passing oxygen into a solution of this complex the color changes to blue with the consumption of 1 mol oxygen per 2 mol Fe(II) pc tetrasulfonate and the reversible formation of a binuclear oxygen adduct. Addition of ethanol to the blue solution completely restores the green color.[52]

"Fe(II) tetrasulfo pc is probably one of the most important pc metal complexes because of its structural similarity to biologically significant iron-porphyrin complex in hemoglobin, myoglobin, and cytochromes. The redox chemistry of various pc metal complexes have been studied previously by using the techniques such as epr, spectroscopy, magnetic measurements, and electrochemistry. A general review can also be found [A. B. P. Lever *Advances in Inorganic and Radiochemistry,* edited by H. J. Emeleus and A. G. Sharpe, 7, pp 27—114 (1965)]. In the present study, polarography, cyclic voltammetry, controlled-potential coulometry coupled with optical measurements were used to deduce the reduction mechanism of the iron complex."* The apparatus is shown in Figure 5. In the electrochemical reductions of Fe(II) 4,4′,4″,4‴ - tetrasulfo pc in DMSO on a mercury electrode, the pc complex undergoes two one-electron reductions to Fe(O) as verified by polarographic measurements.

Photopolarographic and spectrometric investigations of the oxygen-binding properties of Fe(II) pc tetrasulfonate is the subject of a thesis.[54]

* From Li, C.-Y. and Chin, D.-H., *Anal. Lett.,* 8(5), 291, 1975. With permission.

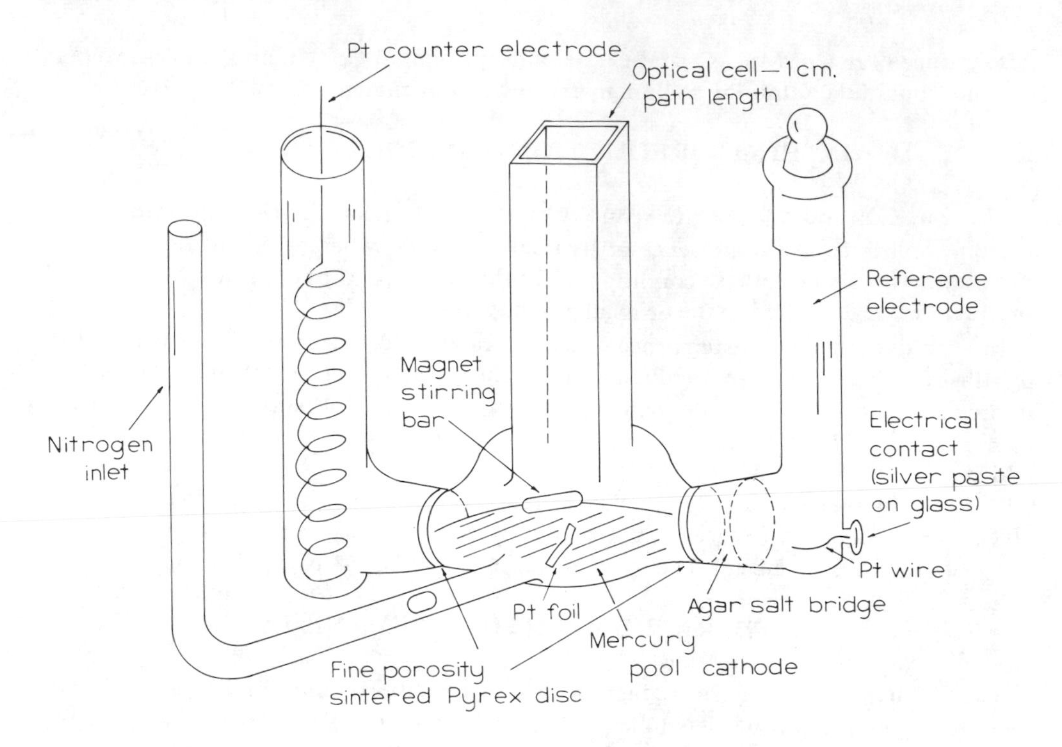

FIGURE 5. Combined optical-electrochemical cell.

REFERENCES

1. Arai, Y. and Sekiguchi, T., Japanese Kokai 69,21,860, September 18, 1969.
2. Sekiguchi, T., Murakami, M., and Banssho, Y., *Kogyo Kagaku Zasshi,* 70(4), 514, 1967.
3. Banssho, Y., Shimura, T., Ueda, O., Tanzaki, H., and Takei, K., *Kogyo Kagaku Zasshi,* 74(9), 1870, 1971.
4. Veprek-Siska, J., Schwertnerova, E., and Wagnerova, D. M., *Chimia,* 26(2), 75, 1972.
5. Abel, E. W., Pratt, J. M., Whelna, R., *J. Chem. Soc. D,* 9, 449, 1971.
6. Busetto, C., Cariati, F., Galizzioli, D., and Morazzoni, F., *Gazz. Chim. Ital.,* 104(1-2), 161, 1974.
7. Wagnerova, D. M., Schwertnerova, E., and Veprek-Siska, J., *Coll. Czech. Chem. Commun.,* 39(8), 1980, 1974.
8. Cariati, F., Morazzoni, F., and Busetto, C., *J. Chem. Soc. Dalton Trans.,* 6, 496, 1976.
9. Islam, A. M., Naser, A. M., El-Mariah, A. A., and Salman, A. A., *J. Oil Colour Chem. Assoc.,* 57(4), 134, 1974.
10. Sekiguchi, T. and Arai, Y., *Tokyo Kogyo Shikensho Hokoku,* 62(6), 202, 1967.
11. Al'yanov, M. I. and Borodkin, V. F., *Izv. Vyssh. Ucheb. Zaved. Khim. Khim. Tekhnol.,* 13(2), 248, 1970.
12. Frank, C. W., Buchacek, R., Dreher, G., Green, E., and Shiffman, S., *2nd Electrophotogr. Int. Conf.,* Soc. Photogr. Sci. Eng., Washington, D.C., 1974, 52.
13. Taube, H. and Arfert, H., *Z. Naturforsch. b,* 22(2), 219, 1967.
14. Stanescu, L., Frunza, E. C., Cohn, M., Wagner, L., and Petcov, R., *Rev. Chim. (Bucharest),* 26(12), 993, 1975.
15. Eigenmann, G., *Helv. Chim. Acta,* 46, 855, 1963.
16. Vogt, L. H., Jr., Zalkin, A., and Templeton, D. H., *Science,* 151(3710), 569, 1966.
17. Yamamoto, A., Phillips, L. K., and Calvin, M., *Inorg. Chem.,* 7(5), 847, 1968.
18. Phillips, L. K., UCRL-17853, U.S. Atomic Energy Commission, Oak Ridge, Tenn., 1967.
19. Glikman, T. S. and Zavgorodnyaya, L. N., *Biofizika,* 15(5), 913, 1970.
20. Przywarska-Boniecka, H., *Roczniki Chem.,* 39(10), 1377, 1965.
21. Canham, G. W. R. and Lever, A. B. P., *Inorg. Nucl. Chem. Lett.,* 9(5), 513, 1973.
22. Engelsma, G., Yamamoto, A., Markham, E., and Calvin, M., *J. Phys. Chem.,* 66, 2517, 1962.
23. Zavgorodnyaya, L. N. and Glikman, T. S., *Zh. Obshch. Khim.,* 39(7), 1443, 1969.
24. Taube, R., Munke, H., and Petersen, J., *Z. Anorg. Allg. Chem.,* 390(3), 257, 1972.

25. Schwertnerova, E., Wagnerova, D. M., and Veprek-Siska, J., *Z. Chem.*, 14(8), 311, 1974.
26. Berezin, B. D. and Sennikova, G. V., *Kinet. Katal.*, 9(3), 528, 1968.
27. Berezin, B. D. and Sennikova, G. V., *Zh. Fiz. Khim.*, 43(10), 2499, 1969.
28. Vul'fson, S. V., Kaliya, O. L., Lebedev, O. L., and Luk'yanets, E. A., *Kinet. Katal.*, 14(6), 1599, 1973.
29. Vul'fson, S. V., Lyubimova, O. I., Kotov, A. G., Kaliya, O. L., Lebedev, O. L., and Luk'yanets, E. A., *Zh. Obshch. Khim.*, 45(8), 1841, 1975.
30. Vul'fson, S. V., Kaliya, O. L., Lebedev, O. L., and Luk'yanets, E. A., *Zh. Org. Khim.*, 10(8), 1757, 1974.
31. Vul'fson, S. V., Kaliya, O. L., Lebedev, O. L., and Luk'yanets, E. A., *Zh. Obshch. Khim.*, 46(1), 179, 1976.
32. Vul'fson, S. V., Kaliya, O. L., Lebedev, O. L., and Luk'yanets, E. A., *Zh. Org. Khim.*, 12(1), 123, 1976.
33. Lever, A. B. P. and Wilshire, J. P., *Can. J. Chem.*, 54(15), 2514, 1976.
34. Sanderson, J. E., One Electron Oxidation and Reduction States of Pcs thesis, avail. Univ. Microfilms, Ann Arbor, Mich., Order No. 71-22,619; *Diss. Abstr. Int. B*, 32(3), 1463, 1971.
35. Lexa, D. and Reix, M., *J. Chim. Phys. Physicochim. Biol.*, 71(4), 511, 1974.
36. Lexa, D. and Reix, M., *J. Chim. Phys. Physicochim. Biol.*, 71(4), 517, 1974.
37. Evstigneev, V. B. and Gavrilova, V. A., *Biokhimiya*, 37(5), 952, 1972.
38. Canham, G. W. R., Myers, J. F., and Lever, A. B. P., *J. Chem. Soc. Chem. Commun.*, 14, 483, 1973.
39. Myers, J. F., Canham, G. W. R., and Lever, A. B. P., *Inorg. Chem.*, 14(3), 461, 1975.
40. Myers, J. F., Lever, A. B. P., and Canham, G. W. R., *16th Proc. Int. Conf. Coord. Chem.*, Univ. College, Dublin, 1.5, 1974.
41. Taube, R., Zach, M., Stauske, K. A., and Heidrich, S., *Z. Chem.*, 3(10), 392, 1963.
42. Taube, R., *Chem. Zvesti*, 19(3), 215, 1965.
43. Taube, R. and Meyer, P., *Angew. Chem. Int. Ed.*, 5(11), 972, 1966.
44. Clack, D. W., Hush, N. S., and Woolsey, I. S., *Inorg. Chim. Acta*, 129, 1976.
45. Clack, D. W. and Hush, N. S., *J. Am. Chem. Soc.*, 87(19), 4238, 1965.
46. Wegmann, J. (to Ciba Ltd.), German Offenlegungsschrift 2,016,365, October 29, 1970.
47. Charman, H. B., *Nature (London)*, 201(4923), 1021, 1964.
48. Savel'ev, D. A., Kotlyar, I. P., and Sidorov, A. N., *Zh. Fiz. Khim.*, 43(7), 1914, 1969.
49. v. Koenig, A., Wolf, W., and Maeder, H. (to Agfa AG), Belgian Patent 637,359, March 13, 1964.
50. v. Koenig, A., Wolf, W., and Maeder, H., German Offenlegungsschrift 1,178,705, September 24, 1964.
51. Fallab, S., *Z. Naturwiss. Med. Grundlagenforsch.*, 2(3), 220, 1965.
52. Trefzer, K. and Fallab, S., *Helv. Chim. Acta*, 48(4), 945, 1965.
53. Li, C.-Y. and Chin, D.-H., *Anal. Lett.*, 8(5), 291, 1975.
54. Joe, F. L., Jr., Photopolarographic and Spectrometric Investigation of the Oxygen Binding Properties of Fe(II) Pc Tetrasulfonate, thesis, avail. Xerox Univ. Microfilms, Ann Arbor, Mich., Order No. 76-14,809; *Diss. Abstr. Int. B*, 37(1), 180, 1976.

Chapter 7

PHTHALOCYANINES IN SOLUTION

I. PHTHALOCYANINES IN SULFURIC ACID SOLUTION

The interaction of Al^{3+}, Sn^{2+}, Sn^{4+}, and V^{4+} pcs and of Cu pentadecachloro pc as a function of the concentration of sulfuric acid is studied.[1] The stability of the bond between the central atom and $HSO_4^{\ominus}$ depends on the solvating capacity of the solvent. In dilute sulfuric acid free water molecules weaken or rupture it.

The solubility of H_2 pc in 17 *M* sulfuric acid solution is determined at 25°C and the equilibrium constants are determined for the reaction

$$H_2\,pc + H_2SO_4 \rightleftharpoons H_2\,pc\,H^{\oplus} + HSO_4^{\ominus}$$

The use of the data obtained in practical calculations in the manufacture of pc dyes is discussed.[2]

A quantitative study is made of the hydrolysis of metal pcs in sulfuric acid at 14 to 17 *M* concentration.[3]

Equilibrium constants of the dissolution of metal pcs in 15 to 17 *M* sulfuric acid are determined according to the equation

$$M\,pc + H_2SO_4 \rightleftharpoons M\,pc\,H^{\oplus} + HSO_4^{\ominus}$$

The dissociation of metal pcs in sulfuric acid solution occurs according to the reaction[4]

$$M\,pc\,H^{\oplus} + 2H^{\oplus} \rightleftharpoons H_2\,pc\,H^{\oplus} + M^{(++)}$$

Solubility determinations are made for Mg pc and Pb pc in 17 *M* sulfuric acid solution.[5]

The solubility of Cu pc, Cu pc Cl_{15}, Zn pc, Zn pc Cl, Co pc Cl, and Al pc Cl (HSO_4) in 15 to 17 *M* sulfuric acid is determined and the equilibrium constants are also determined according to Equation 3 above.[6]

The kinetics of the dissociation of Cu, Pd, Pt, Zn, and VO pcs in sulfuric acid are studied,[7] as well as of Co, Ni, Al, Ga, Os(IV), and Rh (III).[8] The possibility is suggested that the mechanism of dissociation of metal pcs is the same as that of metal porphyrins.[8,9]

Pcs are more soluble in oleum and chlorosulfonic acid than in sulfuric acid.[10] Equilibrium constants for solution in oleum and chlorosulfonic acid are calculated.[11]

The study of stability of polymeric H_2 pc, Cu pc, and Co pc toward dissociation in sulfuric acid reveals that the polymeric metal derivatives are 2 to 4 times more stable than the corresponding monomers.[12]

The protonation equilibria for Zn, Cu, Co, Pt, Pd, and Ni pcs and hydrogen sulfates of Ga, Ru, Rh, Ir, Al, Sn, V, and Os pcs determined spectrally in sulfuric acid give only the monoprotonated species.[13]

The electronic spectra of solutions of Cu pc and of 4,4′,4″,4‴-tetracarboxylic Cu pc in sulfuric acid solutions at 43 to 96% concentration, show several protonation equilibria for both compounds.[14]

II. TETRASULFO PHTHALOCYANINES IN AQUEOUS SOLUTION

The association, dimerization, and polymerization of 4,4′,4″,4‴-tetrasulfo pcs in aqueous solution are studied in terms of the species present, influence of solvent and solute molecules, kinetics of association, equilibrium constant, and ionic strength. Absorption spectrophotometry is a favored method to study these phenomena. Among the tetrasulfo pcs that are known to dimerize in aqueous solution are the Co(II), Cu(II), Zn(II), VO(II), Fe(III), and Mn(III) complexes.

Dimerization and the monomer-dimer equilibrium are represented by the equations, where ts = tetrasulfo,

$$2 \text{ ts pc} \rightleftharpoons (\text{ts pc})_2$$

$$K_D = \frac{[(\text{ts pc})_2]}{[\text{ts pc}]^2}$$

$(Co\ ts\ pc)_2$ and $(Zn\ ts\ pc)_2$ are less stable than $(H_2\ ts\ pc)_2$ and $(Cu\ ts\ pc)_2$, explained by the presence of axially coordinated water molecules in $(Co\ ts\ pc)_2$ and in $(Zn\ ts\ pc)_2$.[15] Typical dimerization $\log K_D$ constants in aqueous solution at 17 to 60°C are in the range 5.4 to 8.2, for Fe(III), H_2, Co(II), Cu(II), and VO(II) ts pc.

Fe(III) ts pc dimerizes and, at $>10^{-5}$ *M* concentration, polymerizes.[15]

Water is the only solvent which promotes formation of the dimeric Co(II) ts pc.[16] The equilibrium between monomer (M) and dimer (D) obeys the equilibrium relationship

$$K = [D] / [M]^2 [H_2O]^n$$

where $n \approx 12$ in both water-methanol and water-ethanol mixtures. It is suggested that the water molecules play a specific role in binding together the monomers by hydrogen bonds.

The addition of molecular oxygen to Co(II) ts pc by bubbling oxygen through an alkaline aqueous solution is reversible.[17] The monomer-dimer system prevails at neutral pH; the oxygen adduct and the dimer predominate in alkaline solution.

The concentration-jump relaxation method is applied to a study of the kinetics of the dimerization of the tetrasodium salt of aqueous Co(II) ts pc.[18] The thermodynamic functions $\Delta H°$, $\Delta G°$, and $\Delta S°$, and activation parameters of dimerization are also calculated, as well as the radius of the activated complex.[19]

Thermodynamic parameters are also determined for Cu(II)-ts pc.[20] "By working at concentrations sufficiently low ($<10^{-5}$ *M*) to eliminate the possibility of aggregates higher than dimers, and by assuming that Beer's Law holds for each component, it is possible to calculate the proportions of monomer (M) and dimer (D). From this information the dissociation constant (K_d) was calculated. . . From the study of K_d at various temperatures, the thermodynamic quantities $\Delta G°$, $\Delta H°$, and $\Delta S°$ were calculated."*

The effect of urea and thiourea on the aggregation of Cu ts pc is studied.[21] "Cu ts pc was chosen as a model dye because it can be prepared in a pure state, it has been carefully studied spectrophotometrically, and the pc chromophore is important in the production of turquoise blue shades on wool. The results. . . show that Cu ts pc in aqueous salt solution is disaggregated by urea and thiourea. The rate of diffusion of the dye in free solution and hence through any hydrodynamic boundary layer of a fiber can be increased or decreased by these compounds."**

* From Blagrove. R. J. and Gruen, L. C., *Aust. J. Chem.*, 26(1), 225, 1973. With permission.

** From Blagrove. R. J., *Aust. J. Chem.*, 26(7), 1545, 1973. With permission.

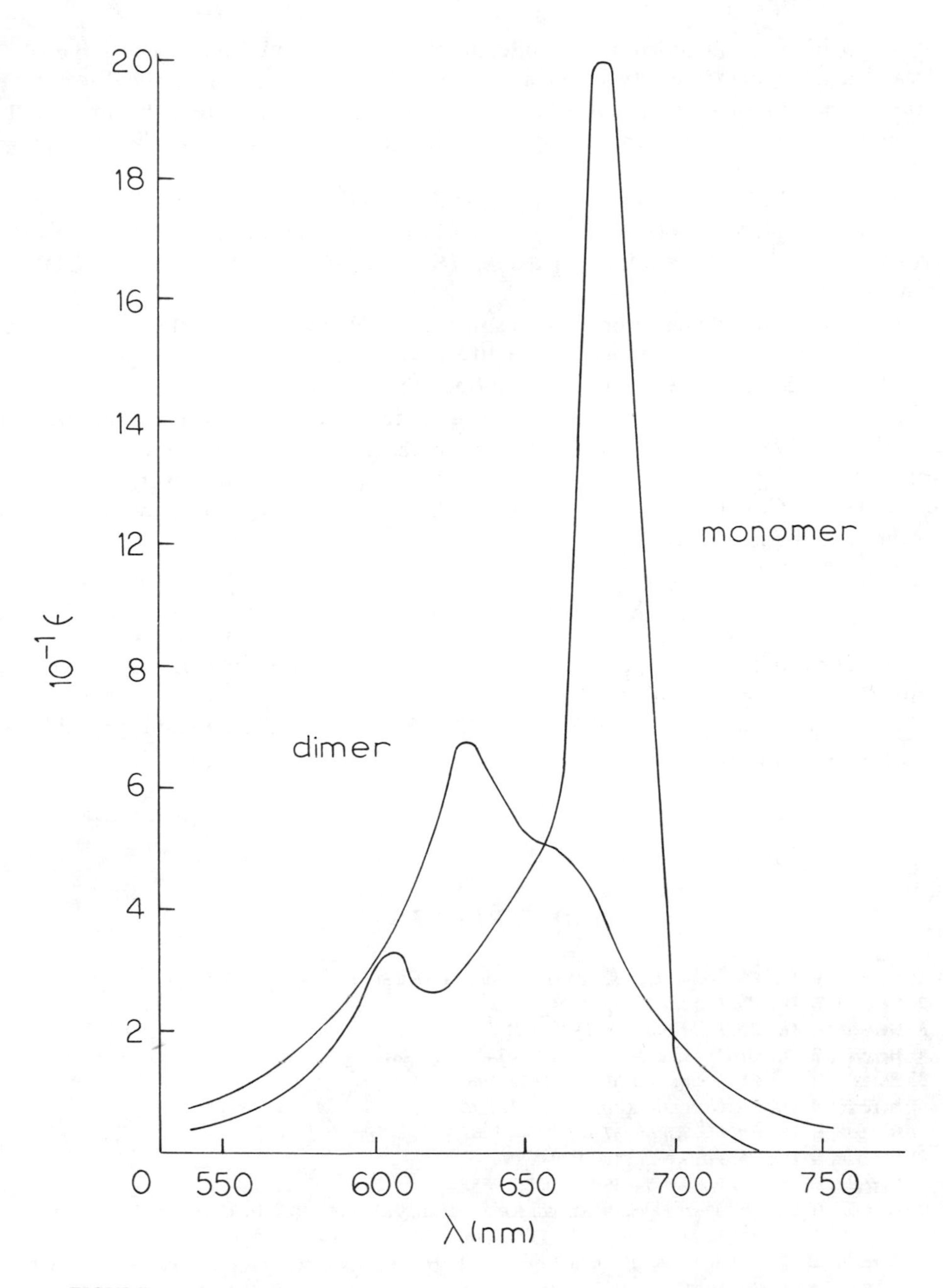

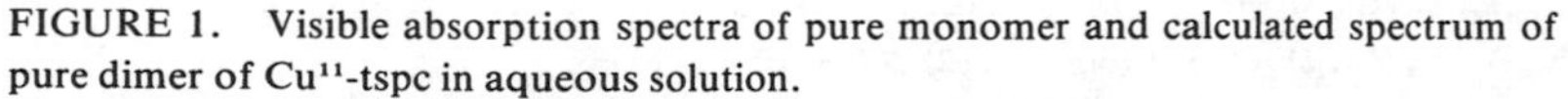
FIGURE 1. Visible absorption spectra of pure monomer and calculated spectrum of pure dimer of Cu^{II}-tspc in aqueous solution.

In water, dimers and monomers and, at high dye concentrations, dimers and tetramers of Cu ts pc are formed.[22] By addition of electrolytes (NaCl, H_2SO_4), the association tendency is increased, and by addition of organic solvents, it is decreased.

The size and shape of the aggregates of sodium Cu pc tetrasulfonate are measured in aqueous solution by small angle X-ray scattering.[23] The smallest aggregates found are dimers and the largest contain 20 molecules.

Visible absorption spectra of pure monomers and the calculated spectrum of pure dimers of Cu(II) ts pc in aqueous solution are shown in Figure 1.[24]

Owing to the strong aggregating tendency of Cu(II) ts pc in water, it "is likely to exist in a highly aggregated form under normal dyeing conditions. The molecule is already a large planar entity in its monomeric form and greater difficulty is to be expected in obtaining even penetration of fibers when it is aggregated. Presumably this problem will be overcome only by using conditions which cause extensive disaggregation in the dye bath.''*

Solutions of Mn(III) ts pc in water contain an equilibrium mixture of Mn ts pc $(OH)(H_2O)^{4-}$ and Mn ts pc $(OH)_2^{5-}$ ions in acidic aqueous solution. At pH ∿ 9, further hydrolysis occurs and a dimer, perhaps (H_2O) Mn ts pc-O-Mn ts pc $(OH)^{9-}$ is formed.[25,26]

Absorption spectra measurements on solutions of H_2 pc sulfonate show that in water at 10^{-4} *M* the pc is mainly dimerized, at 10^{-6} *M* is a mixture of monomers and dimers, and at 10^{-2} *M* is a mixture of dimers and tetramers.[27]

The kinetics and mechanism of formation of metal ts pcs in aqueous solution are investigated.[28-31] Experiments were made with Cu and Zn[28] as well as with Co.[31] The rate of formation of metal pc tetrasulfonic acid depends on the dissociation rate of the N–H bond for Cu^{++}, but not for Zn^{++}, Co^{++}, Mn^{++}, or Ni^{++}. The overall reaction for the acid dissociation of H_2 ts pc is shown to be

$$(H_2\ \text{ts pc})_2 + 2\ OH^{\ominus} \rightleftharpoons 2\ H\ \text{ts pc}^{\ominus} + 2\ H_2O$$

Properties of tetrasulfo pc complexes in solution and bonding in bis(benzene-1,2-dithiolate) Ni(II) are the subject of a thesis.[32]

Studies of VO ts pc at 5×10^{-7} to 5×10^{-5} *M* indicate monomer-dimer equilibrium over this concentration range.[33]

REFERENCES

1. Berezin, B. D., *Izv. Vyssh. Ucheb. Zaved. Khim. Khim. Tekhnol.*, 4, 379, 1961.
2. Berezin, B. D., *Zh. Fiz. Khim.*, 35, 2494, 1961.
3. Berezin, B. D., *Zh. Fiz. Khim.*, 36, 494, 1962.
4. Berezin, B. D., *Dokl. Akad. Nauk S.S.S.R.*, 141, 353, 1961.
5. Berezin, B. D., *Zh. Neorgan. Khim.*, 6, 2672, 1961.
6. Berezin, B. D., *Zh. Neorgan. Khim.*, 7, 2507, 1962.
7. Berezin, B. D., *Zh. Fiz. Khim.*, 37(11), 2474, 1963.
8. Berezin, B. D., *Zh. Fiz. Khim.*, 38(4), 850, 1964.
9. Berezin, B. D., *Zh. Fiz. Khim.*, 39(5), 1082, 1965.
10. Berezin, B. D., *Izv. Vyssh. Ucheb. Zaved. Khim. Khim. Tekhnol.*, 6(6), 1016, 1963.
11. Berezin, B. D., *Zh. Prikl. Khim.*, 36(6), 1181, 1963.
12. Berezin, B. D., Akopov, A. S., and Lomova, T. N., *Tezisy Dokl. Vses. Chugaevskoe Soveshch. Khim. Kompleksn. Soedin., 12th*, 2, 185, 1975.
13. Berezin, B. D., *Zh. Obshch. Khim.*, 43(12), 2738, 1973.
14. Gaspard, S., Verdaguer, M., and Viovy, R., *J. Chim. Phys. Physicochim. Biol.*, 69(11—12), 1740, 1972.
15. Sigel, H., Waldmeier, P., and Prijs, B., *Inorg. Nucl. Chem. Lett.*, 7(2), 161, 1971.
16. Abel, E. W., Pratt, J. M., and Whelan, R., *J. Chem. Soc. Dalton Trans.*, 6, 509, 1976.
17. Gruen, L. C. and Blagrove, R. J., *Aust. J. Chem.*, 26(2), 319, 1973.
18. Schelly, Z. A., Farina, R. D., and Eyring, E. M., *J. Phys. Chem.*, 74(3), 617, 1970.

* From Blagrove, R. J. and Gruen, L. C., *Aust. J. Chem.*, 25(12), 2553, 1972. With permission.

19. Schelly, Z. A., Harward, D. J., Hemmes, P., and Eyring, E. M., *J. Phys. Chem.*, 74(16), 3040, 1970.
20. Blagrove, R. J. and Gruen, L. C., *Aust. J. Chem.*, 26(1), 225, 1973.
21. Blagrove, R. J., *Aust. J. Chem.*, 26(7), 1545, 1973.
22. Ahrens, U. and Kuhn, H., *Z. Physik. Chem.*, 37 1/2, 1, 1963.
23. Kratky, O. and Oelschlager, H., *J. Colloid. Interface Sci.*, 31(4), 490, 1969.
24. Blagrove, R. J. and Gruen, L. C., *Aust. J. Chem.*, 25(12), 2553, 1972.
25. Fenkart, K. and Brubaker, C. H., Jr., *J. Inorg. Nucl. Chem.*, 30(12), 3245, 1968.
26. Fenkart, K. and Brubaker, C. H., Jr., *Inorg. Nucl. Chem. Lett.*, 4(6), 335, 1968.
27. Schnabel, E., Noether, H., and Kuhn, H., *Recent Prog. Chem. Nat. Syn. Colour. Matters*, p.561, 1962.
28. Schiller, I. and Bernauer, K., *Helv. Chim. Acta*, 46(7), 3002, 1963.
29. Schiller, I., Bernauer, K. and Fallab, S., *Experientia*, 17, 540, 1961.
30. Bernauer, K. and Fallab, S., *Helv. Chim. Acta*, 45, 2487, 1962.
31. Bernauer, K. and Fallab, S., *Helv. Chim. Acta*, 44, 1287, 1961.
32. Brooks, H. B., Solvent Interactions and Bonding of Some Delocalized Square-Planar Complexes. Properties of Tetra-Sulfophthalocyanine Complexes in Solution. Bonding in Bis(Benzene-1,2-Dithiolate) Nickel(II), thesis, Order No. 69-395, avail. Univ. Microfilms, Ann Arbor, Mich., *Diss. Abstr. B*, 29(8), 2795, 1969.
33. Farina, R. D., Halko, D. J., and Swinehart, J. H., *J. Phys. Chem.*, 76(17), 2343, 1972.

Chapter 8

DIMERIZATION

Sulfonation of di pcs of Eu, Gd, Tm, Yb, and Lu gives products that are relatively soluble in water and do not colorize in strong NH_4OH, indicating the strength of the metal-organic bond.[1]

Sulfonation of di pcs of Y, Gd, and Lu with gaseous SO_2 in aqueous solution yields sandwich-type structures.[2]

Analysis of the absorption spectra vs. concentration of chloro rare earth pcs in carbon tetrachloride at 10^{-8} to 10^{-4} *M* indicates that chloro rare earth pcs exhibit monomer-dimer equilibria in carbon tetrachloride solution.[3]

Dimerization in carbon tetrachloride, benzene, and in other organic solvents is indicated for 4,4′,4″,4‴-tetraoctadecylsulfonamido pc copper (II).[4]

The thermodynamics of dimerization of 4,4′,4″,4‴-tetraoctadecylsulfonamido Cu (II) pc are determined at 35°C in benzene by adiabatic calorimetry. $\Delta H°$, $\Delta G°$, and $\Delta S°$ for dimerization are calculated.[5] "The calorimetric approach is a useful supplement to the more traditional approach of spectrophotometry."

"Thermodynamics and kinetics of dimerization of 4,4′,4″,4‴-tetraoctadecylsulfonamido metal (II) pc dyes" is the subject of a thesis.[5]

The thermodynamics of dimerization of Cu(II) 4,4′,4″,4‴ tetrakis(octadecylsulfamoyl) pc are also determined at 35°C in benzene by an adiabatic calorimetric technique.[6]

Results of rate studies of the dimerization of several pc dyes in solution are briefly described.[7]

"The absorption spectrum of the Si pc dimer can be satisfactorily interpreted in terms of vibronic coupling."[8]

The slow dedimerization of Fe (II) pc in DMSO is observed when a 5×10^{-4} solution is diluted 100-fold,[9] with all species uncharged. The "rate of dedimerization is of the same order of magnitude as that observed for the dedimerization of the tetrasodium salt of Co (II)-4,4′,4″,4‴-tetrasulfo pc in water, and is 10^5 times slower than the dedimerization rate. . . for a water soluble metal free porphyrin in water."

In "Practical Observations on the Spectrophotometry of Dyes and Aggregation Effects" attention is drawn "to the effectiveness and economy of non-ionic surfactants for disaggregation" and also to promoting aggregation by adding an electrolyte.[10] "When the dyes concerned show the absorption peaks corresponding to monomer and dimer, it is possible to improve the wavelength separation by including agents in the bath which preserve the one dye largely as its dimer while converting the other to the monomer. Various mixtures of such a pair were successfully measured in the presence of a specific quantity of sodium bicarbonate as aggregating electrolyte and a predetermined concentration of Triton-X-100 to disaggregate the dye component that was sufficiently sensitive to it."*

* From Tull, A. G., *J. Soc. Dyers Colour,* 89(4), 132, 1973. With permission.

REFERENCES

1. Moskalev, P. N. and Kirin, I. S., *Zh. Obshch. Khim.*, 38(5), 1191, 1968.
2. Moskalev, P. N. and Kirin, I. S., *Zh. Neorg. Khim.*, 16(1), 110, 1971.
3. Yamana, M., *Shikizai Kyokaishi*, 48(2), 76, 1975.
4. Monahan, A. R., Brado, J. A., and DeLuca, A. F., *J. Phys. Chem.*, 76(3), 446, 1972.
5. Graham, R. C., Thermodynamics and Kinetics of Dimerization of 4,4′,4″,4‴-Tetraoctadecylsulfonamido Metal (II) pc Dyes, thesis, Order No. 73-20, 152, avail. Univ. Microfilms, Ann Arbor, Mich., *Diss. Abstr. Int. B*, 34(3), 1053, 1973.
6. Graham, R. C., Henderson, G. H., Eyring, E. M., and Woolley, E. M., *J. Chem. Eng. Data*, 19(4), 297, 1974.
7. Eyring, E. M., NTIS AD Rep. No. 755867, National Technical Information Service, Springfield, Va., *Govt. Rep. Announce. (U.S.)*, 73(7), 87, 1973.
8. Petelenz, P. and Zgierski, M. Z., *Mol. Physics*, 25(1), 237, 1973.
9. Jones, J. G. and Twigg, M. V., *Inorg. Nucl. Chem. Lett.*, 8(4), 305, 1972.
10. Tull, A. G., *J. Soc. Dyers Colour.*, 89(4), 132, 1973.

Chapter 9

THERMAL STABILITY

I. INTRODUCTION

The thermal stability of pcs is a study of recurring interest because it is one of their elegant and outstanding features. The focus of attention is on their behavior at increasingly elevated temperature in a vacuum, inert atmosphere, oxidizing, or reducing atmosphere. The fact that both sublimation and decomposition with vaporization of some products of decomposition occur simultaneously provides a challenging interplay of factors for the investigator.

II. TECHNIQUES AND APPLICATIONS

Two basic techniques that are popular in the study of thermal properties of pcs are thermogravimetric analysis (TGA), wherein the weight of a sample is followed over a period of time while the temperature is changed at a given rate, and differential thermal analysis (DTA), wherein observation of heat absorbed or liberated of a sample, especially at high temperatures where only a few techniques are applicable, may indicate chemical reactions and phase transitions useful especially in studying structure of solids.

The thermograms of neodymium, cobalt, and nickel pcs are developed.[1] Strong endothermic effects are found in thermograms of Fe^{3+}, Co^{2+}, and Ni^{2+} at about 800°C.

Decomposition of Pt complexes, including pcs, in hydrogen, air, and argon are investigated by TGA and DTA.[2] The Pt complexes are more stable in air than in hydrogen where decomposition is observed in all Pt samples at less than 200°C.

From a study of the temperature dependence of the mass spectra of H_2 pc, VO pc, and Cu pc, it is shown that these pcs are thermally stable at 513 to 673 K, 523 to 623 K, and 448 to 548 K, respectively. These pcs also vaporize as monomeric molecules.[3]

The thermal stability of the following pcs decreases in the sequence Cu pc $>$ VO pc $>$ H_2 pc $>$ Ni pc $>$ $GeCl_2$ pc $>$ $SiCl_2$ pc $>$ $ZrCl_2$ pc $>$ $SnCl_2$ pc. In air the interval of thermal stability of the complexes decreases as a result of the oxidation of the central metal atom. For H_2 pc the interval of thermal stability in air and in an inert atmosphere coincide.[4]

The elemental composition and IR spectra are given for the products of thermal decomposition of $GeCl_2$ pc at 440, 600, and 800°C. The pc ring is completely destroyed at 600°C.[5]

The thermal decomposition of Cu, Mg, Co, VO, $GeCl_2$, $SnCl_2$, $SiCl_2$, $(HO)_2Si$, SO_4Os, Cu tetrachloro, and Cu pentadecachloro pc, Cu tetraamino pc, polymeric Co pc, and H(OSi pc)$_n$ OH is studied by heating in air at $<$ 700°C. The rate of oxidative thermal decomposition of metal pcs depends on the crystal properties of the complexes and on the nature of the metal ligand bond.[6]

To study the carbonization of Si pc, the technique of TGA, elementary analysis, X-ray analysis, IR absorption analysis, and the measurement of electrical resistance are applied[7] to samples of Si pc heated to 1000°C in a nitrogen atmosphere. Si pc is found to be remarkably thermostable. Near 800°C the residual solid contains some orders of molecular arrangement of pc. At 1000°C the sample is carbonaceous.

In an argon atmosphere both Co pc and Ni pc decompose at about 750°C with liberation of nitrogen and the metal, and they change to a carbonaceous structure at 800°C.[8] The pyrolysis pattern studied by TGA to 1000°C of Fe pc is similar to that of

Co pc. On DTA, Fe pc exhibits endothermic reactions at 615, 645, and 690°C, and exothermic reaction at 770°C. The endothermic reaction at 690°C is not observed in Cu pc, Co pc, or Ni pc.[9] The pyrolysis temperatures of $SnCl_2$ pc and VO pc are lower than those of Cu pc, Co pc, and Ni pc, studied by TGA to 1000°C in argon. Between 500 and 530°C, $SnCl_2$ pc changes to Sn pc and at about 550°C the structure breaks down completely with evolution of HCN, benzene, and phenyl cyanide. Crystals of metallic tin deposit at 625°C and carbonization progresses further.[10]

The thermal characteristics of isatin derivatives determined by TGA and DTA are similar to those of Cu pc pigments.[11]

III. POLYMERS

Polymers with a pc ring are more resistant to thermal oxidation than macromolecules with a carborane nucleus with alkyl groups bonded to the carbon atoms of the polyhedron.[12]

The weight losses of pcs and poly pcs in air at 250°C during 1000 hr depend in magnitude upon the metallic component. Copper-containing materials are among the most rapidly degraded under these conditions. The pc polymers formed without metal and those containing Zn or Ni are relatively quite stable.[13]

At the 1974 Spring ACS Meeting a resin system in which polymerization is made to occur through aromatic *o*-dinitrile groups was described.[14] "Indications were that this process involved the formation of the highly stable pc nucleus. The resins were synthesized from 4-aminophthalonitrile and aliphatic diacid chlorides to give diamides of the following structure:

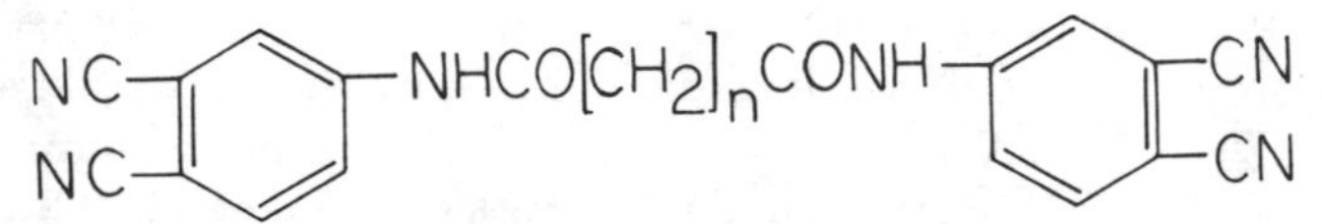

The polymerization of the resin can be represented in the following way:

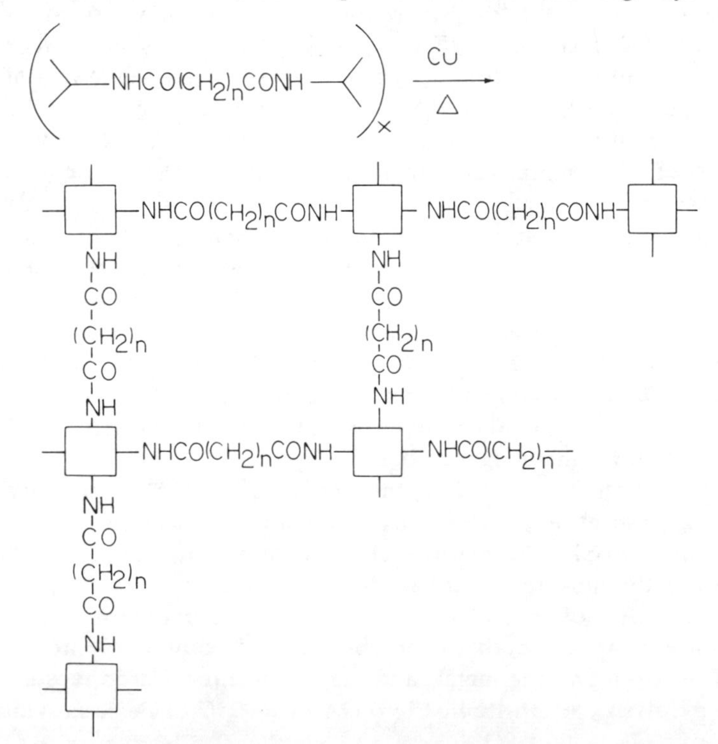

where

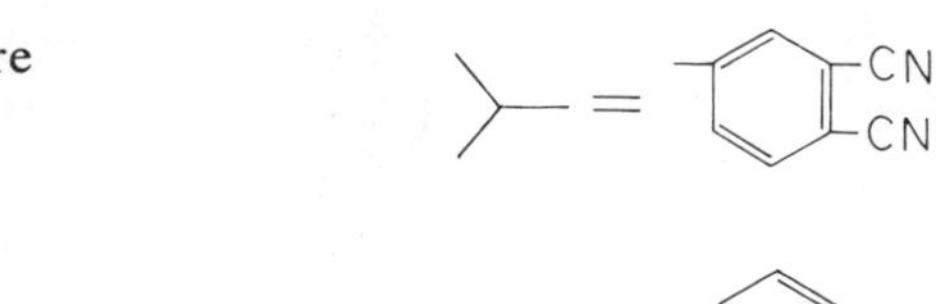

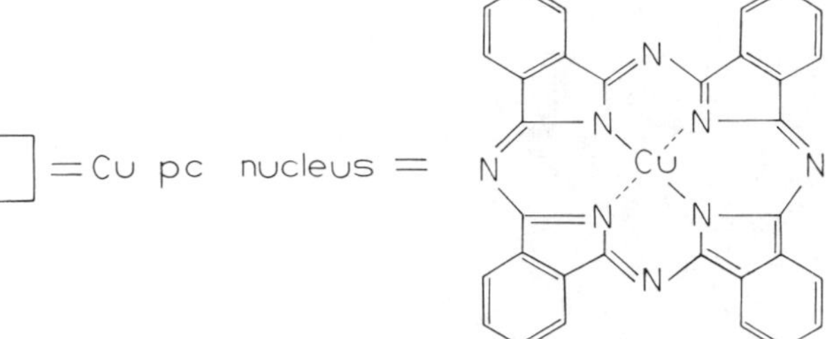

In a subsequent research entitled "Pc Resins: A New Class of Thermally Stable Resins for Adhesives, Coatings, and Plastics"[15] a second class of resins is described where the $(CH_2)_n$ units are replaced by phenyl or phenyl ether groups:

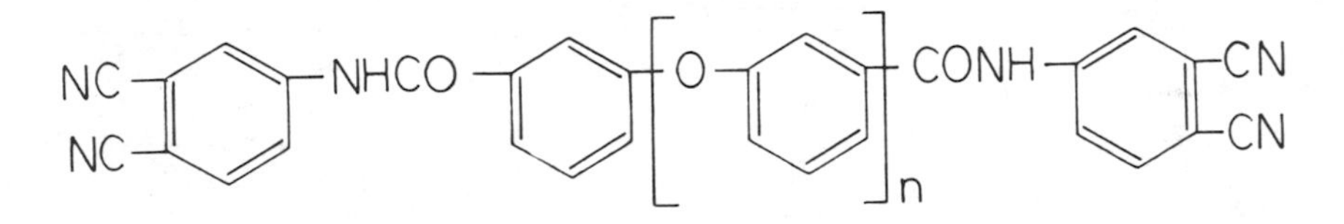

"None of the cured resins showed any weight loss on heating in air at temperatures up to 215°K for as long as 569 hours. Figure 1 shows the weight loss of cured Cu free resins for n = 0, 1, and 4 for the indicated time period (solid line) at 275°C, and then 325°C (dashed line). At the end of the heating period, all three materials still retained their structural integrity and exhibited no gross loss in strength."

IV. FLAMMABILITY

The flammability of dyed cotton fabrics is studied "to investigate the influence of dyestuffs on the combustion and the course of pyrolysis of the dyed materials."[16] The dyes include a Copc and a Cupc dye. This may be the first information obtained on the flammability of dyed cotton fabric as revealed by oxygen index (OI) and by means of TGA following thermal degradation in a nitrogen atmosphere.

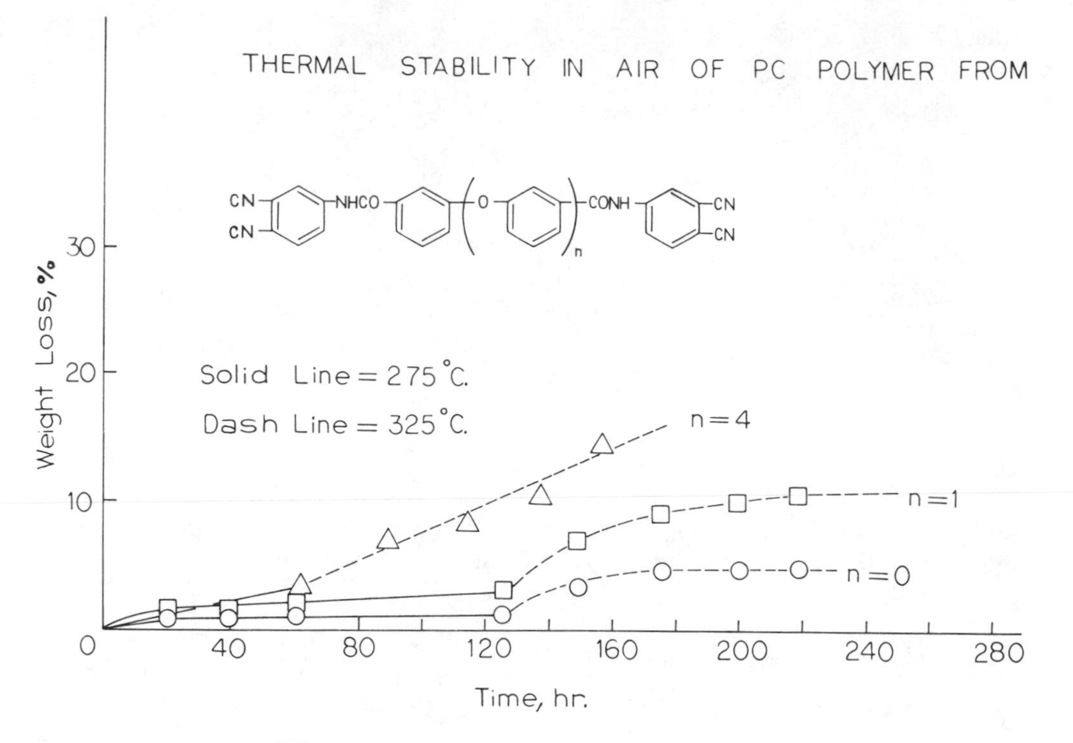

FIGURE 1. Rate of weight loss at high temperatures.

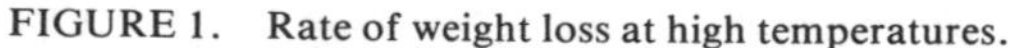

REFERENCES

1. Kirin, I. S., Moskalev, P. N., and Mishin, Ya., *Zh. Obshch. Khim.*, 37(5), 1065, 1967.
2. Kinoshita, K., Routsis, K., and Bett, J. A. S., *Thermochim. Acta*, 10(1), 109, 1974.
3. Belousov, V. I., Kiryukhin, I. A., and Ivanova, G. A., *Zh. Fiz. Khim.*, 50(4), 870, 1976.
4. Markova, I. Ya., Kiryukhin, I. A., Shaulov, Yu. Kh., Benderskii, V. A., and Grigorovich, S. M., *Zh. Neorg. Khim.*, 21(3), 660, 1976.
5. Shaulov, Yu. Kh., Markova, I. Ya., Popov, Yu. A., and Ruchkin, E. D., *Zh. Neorg. Khim.*, 17(3), 634, 1972.
6. Akopov, A. S., Berezin, B. D., Klyuev, V. N., and Morozova, G. G., *Zh. Neorg. Khim.*, 20(5), 1264, 1975.
7. Shindo, A., Souma, I., and Yamaguchi, M., *Kogyo Kagaku Zasshi*, 71(5), 658, 1968.
8. Souma, I. and Kotsuka, K., *Nippon Kagaku Kaishi*, 7, 1249, 1974.
9. Souma, I. and Kotsuka, K., *Nippon Kagaku Kaishi*, 9, 1811, 1974.
10. Souma, I. and Masuda, H., *Shikizai Kyokaishi*, 49(2), 90, 1976.
11. Vilceanu, R., Bader, B. D., Rădulescu, M., and Marinescu, M., *Rev. Roum. Chim.*, 18(7), 1225, 1973.
12. Delman, A. D., Kelly, J. J., Stein, A. A., and Simms, B. B., *Therm. Anal., Proc. 2nd Int. Conf.*, Vol. 1, Academic Press, New York, 1969, 539.
13. Walton, T. R. and Griffith, J. R., *Appl. Polym. Symp.*, 26, 429, 1975.
14. Griffith, J. R., O'Rear, J. G., and Walton, T. R., *Advances in Chemistry Series No. 142, Copolymers, Polyblends, and Composites*, Platzer, N. A. J., Ed., American Chemical Society, Washington, D.C., 1975, 465.
15. Walton, T. R., Griffith, J. R., and O'Rear, J. G., *Polym. Sci. Technol.*, 9B (Adhes. Sci. Technol.), 665, 1975.
16. Segal, L., Timpa, J. D., and Drake, G. L., Jr., *Text. Res. J.*, 44(11), 839, 1974.

Chapter 10

RADIATION

I. RADIATION CHEMISTRY

A. Szilard-Chalmers Process

The Szilard-Chalmers process continues to be examined with respect to the pcs for the preparation of radioactive isotopes of the central metal atom.

In the Szilard-Chalmers process, the target material or compound containing the atom, the radioactive isotope of which is desired, is bombarded with thermal or slow neutrons. For the Szilard-Chalmers process to be effective, upon neutron capture the newly formed atom must separate from the host molecule; it must not recombine with the molecular fragment from which it separates; and there must be a chemical method to separate the isotope from the residual target compound. An effective Szilard-Chalmers separation yields radioactive isotopes of high specific activity, high enrichment or concentration, and high yield of separable activity. The specific activity of a product sample is the ratio of the number of radioactive atoms of the element to the total number of atoms of the element present in the sample. Enrichment or concentration is the ratio of specific activity of a preparation after separation to that of the unseparated target. The yield of separable activity indicates the percentage of the radioactivity that can be separated from the bombarded target.

$$E = S_1/S_2 = Ya/100b$$

where E = enrichment, S_1 = specific activity of separated material, S_2 = specific activity of compound before separation, Y = yield of separable activity, a = total amount of metal in the compound, and b = total amount of metal in the separated material.

"Several metal pcs have been used as target compounds for the preparation of radionuclides via recoil (hot atom) procedures. The typical procedure utilizes concentrated sulfuric acid to dissolve the irradiated pc, followed by precipitation and filtration of the bulk of the parent pc, leaving the recoil- and radiation-decomposition produced species in dilute sulfuric acid solution. Excellent high specific activity ^{64}Cu preparations have been obtained using this procedure. However, relatively low enrichments and specific activities have resulted from production scale experiments with most other metal pcs, as a result of decomposition of target (parent) molecules by the concentrated sulfuric acid, as well as from target radiolysis. We have found the gross decomposition of α ^{65}Zn pc using this procedure to be 1—3%. Higher values of gross decomposition from the dissolution procedure have been reported for this and other metal pcs. Thus, a solvent system other than sulfuric acid would be highly desirable for the purpose of preparing high specific activity radionuclides from metal pcs. . . The promotion of α- to β-transformation of metal pcs by 'crystallizing solvents' can be used to produce radionuclides at high enrichment with satisfactory yields."*

"In the study of solid state hot atom chemistry, the dissolution method for separating the recoil atoms from the original compound involves in some cases a change in the chemical species or an isotope exchange reaction between the recoil and stable atoms. Moreover, the dissolution method, usually applied to metal pcs, has been found to be unsuitable owing to the decomposition of this material in concentrated sulfuric

* From Collins, K. E., Catral, J. C., Toh, W. T., Detera, S. D., and Kapauan, P. A., *Radiochem. Radioanal. Lett.*, 11(5), 303, 1972. With permission.

acid. These factors make it difficult to elucidate the mechanism of recoil reactions. On the other hand, the sublimation method has the advantage of not requiring any reagents in the separation of recoil products from the target material though it may involve some annealing during heating"* . . . A method is proposed for the preparation of enriched ^{231}Th by the hot atom effects of 14 MeV neutron irradiation in Thpc. Sublimotography is successfully applied for the separation of the daughter nuclides and of the target Thp from ^{231}Th[2].

A review studies the Szilard-Chalmers effect on pcs.[3]

In an investigation of the Szilard-Chalmers process in Zr and Hf pcs,[4,5] the authors summarize the process up to the mid 1960s. "The Szilard-Chalmers process in metallopcs has been found particularly suitable for preparation of high specific activity isotopes of the metals. Pcs of Fe, Co, Zn, Cu, Rh, Ga, In, Os, Ir, Pt, U, and Mo have been used as targets. Yields of recoil products of over 90% and enrichment factors of up to 10^4 have been reported in industrial cases."[4]

The dependence of the specific activity or activity yield on the irradiation time by the Szilard-Chalmers process is of recurring interest. Calculations are compared with experimental results of the Szilard-Chalmers enrichment for Cu, Ga, Co, Fe, Al, and In pcs.[6] The relation between the specific activity given by Adamson and Williams is extended by a term for decomposition of irradiated targets during the chemical separation.[7]

The accumulation of $^{114}In^m$ by the Szilard-Chalmers process is studied by using In pc as a target. Reactor radiation results in very high specific activities which otherwise can be obtained only with very long irradiation periods.[8]

The Szilard-Chalmers process is observed with Cd pc.[9] The daughter tracing technique is used. The principle of this technique is based on the fact that the behavior of a recoil ^{115}Cd atom can be traced by the behavior of its daughter, $^{115}In^m$.

In a Szilard-Chalmers study, new short-lived activities of Y-163, Tm-165m, Tm-163m, Tm-161m, and Tm-159m are observed.[10]

A procedure for the Szilard-Chalmers reaction with Co pc is outlined as follows.[11] Co pc is bombarded with a flux of 3.6×10^{16} thermal neutrons per cm^2 and is dissolved in concentrated sulfuric acid yielding a dark green solution; dilution of this solution with water precipitates the Co pc while the ^{60}Co atoms formed in the neutron bombardment remain in solution. The yield of ^{60}Co is 94.4% and the enrichment coefficient is about 200.

The Szilard-Chalmers effect is studied on Cu pc preirradiated with γ-rays from ^{60}Co source, followed by irradiation with 10^{12} neutrons per cm^2 per second for 30 min. Activity is determined with a NaI γ-ray scintillator.[12] Na pc is irradiated with 10^{10} neutrons per cm^2 per second for 40 hr. The yield is 95%.[12]

The Szilard-Chalmers process is studied at neutron fluxes of $\leqslant 2 \times 10^{13}$ neutrons per cm^2 per second[13] and is studied on vanadium pc.[14] On vanadium pc the retention value is 93%.

A considerable interest in hot atom chemistry of pcs has been demonstrated by workers at the Japanese Atomic Energy Research Institute at Tokai,[15-30] as well as at Japanese universities. The following discussion relates to their efforts.

The effect of crystal structure on annealing of ^{64}Cu recoil atoms in Cu pc is considered.[15] Annealing proceeds more rapidly and with greater retention in β crystals than in α crystals probably because the β form has a structure with closer packing than the α form.[16]

"Crystal structure effects on the initial recoil reaction and on the annealing reactions were studied in neutron irradiated α and β Zn pcs, and the initial retention ratio is

* From Endo, K. and Sakanoue, M., *Radiochem. Radioanal. Lett.*, 9(4), 255, 1972. With permission.

compared with the annealing rate ratio. The initial retention in the β crystals is greater than in the α crystal. As in the case of Cu pc, the annealing processes can be divided into two stages for both crystals.''*

''A definite dependence has been found of the initial retention on recoil energy in Zn pc after a systematic investigation of the chemical effects of different types of nuclear transformations. High energy γ-rays and neutrons of 14 MeV were employed.''[18] The α crystals of Zn pc have a narrower annealing domain (18 to 20% for $^{69}Zn^{m}$) than the β crystals, while the annealing reaction of Stage II in the α crystals has a wider annealing domain than in the β crystals below their decomposition temperature of about 400°C.[19]

''The effects of bulk properties of the crystals on the behavior of recoil atoms produced in Cu pc were studied. The effect of crystal size on the initial retention and the annealing reaction of Stage I are only slight, whereas the effects on Stage II are very prominent. The crystals of smaller size which have a higher defect density in the crystal lattice are found to anneal much more easily in Stage II, but no such prominent effect can be observed in the initial retention nor in the annealing reaction of Stage I. Stage I is considered to be a short range process whereas Stage II is a very long range process.''*

The initial retentions of ^{64}Cu recoil atoms in β, α, γ, and δ crystals of Cu pc are 23.5 ± 0.7, 6.1 ± 0.2, 6.1 ± 0.5, and 5.5 ± 0.2%, respectively.[21] The isothermal annealing reaction rate at $\leqslant$200°C is greatest in the β crystals.

The chemical behavior of ^{115}Cd recoil atoms in Cd pc irradiated at dry ice temperature in a nuclear reactor, is studied by the use of the daughter tracing technique. The thermal annealing process can be divided into two stages, the activation energies being 0.12 ± 0.05 eV for Stage I and 0.40 ± 0.05 eV for Stage II, respectively.[22] By comparing with Cu and Zn pcs, it is suggested that the nature of the bonding between the central metal atom and the ligand in the complex is one of the most important factors in determining the chemical fate of the recoil atoms.

''The relative importance of recoil energy and of charge in determining the chemical fate of the recoil atom has not been established in most cases. Therefore the nature of the 'isotope effect' in hot atom chemistry is not normally understood. . . the nature of the isotope effect is examined. A wide range of the recoil energies is covered by using a variety of nuclear processes. As a result, it is found that the influence of the recoil energy on the chemical fate of the recoil atom is very important in all of the metal pc systems studied, including Zn, Ni, and Co pcs.''**

Chloroindium chloro pc, (InCl)pcCl, is irradiated at dry ice temperature in a nuclear reactor and the Szilard-Chalmers process is investigated.[24] Neutron dose dependence of the apparent initial retention is observed in the range of neutron dose of 3×10^{14} n/cm^2 to 6.4×10^{15} n/cm^2. The apparent initial retention of $^{116}In^{m}$ produced in chloroindium chloro pc by neutron irradiation increases with the increasing neutron dose, probably due to the recombination of the recoil fragments during neutron irradiation under the influence of local heating.

''Pre-irradiation treatments such as crushing, quenching from an elevated temperature and irradiation with ionizing radiation, sensitize the material to thermal annealing. These effects are, in many cases, explained by the interaction of the recoil fragments with the defects introduced in the crystals. . . In order to assess the role of the defects in the molecular crystals . . . study is extended to the case in which Cd pc has been quenched, irradiated with Co - 60 γ-rays, or subjected to previous irradiation in a reactor prior to the recoil processes due to the radiative neutron capture.''***

* From Yoshihara, K. and Yang, M., *Radiochim. Acta,* 9(2—3), 168, 1968. With permission.

** From Yang, M., Yoshihara, K., and Shibata, N., *Radiochim. Acta,* 14(1), 16, 1970. With permission.

*** From Kudo, H., *J. Inorg. Nucl. Chem.,* 34(2), 453, 1972. With permission.

To further elucidate the effect of the crystal structure on the chemical behavior of ^{64}Cu recoil atoms in Cu pc, the mixed system of Cu pc and H_2 pc is studied.[26] Initial retention increases in the mixed crystal. The addition of metal-free pc sensitizes the target to the thermal annealing reaction and the rate of the reaction in the mixed crystal is faster than that in the unmixed crystal.

The effects of preirradiation treatments with ^{60}Co γ-rays on the chemical behavior of ^{64}Cu recoil atoms in neutron irradiated Cu pc are studied and compared with those for Cd pc.[27] Although the variations are small, irradiation with ^{60}Co γ-rays causes a change in the chemical behavior of the recoil atom.

The Stage II annealing process of ^{64}Cu recoil atoms in neutron irradiated α Cu pc is examined.[28] The isothermal annealing curve above 270°C shows an anomalous inflection point, probably caused by crystal structural transformation.

A study of the hot atom chemistry of Cu pc by the (n,2n) reaction at Kanazawa University, Kanazawa, is made of the chemical effects of the ^{65}Cu (n,2n) ^{64}Cu reaction in α and β Cu pcs and a comparison of the results with those obtained for the $^{63}Cu(n,\gamma)^{64}Cu$ and $^{65}Cu(\gamma,n)^{64}Cu$ reactions indicate that the purity of the sample affects the retention values.[29] In another study at this university the initial retention, R, of ^{60}Co recoil atoms in β Co pc irradiated with neutrons at dry ice temperature shows an initial retention time, t, dependence of the type R % = 2.75t + 5.40, where t is in minutes.[30] The initial retention in α crystals is independent of the irradiation time at dry ice temperature.

Work on hot atom chemistry of pcs has also been conducted at Tokyo Kyoiku University.[31-34]

The chemical behavior of recoil atoms in solid solutions of Co pc and Cu pc mixed crystals is studied.[31] "The initial retention values of ^{60}Co and ^{64}Cu change rather markedly upon dilution with the opposite component and show different trends from each other."*

The behavior of ^{60}Co hot atoms in the mixed crystal of Co pc and metal-free pc is studied.[32] The initial retention and the thermal annealing behavior are different between α and β modifications of mixed crystals.

The initial retention of ^{65}Zn and $^{69}Zn^m$ are investigated in α and β mixed crystals of Zn pc and metal-free pc. Initial retention increases with the increasing mole fraction of metal-free pc.[33]

The chemical behavior of ^{58}Co and ^{64}Cu hot atoms produced by (γ,n) reactions is investigated in the mixed crystals of Co and Cu pcs.[34] Initial retention of ^{58}Co and ^{64}Cu are compared with those of ^{60}Co and ^{64}Cu produced by (n,γ) reactions. In mixed crystals the initial retention for (γ,n) reaction is lower in general than that for the corresponding (n,γ) reaction.

A review with 25 references discusses the hot atom chemistry of the pcs.[35]

The recoil effect with ^{58}Co after a (γ,n) reaction in Co pc is measured. Initial retention is 3.5% and the retention is 4.7% after heating 120 hr at 200°C.[36] The retention of Co upon (n,γ) and/or (γ,n) irradiation is determined for Co pc.[37] Irradiation with neutrons was in the range (1 to 5) × 10^{15} n/cm^2 and irradiation with γ rays was carried out with 26 MeV bremsstrahlung. Thermal annealing is studied in neutron irradiated α and β Co pc. The molecular symmetry has a considerable influence on the initial retention (α form 2% and β form 4%); annealing is faster in the β form than in the α form.[38]

Thermal annealing is observed on α and β Co pc in oxygen, nitrogen, hydrogen, and *in vacuo* atmospheres.[39] For example, oxygen atmosphere inhibits annealing and pro-

* From Kujirai, O. and Ikeda, N., *Radiochem. Radioanal. Lett.*, 15(2), 67, 1973. With permission.

motes decomposition of the parent compound. Hydrogen atmosphere slightly enhances the annealing of α crystals and greatly enhances that of the β crystal. Annealing under nitrogen gives about the same retention as in air and vacuum gives higher retention than in air.

An enrichment procedure of radioactive antimony preparations is described.[40] Sb pc powder is irradiated with neutrons until a sufficient amount of ^{124}Sb accumulates, then is stored 10 to 20 days. A 0.3-g sample is dissolved in 5 to 10 mℓ quinoline. The ^{124}Sb atoms that break free from the complex through recoil effects are coprecipitated with MnO_2 formed *in situ* by the addition of 2 mℓ distilled water, 2 mℓ of alcoholic $MnCl_2$ (0.6 g/mℓ), and 2 mℓ of a solution of $KMnO_4$ in acetone (5 g/100 mℓ). The precipitate is dissolved in 6 *N* HCl; the Sb is extracted with isoamyl acetate and reextracted with *N* sulfuric acid in the presence of a few drops of bromine. The precipitation of ^{124}Sb with the MnO_2 is quantitative and the total yield of enriched ^{124}Sb is 90%; the enrichment factor is (2 to 8) $\times 10^5$.[40]

Chemical consequences of nuclear recoil in pc solids — isotopic, steric, and annealing effects in Co, Cu, and Zn pcs — are studied.[41,42]

The pcs of Cu, Zn, Ga, Ce, Pr, Ge, Zr, As, Te, Mo, W, Re, Ru, Pd, Os, Ir, and Pt are irradiated with slow and fast neutrons and with the quantum radiation of UV, X-rays, and γ-rays, to induce (n,γ), (n,2n), (n,p) reactions, β decay, isomeric transition, and K capture.[43] The percentage retention of these pcs are tabulated for the various reactions.

In a study of the influence of atmospheres and crystal phase on the isochromal annealing of Cu pc,[44] the initial retentions of α and β Cu pc isochronally annealed in air agree well with prior results.[15,16] "All samples present however several new and well characterized annealing stages: six for the α crystals, and four and three, respectively, for the β crystals."[44]

Using a differential counting technique, the radioactive decay rates of ^{64}Cu atoms in recoil and in lattice sites of β Cu pc, which have been irradiated with neutrons, are compared.[45] "We see no chemical effect on the rate of decay." "A few analogous experiments with α Cu pc likewise show no chemical effect on decay, though with somewhat larger limits of error."

The lifetime variation of ^{64}Cu and the chemical state of (n,γ) recoil copper in α Cu pc crystals is examined.[46] "The decay constant, λ, of ^{64}Cu in α Cu pc is changed by $\Delta\lambda/\lambda = (10 \pm 1.6) \times 10^{-4}$ when the electron capture decay rate in the labelled crystal is compared to that of samples in which the ^{64}Cu activity is introduced by neutron irradiation."

B. Separation Procedures

"An extraction method was developed for the separation of shortlived isomeric states of lanthanides. The method is based on the chemical decomposition of lanthanide pc molecules following nuclear transformation of the central metal atom, and allowed the separation of 11 min ^{158g}Ho from 27 min ^{158m}Ho, 26 min ^{160g}Ho from 5.0 h ^{160m}Ho, 6.8 s ^{161m}Ho from 3.1 h ^{161}Er and 15 min ^{162g}Ho from 68 min ^{162m}Ho. Due to the chemical similarity of the lanthanides, there are reasons to believe that the method should work also for other lanthanide nuclides. A continuous separation in which the transformed nuclei migrate from a finely dispersed solid pc phase to a moving liquid phase was also studied. This method appears to allow separation of nuclides with half-lives of the order of 1 s."*

An attempt is made to develop separations of Np and of fission products from neu-

* From Stenstrom, T. and Jung, B., *Radiochim. Acta*, 4(1), 3, 1965. With permission.

tron irradiated U and U compounds by sublimation or extraction, aided by complexation, and optionally aided by oxidation, reduction, or phase change.[48] For example, oxide samples are treated with $C_6H_4(CN)_2$ vapor, with molten $C_6H_4(CN)_2$, or with mixed $C_6H_4(CN)_2$-pc melt, in order to form in each case the pc complexes. Results are disappointing, but some separation of Zr, Ce, Nd, and Ru is obtained (10^{-2} to 10^{-1} % yield).[48]

A microsynthesis for the rapid transfer of radioactive rare-earth reactions as pc complexes and an extraction chromatographic column for fast mother-daughter separation are described.[49]

C. Radiolysis by γ-Rays

Solutions of pc and Cu pc in concentrated sulfuric acid and in cyclohexanol:ethanol 10:1 in 2×10^{-8} to 6×10^{-8} *M* concentrations are exposed to 15×10^3 to 55×10^3 R. from a ^{60}Co source. With this exposure, their color changes and the absorbance decreases proportionally with the γ dose. The absorbance of a 6×10^{-8} *M* solution of Cu pc in sulfuric acid in the 430 mμ region changes from 1.5 before γ-radiation to 1.0 and 0.14 after receiving 15×10^3 and 45×10^3 roentgens, respectively. A similar decrease in absorbance is noted for pc. The change of the absorption spectra "indicates that the decomposition of pc proceeds mainly accordingly to the tetraazaporphyrin cycle."[50] Aqueous alkaline Cu pc solutions are irradiated by a ^{60}Co γ source.[51] Tetrasulfonated Cu pc coupled with *p*-aminoacetoacetanilide in alkaline solutions is irradiated and an irreversible discoloration, observed by the decrease in the absorption at λ_{max} = 340 nm, takes place. The strong role of the OH radical is indicated.[51]

D. Element Transformation, Ion Implantation, and Recoil Implantation

The protactinium and neptunium bis pc complexes, $^{233}Papc_2$ and $^{239}Nppc_2$, are prepared from the corresponding thorium and uranium complexes $Thpc_2$ and Upc_2, by the element transformations $(n,\gamma)(\beta^-)$ of ^{232}Th and ^{238}U, respectively.[52] Their existence is proved by repeated sublimation of the irradiated parent compounds, showing that the specific activities of ^{233}Pa and ^{239}Np remain constant in the sublimates.

The chemical effects of ion implantation in molecular solids are studied. α and β Cu pc are irradiated with 30 and 60 keV $^{64}Cu^+$.[53] The targets are dissolved and radioactive products are separated and analyzed. "The fraction of labeled Cu complex decreases with increasing energy and is greater for the β form. An increased energy deposit per unit volume will initiate annealing reactions able to change the chemical distribution of the implanted ions by an electronic mechanism."

The synthesis of technetium and rhenium pcs, Tc pc and Re pc, is undertaken by recoil implantation of isotopes generated by (d, xn) reactions in a mixture of Mo or W, or Mo or W oxides, with Cu pc.[54] Radioactive isotopes of technetium and rhenium pcs are "found unambiguously in the Cu pc sublimate and the fraction, which is dissolved in concentrated sulfuric acid and coprecipitated with Cu pc by dilution of the solution with water. . . The most probable valence state of technetium and rhenium in the complex seems to be +4."

E. Interaction Between Nuclear Fission Products and Solids

"The interaction between nuclear fission products and solids can be detected directly as tracks by the use of an electron microscope with high magnification. Pt pc is among the materials examined.[55] Tracks are observed in polycrystalline metal films and are analyzed by etching the materials with sodium hydroxide, 20% HF, and aqua regia in order to prevent the fading of tracks by heat effects of the beam in the electron microscope. An ion explosion model is proposed as the mechanism of the formation of tracks.[55]

II. RADIATION DAMAGE

A. Electrons (Electron Microscope and Electron Accelerator)

"In order for the single atoms in organic molecules to be resolved, considerably high magnification is needed up to 100 to 200 thousand times. A highly bright illumination of electron beam is an essential requirement for this purpose, consequently giving rise to radiation damage. . ."[56,57] $CuCl_{16}$ pc "is remarkably resistant to radiation damage." Its resistivity to the electron beam at various accelerating voltages and different specimen temperatures is measured semiquantitatively.

Experimental values of the critical exposure for a given damage endpoint are measured for several organic compounds including pc and hexadecachloro pc.[58]

Limits in the high resolution electron microscopy of halogen substituted organic molecule single crystals, including $CuCl_{16}$ pc, caused by radiation damage are observed in terms of the decrease in the electron diffraction intensities.[59]

Low energy electron beam irradiation of liquids using a 500 keV electron accelerator is tested.[60] This operation is piloted particularly for irradiation of wastewaters containing dye substances. Decolorization effectiveness is measured spectrophotometrically on dyes including Cu pc sodium disulfonate solutions, Direct Blue 86, C. I. 74180, which decrease in concentration from 0.5 mg/ℓ to less than 0.1 g/ℓ with an electron dosage of 2 Mrad.[60]

B. γ-Rays and Neutrons

Radiation damage to pcs is studied in terms of irradiation by neutrons and γ-rays.[61] "Most samples were irradiated at temperatures less than 50°C in the thermal neutron column of the HIFAR reactor at Lucas Heights, where the thermal neutron flux was 3×10^{12} $cm^{-2}sec^{-1}$ and γ-ray dose rate of 4—20 rad hr^{-1}. Different sections of the same crystal were irradiated for times ranging between 3 and 96 hours, the plates were irradiated for up to 40 hours while the films were only irradiated for 8 hours.

"Samples irradiated with a ^{60}Co source or in a spent fuel facility gave similar results to those obtained from samples irradiated in HIFAR. The spectral changes on irradiation described below also appeared when a H_2(pc) crystal was X-irradiated for eight hours using a Philips X-ray tube with a copper target at 30 kV. Unduly long times would have been required to produce observable changes in the spectra of the metal pc crystals. These observations indicate that most of the damage results from the γ-rays. The thermal neutron capture cross-sections are such that any spectral changes produced by them are probably too small to be observed."*

REFERENCES

1. Collins, K. E., Catral, J. C., Toh, W. T., Detera, S. D., and Kapauan, P. A., *Radiochem. Radioanal. Lett.*, 11(5), 303, 1972.
2. Endo, K. and Sakanoue, M., *Radiochem. Radioanal. Lett.*, 9(4), 255, 1972.
3. Pertessis, M., *Chim. Chronika*, 28(5-6), 54, 1963.
4. Hillman, M. and Weiss, A. J., AEC Accession No. 26921, Report No. BNL-10208 1966.
5. Hillman, M., Kim, C. K., Shikata, E., and Weiss, A. J., *Radiochim. Acta*, 9(4), 212, 1968.
6. Grossmann, G., Muehl, P., Grosse-Ruyken, H., and Knoefel, S., *Isotopenpraxis*, 4(1), 23, 1968.
7. Muehl, P. and Grosse-Ruyken, H., *Radiochem. Conf., Abstr. Pap., Bratislava*, 58, 1966.
8. Muehl, P. and Grosse-Ruyken, H., *Isotopenpraxis*, 3(12), 486, 1967.

* From Boas, J. F. Fielding, P. E., and MacKay, A. G., *Aust. J. Chem.*, 27(1), 7, 1974. With permission.

9. Yoshihara, K. and Kudo, H., *Nature (London)*, 222(5198), 1060, 1969.
10. Gromov, K. Ya., Zhelev, Zh. T., Kalinnikov, V. G., Malek, Z., Nenov, N., Pfrepper, G., and Strushnyi, Kh., Joint Inst. Nucl. Invest., Rep. JINR-P6-3945, 1968.
11. Kirin, I. S., Ivanchenko, A. F., and Moskalev, P. N., *Radiokhimiya*, 9(3), 346, 1967.
12. Pertessis, M. and Henry, R., *Radiochim. Acta*, 1, 58, 1963.
13. Schwartz, A., Rafaeloff, R., and Yellin, E., *Int. J. Appl. Radiat. Isotop.*, 20(12), 853, 1969.
14. Fucugauchi de Crowley, L. A. and Cruset, A., *Rev. Soc. Quim. Mex.*, 19(6), 279, 1975.
15. Yoshihara, K. and Ebihara, H., *Sonderdruck Radiochim. Acta*, 2, 219, 1964.
16. Yoshihara, K. and Ebihara, H., *Nature (London)*, 208, 482, 1965.
17. Yoshihara, K. and Yang, M., *Radiochim. Acta*, 9(2—3), 168, 1968.
18. Yoshihara, K. and Yang, M., *Inorg. Nucl. Chem. Lett.*, 5(5), 389, 1969.
19. Yang, M., *Radiochim. Acta*, 12(3), 167, 1969.
20. Yang, M., Yoshihara, K., and Shibata, N., *Radiochim. Acta*, 14(1), 16, 1970.
21. Yang, M., Kudo, H., and Yoshihara, K., *Radiochim. Acta*, 14(1), 52, 1970.
22. Kudo, H. and Yoshihara, K., *J. Inorg. Nucl. Chem.*, 32(9), 2845, 1970.
23. Yang, M., Yoshihara, K., and Shibata, N., *Radiochim. Acta*, 15(1), 17, 1971.
24. Kudo, H. and Yoshihara, K., *Radiochim. Acta*, 15(4), 167, 1971.
25. Kudo, H., *J. Inorg. Nucl. Chem.*, 34(2), 453, 1972.
26. Kudo, H., *Bull. Chem. Soc. Jpn.*, 45(2), 392, 1972.
27. Kudo, H., *Bull. Chem. Soc. Jpn.*, 45(5), 1311, 1972.
28. Kudo, H., *Bull. Chem. Soc. Jpn.*, 45(2), 389, 1972.
29. Sakanoue, M. and Endo, K., *Radiochem. Radioanal. Lett.*, 4(3), 99, 1970.
30. Endo, K. and Sakanoue, M., *Radiochim. Acta*, 17(1), 7, 1972.
31. Kujirai, O. and Ikeda, N., *Radiochem. Radioanal. Lett.*, 15(2), 67, 1973.
32. Kujirai, O. and Ikeda, N., *Radiochem. Radioanal. Lett.*, 18(4), 197, 1974.
33. Ikeda, N. and Kujirai, O., *Radiochem. Radioanal. Lett.*, 23(3), 125, 1975.
34. Kujirai, O., Ikeda, N., and Shoji, H., *Radiochem. Radioanal. Lett.*, 26(1), 5, 1976.
35. Yoshihara, K., *Kagaku No Ryoiki*, 27(9), 762, 1973.
36. Nath, A. and Nesmeyanov, A. N., *Radiokhimiya*, 5(1), 125, 1963.
37. Shankar, J., Nath, A., and Rao, M. H., *Radiochim. Acta*, 3(1—2), 26, 1964.
38. Mathur, P. K., *Indian J. Chem.*, 7(8), 820, 1969.
39. Scanlon, M. D. and Collins, K. E., *Radiochim. Acta*, 15(3), 141, 1971.
40. Ziv, D. M., Kirin, I. S., Ivanchenko, A. F., and Ishina, V. A., *Radiokhimiya*, 5(5), 632, 1963.
41. Apers, D. J. and Capron, P. C., *Chem. Effects Nucl. Transforms, Proc. Symp., Prague*, 429, 1961.
42. Apers, D. J., Dejehet, F. G., van Outryve d'Ydewalle, B. S., and Capron, P. C., *J. Inorg. Nucl. Chem.*, 24, 927, 1962.
43. Merz, E., *Nukleonik*, 8(5), 248, 1966.
44. Odru, P. and Vargas, J. I., *Inorg. Nucl. Chem. Lett.*, 7(4), 379, 1971.
45. Johnson, J. A., Dema, I., and Harbottle, G., *Radiochim. Acta*, 21(3—4), 196, 1974.
46. Odru, P. and Vargas, J. I., *Chem. Phys. Lett.*, 15(3), 366, 1972.
47. Stenstrom, T. and Jung, B., *Radiochim. Acta*, 4(1), 3, 1965.
48. Lux, F. and Ammentorp-Schmidt, F., *Radiochim. Acta*, 4(2), 112, 1965.
49. Pfrepper, G., Herrmann, E., and Khristov, D., *Radiochim. Acta*, 13(4), 196, 1970.
50. Starodubtsev, S. V., Tikhomolova, M. P., and Mzhel'skaya, L. G., *Vopr. Sovrem. Fiz. Mat., Akad. Nauk Uz. S.S.R.*, 28, 1962.
51. Tarabasanu-Mihaila, E. and Sofronie, E., *Rev. Roum. Chim.*, 14(11), 1467, 1969.
52. Lux, F., Ammentorp-Schmidt, F., Dempf, D., Graw, D., and Hagenberg, W., *Radiochim. Acta*, 14(2), 57, 1970.
53. Andersen, T., Langvad, T., and Soerensen, G., *Nature (London)*, 218(5147), 1158, 1968.
54. Yoshihara, K., Wolf, G. K., and Baumgaertner, F., *Radiochim. Acta*, 21(1—2), 96, 1974.
55. Sakanoue, M., *Radioisotopes (Tokyo)*, 17(5), 212, 1968.
56. Harada, Y., Taoka, T., Watanabe, M., Ohara, M., Kobayashi, T., and Uyeda, N., *Proc. Electron Microsc. Soc. Am.*, 30, 686, 1972.
57. Uyeda, N., Kobayashi T., Ohara, M., Watanabe, M., Taoka, T., and Harada, Y., *Electron Microsc., 5th Proc. Eur. Congr. Electron Microsc.*, Inst. of Physics, London, 566, 1972.
58. Glaeser, R. M., *Proc. Electron Microsc. Soc. Am.*, 31, 226, 1973.
59. Kobayashi, T. and Reimer, L., *Bull. Inst. Chem. Res. Kyoto Univ.*, 53(2), 105, 1975.
60. Wiesboeck, R. and Proksch, E., *Kerntechnik*, 18(1), 20, 1976.
61. Boas, J. F., Fielding, P. E., and MacKay, A. G., *Aust. J. Chem.*, 27(1), 7, 1974.

Chapter 11

CATALYSIS

I. INTRODUCTION

The focus of research on pcs as catalysts is well expressed as follows: "Metal pcs and their derivatives have received a great deal of attention, partly because of their relevance to various biological systems. These compounds have characteristic features that various kinds of metals can be introduced in the center of the same porphyrin ligand and the surface of a solid metal pc may be regarded as one of the most well defined systems for studying chemisorption and catalysis in correlation with the electronic properties of various metal atoms."*

Catalysis involving pcs was the subject of a symposium held in Hamburg on May 10, 1972.[2]

Pcs as heterogeneous catalysts in dehydrogenation, oxidation, and electrocatalysis are the purpose of a review with 35 references.[3] Five reasons are given as advantages in using tetraphenylporphyrins and pcs as catalysts: "(1) They form complexes which are of significance in catalysis with most metal ions; (2) These complexes are thermostable; (3) Because of the highly conjugated nature of the ligand, all complexes are square planar, leaving two octahedral sites open for additional ligands; (4) The phenyl groups give the opportunity for introducing chemical substituents on the periphery of the π system, by which electrons can be withdrawn from or donated to the system; (5) The complex can be dissolved in an organic solvent, and the physical properties of the metal ions can be studied in solution in the same surroundings as they occur on the surface."**

Heterogeneous catalysis in gas phase reactions by metal pcs is the subject of a review with 51 references.[4] According to the authors, "adsorption of gases and catalytic reactions of gaseous substrates on catalysts of monomeric β metal pcs of the first transition period are governed by two characteristic properties of monomeric β metal pcs, i.e., the basically uniform crystal structure together with the different number of electrons in the orbitals of the central metal ion."

II. CATALYSTS IN THE REDUCTION OF OXYGEN ON γ-ALUMINUM OXIDE SURFACES

ESR direct spectroscopic investigation of the Co pc catalyst complex -γ-Al_2O_3 surface during oxygen reduction on γ-Al_2O_3 surfaces with Co pc catalyst shows a large amount of chemisorbed O_2^-. "A reasonable hypothesis is that the superoxide anion is bonded to Al^{3+} centers. . . It may be concluded that in γ-Al_2O_3-Co pc samples there is no evidence that the O_2^- species is bonded to cobalt; however, the presence of the metal complex on the surface is indispensable in forming the chemisorbed anion." "An electron transfer from the transition metal complex to oxygen is the origin of the O_2^- which is then fixed to Al^{3+} centers. . . The strength of the chemisorption is very high. . . This is probably the first example of an electron transfer process from a porphyrin-like complex to be confirmed with direct evidence."***

* From Kawai, T., Soma, M., Matsumoto, Y., Onishi, T., and Tamaru, K., *Chem. Phys. Lett.*, 37, 2, 378, 1976. With permission.

** From Manassen, J., *Catal. Rev.*, 9(2), 223, 1974. With permission.

*** From Campadelli, F., Cariati, F., Carniti, P., Morazzoni, F., and Ragaini, V., *J. Catal.*, 44(1), 167, 1976. With permission.

III. CATALYSTS IN THE REDUCTION OF OXYGEN IN BIOLOGICAL SYSTEMS

In a discussion of in vivo electrochemical power generation, an in vivo electrode system is described. It consists of an Fe metal anode and a catalytically acting cathode (pc, C. I. Pigment Blue 16 [574-93-6]) which reduces molecular oxygen.[6]

IV. CATALYSTS IN THE ELECTROREDUCTION OF OXYGEN

An underlying theme in the reduction of oxygen using pc catalysts lies in the development of useful fuel cells. For example, a hydrogen-oxygen fuel cell has been used in manned spacecraft.[7] In a typical fuel cell, "the two gases, hydrogen and oxygen, are led into the cell, where each comes in contact with a porous electrode of either nickel or graphite. The two electrodes are separated by an electrolyte such as potassium hydroxide. A theoretical reaction at the cathode is

$$O_2 + 2H_2O + 4e^{\ominus} \rightarrow 4\ OH^{\ominus}$$

However, the actual reaction results in the formation of hydrogen peroxide ions:

$$O_2 + H_2O + 2e^{\ominus} \rightarrow HO_2^{\ominus} + OH^{\ominus}$$

Catalysts are embedded in the porous electrodes to hasten the decomposition of the peroxide ions:

$$2\ HO_2^{\ominus} \xrightarrow{\text{catalyst}} O_2 + 2\ OH^{\ominus}$$

Hydroxide ions then diffuse from the cathode to the anode, where they participate in the oxidation of hydrogen:

$$H_2 + 2\ OH^{\ominus} \rightarrow 2\ H_2O + 2e^{\ominus}$$

The overall reaction is

$$O_2 + 2\ H_2 \rightarrow 2\ H_2O$$

. . . Fuel cells in which acid electrolytes are used produce hydrogen ions at the anode and consume hydrogen ions at the cathode."[7]

The recognition of pcs as oxygen reduction catalysts relates to the work of Jasinski, Jahnke, and Schoenborn, who studied catalysts for electrochemical reactions taking place in fuel cells.[8-11] Jasinski[8] used Co pc in alkaline electrolyte; Jahnke,[9] and Jahnke and Schoenborn[10] discovered the catalytic activity of pc-carbon support combinations for the electroreduction of oxygen in acid electrolyte."[12]

In a study published in 1973, the catalytic activity for the electroreduction of oxygen in acid electrolyte, it was found that the catalysts were not completely stable when supported by active carbon or carbon black. "Especially the pcs readily disintegrate in acid."[12] The mechanism of the activation of the oxygen molecule is explained on the basis of simple MO considerations, "which also provide an explanation for the order of activity of central metal ions — Co > Fe > Ni — and for the effect of the support and of substituents of the ligand on the activity and the stability of the chelate."[12]

A comparison is made between the chemical and electrochemical catalysis by tetraphenylporphyrin and pc complexes.[13] Catalytic electrodes are prepared by mixing carbon black on which the catalyst is precipitated, with a Teflon® emulsion and painting the paste obtained on a 100-mesh metal gauze. By heating at 250°C for 1 hr in an argon stream, good hydrophobic electrodes of possible use in fuel cells are obtained. Oxygen reduction with Fe^{2+}, Fe^{3+}, Co^{2+}, and Cu^{2+} are studied, as well as with various degrees of fluorination of the macrocycles.

A ferric pc catalyzed porous flow-through graphite electrode for oxygen reduction is prepared and is shown to be effective in the oxidation of whole blood and saline solution.[14] A mathematical model is developed that describes the experimental behavior of porous electrodes.

It is pointed out that the central iron atom, the aromatic rings, and the nitrogen atom in ferric pc are candidates as adsorption sites for additives when graphite electrodes are catalyzed with Fe^{3+} pc for oxidation reduction reactions. It is shown in sodium acetate solutions at pH 6 that most of the oxygen is reduced in two steps, to H_2O_2, then to H_2O; at pH > 6 the first step disappears. Addition of PO_4^{3-}, $CN^{\ominus}$, and EDTA to the sodium acetate solution decreases catalytic activity; the addition of SO_4^{2-}, $Cl^{\ominus}$, and $Br^{\ominus}$ has only a slight effect; Pb^{2+} has a beneficial effect.[15]

The adsorption of oxygen by pcs is studied by temperature programmed desorption between −160 and 400°C, providing "the first direct evidence on the lack of rupture of the bond between two oxygen atoms in the molecule adsorbed on pcs."[16] "This would tend to explain why electroreduction on pc leads principally to the peroxide formation, instead of four electron to the hydroxyl ions on water molecules. . ." and "since electroactivity is observed on all pcs one must conclude the presence of adsorbed oxygen on all of them." In another experimental study, the stability of the Fe $pc{-}O_2$ bond is emphasized.[17]

"The electrocatalytic activity of metal pcs has been found to depend considerably on the nature of the central metal ion. The order of decreasing electrocatalytic activity is Fe > Co > Ni > Cu > H_2. The ligand with no metal ion (H_2 pc) has a very low activity. The electrocatalytic activity also depends to some extent on the degree of polymerization, on the substitution of the ligand, and on the nature of the support on which the catalyst is carried."* There is also a general correlation between electrocatalytic activity and the first oxidation potential of metal pcs, between electrocatalytic activity and the magnetic properties of metal pcs.

"In addition to the N_4-chelates. . . there are numerous complexes with other ligands capable of reducing oxygen cathodically. The activity decreases in the order $N_4 > N_2O_2 > N_2S_2 = O_4 > S_4$ which corresponds to that of the field strengths of the ligands."[19] It is found that the N_4 catalysts such as Co pc can be stabilized for long-term operation as a catalyst by thermal treatment. The catalyst is mixed with powdered carbon and heated in a current of nitrogen preferably at 800 to 900°C. . . Ideas as to the nature of the heat treatment reaction are discussed and a model for the catalysis of cathodic reduction of oxygen is presented on the basis of the MO theory."[19]

Light also plays a role in pc catalysis in the electrocatalytic reduction of oxygen on Fe pc, Cu pc, and H_2 pc coated on graphite.[20-23] On exposure to light a significant increase in the rate of oxygen reduction occurs in the order H_2 pc > Cu pc > Fe pc.

The crystalline phase of the metal pc electrocatalyst also may play a role in the effectiveness of the catalyst.[24]

"In the electrochemical reduction of oxygen in aqueous 0.2 M H_3PO_4 (pH 1.3) at 25° on electrodes comprised of a 4000 Å thick layer of monomeric Fe pc vacuum deposited onto an Au support (at ambient temperature), the electrochemical activity of the Fe pc was greater for the α phase structure (obtained by fast vacuum deposition)

* From Randin, J. P., *Electrochim. Acta*, 19(2), 83, 1974. With permission.

than for the β phase (obtained by slow vacuum deposition), owing to the greater rate of oxygen chemisorption in the α phase." It is also found, in a study of the reduction of oxygen on thin layers of pc monomers of Cu and Fe on gold in the presence of ethanol by the potentiokinetic method that the radicals formed during oxygen reduction can recombine with the ethanol and thus influence the kinetics of the electrochemical reaction.[25] Another related study compares electroreduction of oxygen on electrodes of Fe pc deposited on gold by rapid vacuum deposition and by slow thermal deposition at 400°C. The physical properties are studied by visible, X-ray, and electron spectroscopy for chemical analysis (ESCA) spectroscopy.[26]

The effect of Fe pc as electrocatalyst at the oxygen reduction cathode is postulated to be as follows:[27,28]

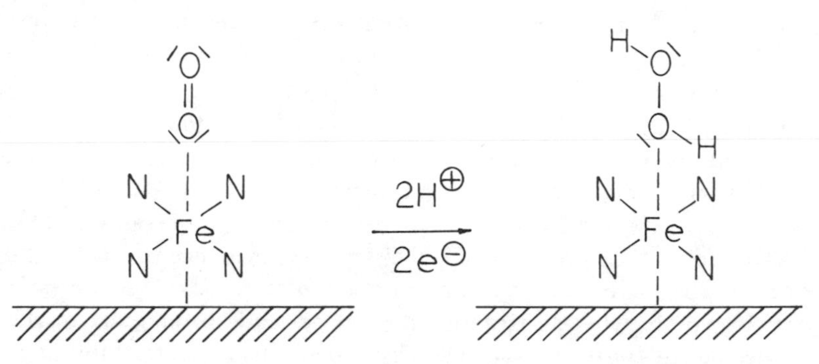

An explanation for the electrocatalytic activity of pcs is given, in terms of a mechanism called "redox catalysis". In a chemical step, the electrocatalyst is oxidized by the oxygen to the cation, which is reduced in an electrochemical follow up step. This gives a satisfactory explanation for the complementarity of activity and stability of these electrocatalysts.[27,28]

The superoxide ion as a crucial oxygen electroreduction intermediate that relates to the catalytic activity of metal pcs by adsorption on the metal cation, is the focus of experimental study.[29,30] In the autoxidation of 9,10-dihydroanthracene to anthraquinone and anthracene by the superoxide ion in pyridine, addition of Fe^{2+} pc and Co pc doubles the reaction rate.

The addition of $CN^{\ominus}$ ions ($> 10^{-6} M$) in the electroreduction of oxygen in 0.1 *M* LiOH in methanol "deactivates the Fe pc complex. This poisoning can be understood in terms of the bridging complex as a displacement of $O_2^{\ominus}$ by $CN^{\ominus}$ from the sixth coordination position."[31]

Oxygen reduction is also determined on electrodes composed of Fe, Co, Ni, and Cu polypcs impregnated in acetylene black deposited in a thin layer on a graphite support.[32,33] Acetylene black is chosen because of its low activity toward oxygen reduction. Current-voltage characteristics at various pH in aqueous phosphate solutions indicate that Fe poly pc is the most effective of this catalytic group. The electrocatalysis of oxygen reduction in acidic solutions with Fe poly pc and Co poly pc catalysts on porous carbon-teflon electrodes is the subject of another study.[34-36] Current-voltage curves and time dependence of the voltage diagram are given. Co poly pc and Fe poly pc show correlations between electrocatalytic activity of oxygen reduction and conductivity. A mechanism is proposed for the high electrocatalytic activity of Fe poly pc based on the observation of a disturbance of the delocalization of π electrons (rubiconjugation) in Fe poly pcs in addition to the correlation between conductivity and electrocatalysis and the catalytic decomposition of hydrogen peroxide.[37]

It is pointed out that, in addition to use as fuel cell catalysts, "the use of pc derivatives and similar organic N_4 macrocycles in the electrocatalysis of oxygen reduction, especially in acid media, was no doubt motivated by biochemical analogies arising from the necessity to completely eliminate hydrogen peroxide whose presence is noxious both in living organisms and in the pores of oxygen electrodes."*

Another study[38] is also directed at oxygen reduction on carbon activated by Fe and Co pcs and poly pcs. The active electrode area contains carbon black of 500 m^2/g surface area, the pc catalysts, and a wetproofing agent. Tests are made in 12 *M* KOH at 20 to 90°C. Increase in metal pc content in the active layer from 5 to 20 wt% results in increased electrochemical activity. Other studies relate to the cathodic reduction of oxygen on monomeric pcs of Cu, Co, Ni, Mn, and V and of polymeric pcs of Co, Fe^{2+}, and Mn in alkali solutions.[39-41]

Current densities of 130, 120, and 140 mA/cm^2, respectively, are obtained on cathodes containing Co pc, Fe pc, and Mn pc in 12 *M* KOH at 90°C at 0.8 V.[42] Fe^{2+} pc and Co pc polymers are also effective in the reduction of oxygen on a fuel cell electrode made of FP-4D Floroplast wet proofed carbon black-nickel grid.[43] Still another study also relates to electrochemical reduction of oxygen on carbon modified by Co pc in alkaline solution.[44]

In the reduction of oxygen on activated carbon and carbon black, there is intermediate formation of H_2O_2. The proportion of H_2O_2 is only 20% on carbon black activated with $NiCo_2O_4$ and 10, 30, and 40% for Co, Mn, and V pcs.[45]

The properties of Co octamethoxy pc, and Cu and Co octahydroxy pc are compared with those of Co pc as catalysts in the electroreduction of oxygen in KOH solution, with acetylene black cathode with 5% by weight of pc. The electrode voltage vs. current density (50 mA/cm^2) and time are described. The catalytic activity is in the sequence Co octahydroxy pc $>$ Co octamethoxy pc $>$ Co pc $\approx$ Cu octamethoxy pc. In other words, the "electrodonor substituents have a considerable influence on the electrocatalytic properties of the pcs."[46] The electrocatalytic properties of mixed Fe-Co and Fe-Cu pcs are also compared and are sequenced in the order Fe-Fe poly pc $>$ Fe-Co poly pc $>$ Fe-Cu pc.[47] Fe poly pcs are used successfully as cathodic carbon oxygen reduction catalysts in 4.5 *N* H_2SO_4 which are active for more than 1000 hr at a current density of 20 mA/cm^2 at a voltage or potential of 600 to 700 mV. It is also shown that there is a relationship between catalytic activity and electrical conductivity.[48] Co pc and Co poly pc, Fe pc and Fe poly pc, Cu pc, and Ni pc are also studied as oxygen reduction catalysts, mixed with carbon electrodes in a 1:1 ratio by weight, with respect to their effectiveness in cathodic reduction plotted as potential in mV vs. current density in 4.5 *N* H_2SO_4. By comparison with other N_4 chelates, tetradithiacyclohexenotetraazaporphyrin, dibenzotetraazaannulene, and tetraphenylporphyrin, it is concluded that, along with other investigations, oxygen reduction depends on the central metal atom of the chelate.[49]

Pcs are catalysts in cathodic reduction with carbon cathodes of substances other than oxygen.[50,51] Carbon electrodes are coated or impregnated with transition metal pc and porphyrin complexes which significantly accelerates the cathodic reduction $ClO_3^{\ominus} + 6\ H^{\oplus} + 6\ e^{\ominus} \rightarrow Cl^{\ominus} + 3\ H_2O$. Co pc is applied on finely divided carbon (particle size 40 μm) on a 30% catalyst by weight, and the compressed mixture is used as indicator electrode in the electrochemical determination of $ClO_3^{\ominus}$.[50] Co and Ni pcs are catalysts for the electrode reduction of carbon dioxide.[51] The electrodes are prepared by immersing a suspension of a metal pc in benzene and then drying in air.

Electroreduction and electrooxidation of hydrogen peroxide takes place on iron and cobalt pcs.[52] The mechanism of the process is studied electrochemically, gasometrically, and by mass spectrometer. Fe^{3+} pc and Co^{2+} pc and a polymer of Co^{2+} pc are deposited on a graphite rotating disk electrode. Labeled H_2O_2, $H_2O_2^{18}$, is used to explain the reaction kinetics.

* Musilová, M., Mrha, J., and Jindra, J., *J. Appl. Electrochem.*, 3(3), 213, 1973. With permission.

Electrocatalysis by Fe, Co, Ni, and Cu pcs in concentrated sulfuric acid solution is studied.[53] Their oxidation behavior is observed in 96% sulfuric acid at a rotating Pt(Au) disk electrode.

Electrochemical evolution of oxygen takes place from alkaline solutions in the presence of metal pcs.[54]

The reduction of oxygen takes place in alkali solutions on carbon-graphite substrates modified by Fe(II), Mn(II), Cu, and Ni pcs.[55]

V. CATALYSTS ON SEMICONDUCTOR ELECTRODES FOR THE ELECTROREDUCTION OF OXYGEN

Electrochemical reactions can be catalyzed at highly doped semiconductor electrodes such as ZnO or CdSe by precipitation of the metal pc on the surface of the electrode, from concentrated sulfonic acid by dilution with water.[56] Co pc is more active than Fe pc.

VI. CATALYSTS IN THE REDUCTION OF NITRIC OXIDE

Metal pcs catalyze the reduction of nitric oxide, NO, at 200 to 340°C and 1 atm, with hydrogen, to the following reduction products: N_2, N_2O, and NH_3.[57] The catalytic activity is in the order Co pc > Cu pc > Fe pc > Mn pc.

VII. CATALYSTS IN THE REDUCTION OF ACETYLENE

Co(II) pc sodium tetrasulfonate is a catalyst for ethylene formation in the reduction of acetylene with $NaBH_4$.[58] A possible mechanism is hydrogen transfer from a proton adduct of Co pc sodium tetrasulfonate formed with $NaBH_4$ in alkaline solution, according to the scheme

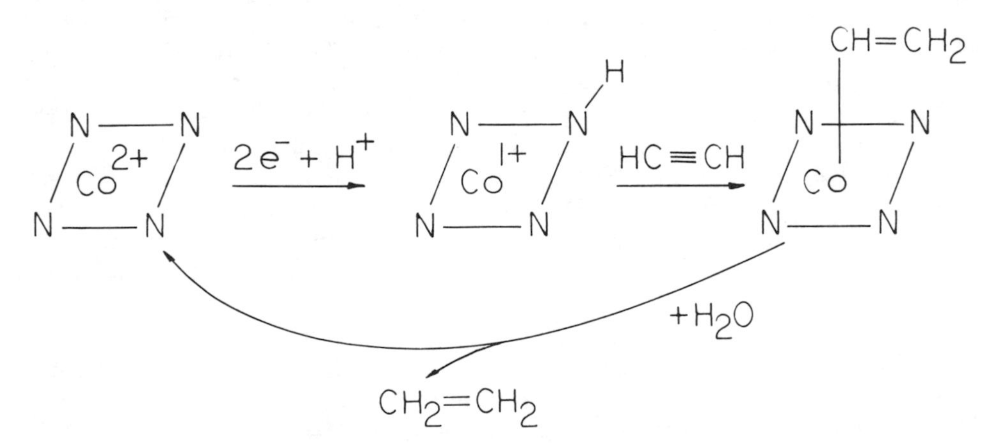

Also, the catalytic activity of the molybdenum (V)-cysteine complex for acetylene reduction by $NaBH_4$ or $Na_2S_2O_4$ is increased by addition of charge carriers such as pcs.[59]

VIII. POLY PHTHALOCYANINES AS CATALYSTS IN REDUCTIVE DEHALOGENATION

Fe poly pcs catalyze the reductive dehalogenation of α chlorodeoxybenzoin with 1-benzyl-1,4-dihydronicotinamide in benzene or aqueous methanol.[60] The Fe poly pc acts as an electron transfer carrier between 1-benzyl-1,4-dihydronicotinamide and α chlorodeoxybenzoin.

Also, α halo ketones are reduced with benzenethiol in the presence of Fe poly pc to the parent ketones although they are not reduced in the absence of Fe poly pc. The Fe poly pc appears to serve as an electron transfer catalyst for the reductive dehalogenation.[61]

IX. HYDROGENATION CATALYSTS

Pcs are studied as hydrogenation catalysts.[62-64] Ethylene is prepared by hydrogenation of acetylene in alkaline solution with reducing agents such as $NaBH_4$ and with water soluble Co pc, namely, Co pc tetrasulfonic acid sodium salt at 25°C.[62] Similarly prepared is *trans*-2-butene from butadiene.[63] The hydrogenation and dehydroformylation of hydrotropic aldehydes are catalyzed on Ni pc.[64]

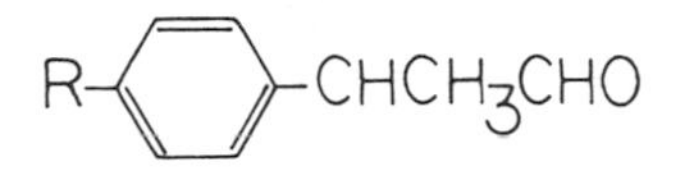

(R=H, CH_3, Cl, F, CH_3O)

are hydrogenated on Ni pc at 120 to 300°C to hydrotropic alcohols and dehydroformylated to the corresponding styrenes.

X. CATALYSTS IN HYDROGENATIVE THERMAL CRACKING

Copper pc is used as a catalyst in the hydrogenative thermal cracking of heavy feedstocks.[65] For example, black oil is cracked at 750°F with hydrogen at 2050 psig and 950°F in the presence of copper pc catalyst.

XI. HYDROGEN EXCHANGE REACTIONS

A series of studies have been published relating to the hydrogen exchange reactions of the pcs,[66-70] which are either participants or catalysts in hydrogen exchange. The pcs are treated as EDA (electron donor acceptor) complexes. The H exchange reaction takes place at room temperature between C_2H_2 and the EDA complexes of pcs with sodium as follows:

$$C_2H_2 + DZ = C_2HD + HZ$$

$$C_2D_2 + HZ = C_2HD + DZ$$

where HZ = the EDA pc complex.[66]

The catalytic activity and electron configuration of the stoichiometric EDA pc complexes with alkali metals and the H adsorption and H_2-D_2 exchange reactions are the object of investigation.[67] Hydrogen exchange reactions have also been studied over such stoichiometric EDA complexes as $Nipc^{4-}$ $4Na^+$ using mass and microwave spectroscopy.[68] The H_2-D_2 and propene-D_2 exchange reaction catalyzed by stoichiometric EDA pc complexes with sodium in solution are also studied by microwave spectroscopy.[69]

The hydrogen exchange reaction between aromatic hydrocarbons and deuterium is also catalyzed by EDA pc complexes with sodium such as $Nipc^{4-}4Na^+$, Ni pc^{2-} $2Na^+$, Co $pc^{5-}5Na^+$, Ni pc^{3-} $3Na^+$, and Co pc^{4-} $4Na^+$ with such reactant pairs as $C_6H_6 + D_2$, $C_6H_5CH_3 + D_2$, $C_6H_5C_2H_5 + D_2$, $C_6H_6 + C_6D_6$, or $C_6H_6 + D_2$. "It is concluded that the mechanism of the exchange reaction of aromatic compounds is quite different from that of aliphatic olefins, in which the reaction intermediate is an associative one. Aromatic compounds are adsorbed on these EDA complexes dissociatively and the reaction proceeds via the multiple exchange mechanism."*

EDA pc complexes also express catalytic activity in isomerization reactions, such as the isomerization of *cis*-2-butene to 1-butene and *trans*-2-butene.[71] This isomerization is closely related to the H_2-D_2 exchange reaction.

XII. HETEROGENEOUS CATALYSTS IN RELATION TO THE ELECTRON EXCHANGE OF SODIUM

"Transfer of electrons from metals to adsorbed molecules is often postulated as the first step in heterogeneous catalysis."[72] A thin film of sodium is used as the source of free electrons, a quantitative measure of which is furnished by their ESR signal. The film "transferred electrons to naphthalene, anthracene, and pc. While it did not react with methyl iodide at room temperature, the reaction proceeded at 40°C. However no methyl radicals were produced. Attempts to catalyze the reaction using pc or palladium films were unsuccessful. However irradiation with 2537 Å or with visible light when pc was present produced a strong methyl radical signal accompanied by a marked decrease in the sodium signal."

XIII. DEHYDROGENATION CATALYSTS

The use of pcs as catalysts in dehydrogenation reactions is the subject of several studies.[73-80]

The catalytic activity of pcs in the heterogeneous dehydrogenation of cyclohexadiene in the gas phase, using nitrobenzene as the oxidizing agent, is measured.[73] By polarographic measurements, it is shown that the pcs are capable of two types of redox modes, either valency change of the central metal atom (Fe^{2+}, Co^{2+}), or ligand oxidation-reduction (Cu^{2+}, Zn^{2+}), or both modes (Ni^{2+}). When 1,4-cyclohexadiene is reacted with nitrobenzene in the gas phase over Fe pc or Co pc crystals at 240°C, benzene, aniline, and water are formed.[74] "The suggested mechanism is a rate determining electron transfer from catalyst to the nitrobenzene. This is in accordance with the fact that catalytic activity decreases when four electron attracting fluorine substituents are attached to the pc ring. If, however, sixteen fluorine substituents are attached to the ring, activity increases again. It can be shown that this is due to a change of mechanism: because of the great number of electron attracting substituents, the catalyst is unable to donate electrons but has become an electron acceptor instead. The mechanism has changed from that of oxidative dehydrogenation to that of direct dehydrogenation and no nitrobenzene is necessary for the reaction to proceed. The rate determining step is now an electron transfer from the cyclohexadiene to the catalyst and the products of reaction are benzene and molecular hydrogen."**

Pcs also exhibit catalytic activity in the dehydrogenation of isopropyl alcohol.[75-77] Cu^{2+} and Fe^{3+} pcs on $BaSO_4$ catalyze the oxidative dehydrogenation of isopropyl alcohol to dimethyl ketone at 260 to 320°C in air. In an inert atmosphere Cu^{2+} pc catalyzes the dehydrogenation of isopropyl alcohol, but at a substantially slower rate, and also catalyzes its dehydration to propylene.[76] AlClpc also catalyzes the dehydrogenation of isopropyl alcohol in oxygen containing and inert atmospheres, but some dehydration to C_3H_6 occurs.[75]

* From Naito, S. and Tamaru, K., *Z. Phys. Chem. (Frankfurt am Main)*, 94(1—3), 156, 1975. With permission.

** From Manassen, J. and Bar-Ilan, A., *J. Catal.*, 17(1), 86, 1970. With permission.

Co pc catalyzes the dehydrogenation of isopropyl alcohol, but only when oxygen is present, suggesting a pc-O complex. Metal free pc is inactive.[77]

"There have been several studies of the catalytic activity of metal poly pc on the oxidation of cumene[1,2], cyclohexene[1], acetaldehyde, ethylene acetal[3] and some aldehydes[4,5]. But all of them concern the liquid phase oxidation. In this report, heterogeneous (gas-solid) catalytic activities of metal poly pc for the oxidative dehydrogenation of methanol, ethanol, isopropanol and aryl alcohol are studied."* [The references 1 to 5 are: (1) S. Z. Roginskii, A. A. Berlin, L. N. Kutseva, R. M. Aseeva, L. G. Cherkashina, A. I. Sherle, and N. G. Matveeva, *Dokl. Akad. Nauk SSSR,* 1963, 148, 18; (2) T. Hara, Y. Ohkatsu, and T. Osa, *Chem. Lett.,* 103, 1973; (3) H. Inoue, Y. Kida, and E. Imoto, *Bull. Chem. Soc. Jpn.,* 40, 184, 1967; (4) H. Inoue, Y. Kida, and E. Imoto, *Bull. Chem. Soc. Jpn.,* 684, 1968; (5) T. Hara, Y. Ohkatsu, and T. Osa, *Chem. Lett.,* 953, 1973.]

The reactivity of various metal poly pc is Cu > Fe > Fe-Cu > Fe-Pd ≅ Fe-Mn. "It is notable that metal free poly pc exhibited no activity which corresponds well to the fact that oxygen is not adsorbed on it." It is concluded "that the activation of oxygen in the metal-poly pc occurs on the central metal ion and the activated oxygen has the ability of hydrogen abstraction from alcohol. The large π-electron system of the polymerized ligand has a rather marked influence on the central metal ion, which results in the so-called 'polymer effect'."*

The gas phase hydrogenation and dehydrogenation of pcs in an allyl alcohol-propanol-acrolein system is also studied[79] with Fe, Co, Ni, Zn, and Pt pcs. It is postulated that active centers involve not only the central metal atom of the pc molecule, but the nitrogen atom ring systems within the pc molecule.

The participation of singlet oxygen in the dehydrogenation photoreactions sensitized by tetrapyrrole pigments including porphyrin, pc, and chlorophyll, is studied[80] by irradiation using a Hg vapor lamp, in the photolysis of

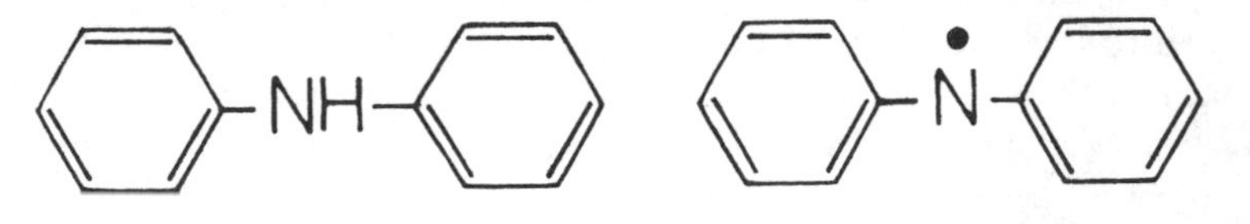

and

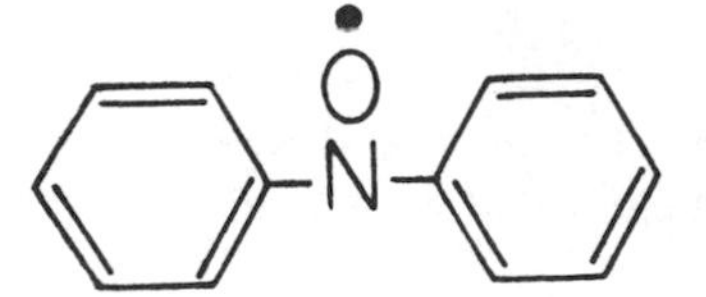

XIV. POLY PHTHALOCYANINES AS OXIDATION CATALYSTS

In a patent relating to the use of pcs as catalysts in fuel cells, a discussion of the role of pcs in electrodes is given.[81] It is pointed out that the use of organic catalysts such as the pcs as electrocatalysts is advantageous in that they are relatively inexpensive compared to precious metal electrocatalysts. However, pcs possess an extremely low conductivity and can not themselves be readily used as the electrode. They must be applied as a thin film or mixed as a powder with other conductive material. According to the discussion of the invention, therefore, it is advantageous to use polymeric pcs because they have a specific conductivity "several powers of ten higher than that of monomeric pcs," have increased [catalytic] activity whereby "the chemical yield at the electrodes is additionally increased," "they may be employed both in anodes and in

* From Naito, S. and Tamaru, K., *Z. Phys. Chem. Neue Folge,* 94, 150, 1975. With permission.

cathodes of fuel cells," and "a number of polymeric pcs, more especially Cu poly pcs, are stable in acids." According to the patent discussion, "poly pcs of copper, silver, nickel, iron, manganese, cobalt and chromium have proved particularly suitable as catalytically active cathodes, while copper poly pc and metal free poly pc have proved particularly suitable as catalytically active anodes."

An experimental study of liquid phase oxidation catalyzed by metal poly pcs[82] includes the poly pcs of Cu, Co, Cu–Fe, Fe, Mn, Cr, Cu–Co, Cu–Mn, Co–Fe, Co–Mn, and Fe–Mn.

XV. CATALYSTS IN THE OXIDATION OF ALDEHYDES

Cu-Fe poly pcs are used as catalysts in the oxidation of acetaldehyde ethylene acetal, benzaldehyde ethylene acetal, and cumene.[83] "The oxygen absorption rate is dependent, among other factors, on the atomic ratio of Cu to Fe ions in the poly pc lattice."

Aldehydes are also oxidized to carboxylate esters with a Pd pc catalyst.[84] For example, passage of an oxygen-nitrogen gas mixture into 10 g ethyl aldehyde, 40 g methyl alcohol, and 0.3 g Pd pc in an autoclave at 100°C for 4 hr gives 4.1 g methyl propionate, 1.5 g methyl formate, and 1.2 g unreacted ethyl aldehyde. In a like manner, methacrolein is converted to methyl methacrylate in the presence of Pd poly pc.

XVI. CATALYSTS IN THE OXIDATION OF PHENOLS AND ALCOHOLS

A process is described for oxidizing alkyl substituted phenols to *p*-benzoquinones with oxygen in the presence of Cu^{2+} and Co^{2+} pcs as catalysts. Preferred reaction conditions include 15 to 50°C at a pressure of oxygen bearing gas of 10 to 1000 psi.[85]

Co pc also catalyzes the oxidation of phenols to quinones in another similar process.[86] For example, 2,3,6-trimethylphenol is oxidized to 2,3,6-trimethylbenzoquinone.

Co pc tetrasulfonate and Ni pc tetrasulfonate are catalysts for the oxidation of polyhydric phenols such as catechol, hydroquinone, and pyrogallol.[87]

The vapor phase oxidation of 2-propanol to CH_3COCH_3 and H_2O takes place with heterogeneous catalysis over Cr, Mn, Fe, Co, Ni, and Cu pc.[88,89] "The oxidation is shown to occur on the central metal atom of the pc without any bonding interaction with the organic environment of the metal atom."

The vapor phase oxidation with oxygen of methanol to CH_2O with Cu pc catalyst[90] indicates that the catalytic effect is due to the central metal atom –Cu– but that the organic ring surrounding the central atom may also exert an influence on the catalysis.

XVII. CATALYSTS IN THE OXIDATION OF SUGARS

Monosaccharides such as glucose, arabinose, and threose can be oxidized with oxygen in the presence of Cu pc catalyst.[91,92] In Example 1, "Into a creased flask equipped with a high speed, shearing action stirrer, there was added 27 grams (150 millimoles) of glucose dissolved in 220 cc of water, 1.5 grams of Cu pc and 12.6 grams (150 millimoles) of sodium bicarbonate. The resulting reaction mixture was stirred and heated to a temperature of 70°C for 2½ hours during which time air was introduced into the reaction mixture. At the end of this time, 30 percent of the glucose had been oxidized to a material having a calcium ion sequestering index of 0.95."[91]

XVIII. CATALYSTS IN THE OXIDATION OF OLEFINS AND AROMATIC COMPOUNDS

The air oxidation of propylene to propylene oxide takes place with a Cu pc catalyst at 40 atm and 185°C on a variety of supports such as K_2CO_3, MgO, SiO_2, and SiO_2-Al_2O_3.[93] The selectivity of the catalyst is a function of the acidity of the catalyst support. The air oxidation of propylene is also catalyzed by Co pc.[94]

Mechanisms of inhibition against the copper catalyzed oxidation of polyethylene are postulated.[95,96] Cu pc is one of the catalysts. The inhibitors are mostly derivatives of *N,N'*-diphenyloxamide.

The catalytic properties of pcs in the oxidation of aromatic hydrocarbons also elicit interest.[97-99] They include the oxidation of isopropylbenzene with Ru and Os pc catalysts.[97]

The superoxide ion formed by the electrolysis of the oxygen molecule dissolved in pyridine acts as a base catalyst in the autoxidation of 9,10-dihydroanthracene. Fe^{2+} and Co^{2+} pcs accelerate the reaction[98] presumably by forming a complex with the superoxide ion.

In the catalysis of the homogeneous oxidation of mesitylene by cobalt and bromide ions[99] it is shown that Mn pc acts as a cocatalyst if it is added to Co^{2+} acetate and lithium bromide.

XIX. CATALYSTS IN THE OXIDATION OF ALKANES

The use of pcs as catalysts in the oxidation of alkanes has elicited interest.[100-107]

Cu pc accelerates the oxidation of *n*-tetradecane to a certain concentration above which it acts as an inhibitor. The addition of sodium stearate as promoter further accelerates the oxidation.[100]

Cu pc also catalyzes the autoxidation of phenylcycloalkanes.[101] Included are diphenylcyclopentanes and diphenylcyclohexanes.[102]

"In the autoxidation of phenylalkanes and phenylcycloalkanes in the presence of the pc complexes of Cu, Zn, Pd, Pt, Ag, and Hg, the metal-pc-O_2 complex reacts with dimeric hydroperoxide to form alkylperoxy radicals; at higher temperatures, the pc complex itself reacts with dimeric hydroperoxide."[103]

"The autoxidation of 1,1-diphenylethane at 85—112°C is accelerated by the pcs of copper and nickel, the initial reaction being catalyzed by activation of molecular oxygen. When hydroperoxide is accumulated it reacts with the pc-oxygen complex forming chain initiating alkylperoxyl radicals."[104]

A mathematical model is developed for the catalytic oxidation of isobutane, with Co pc.[105]

The oxidation of isoparaffins such as isobutane at 100 to 150°C and 400 to 1000 psi in the presence of Co, Fe, or Cu to alcohols as the main product is the subject of a patent.[106]

A process is provided "for the catalytic oxidation of *p*-amino substituted di-(hetero)-arylmethane and tri-(hetero)-arylmethane compounds with oxygen in the presence of a quinone and a catalyst containing a heavy metal in complex form." Preferred catalysts include Fe^{2+} and Fe^{3+} pcs and desired products are *p*-amino-substituted diarylketones.[107]

XX. CATALYSTS IN THE OXIDATION OF LEUCO COMPOUNDS

Certain basic dyes containing $=N^{\oplus}R\,R'$ groups, where R and R′ are alkyl or phenyl moieties, are prepared by oxidation of the corresponding leuco compounds with oxygen in the presence of halogenated benzoquinones, e.g., tetrabromo-*p*-benzoquinone and metal complex catalysts such as Fe(III) pc.[108]

XXI. CATALYSTS IN THE OXIDATION OF POLYMERS

Cu pc and other copper complexes are examined as catalysts in the oxidation of polyethylene.[109] The surface areas of the powdered catalyst compounds are determined by the BET method and the powders are dispersed in polyethylene by milling. The rates of oxidation of the polyethylene samples at 100°C are determined by oxygen absorption. Cu_2O is extremely active with a relative specific rate constant of 19.6 although its specific surface area is only 0.4 m^2/g. Cu pc exhibits little or no activity although its specific area is 50.0 m^2/g.

XXII. CATALYSTS IN OXIDATION ON ALUMINUM OXIDE SURFACES

Pigments deposited on various supporting surfaces behave similarly to natural pigment systems of plants, yielding a photoresponse, ascribed to the formation of oxygen, which is promoted by the addition of Fe(III) salts.[110] A system of pc on Al_2O_3 is photoactive in this sense, and the dynamics of the response are similar to those of a *Chlorella* culture. A sequence of six reaction steps in generation of oxygen from adsorbed HO and H ions is suggested, three of which require photostimulation.

XXIII. CATALYSTS IN THE OXIDATION OF CUMENE

Pcs elicit attention as catalysts in the oxidation of cumene.[111-119]

Cu pc and its octahydroxy, octamethoxy octahydroxy tetrachloro, octahydroxy-octachloro, octaoxo, and octaoxotetrachloro derivatives are used as catalysts for the oxidation of cumene to give cumene hydroperoxide. The reactivity of the catalysts increases with their increasing electrical conductivity.[112] The catalytic activity of Cu, Co, and Ni pcs is the subject of another study.[113]

The oxidation of cumene using Cu-Fe poly pc, Fe poly pc, and Cu poly pc catalysts indicates that the catalysts activate an oxygen molecule to form a pc-O_2 complex which abstracts the tertiary hydrogen of cumene to initiate the reaction.[114]

A process is described for the preparation of halo-substituted derivatives of acetophenone by oxidizing halocumenes by air or oxygen with a metal pc catalyst.[115,116] The preferred halocumenes are *p*-chlorocumene and *p*-bromocumene and the metal pcs include Cu, Co, and Fe.

Autoxidation of cumene with pc catalysts is the subject of several studies.[117-119] The catalytic efficiency of alkali pcs in the autoxidation of cumene appears to be affected by the metal hydrogen interchange of the pc with cumene hydroperoxide, such that the actual catalyst is the alkali salt of the hydroperoxide.[117,118]

In the initial phase of the autoxidation of cumene with Pb pc catalyst, the complex is destroyed by alkylperoxyl radicals. The resulting Pb compounds then catalyze the rapid adsorption of oxygen.[119]

XXIV. CATALYSTS IN THE OXIDATION OF AMINES

Pcs exhibit catalytic behavior in the oxidation of amines.[120-123]

Co pc sulfonic acids are used in the catalytic oxidation of aqueous *p*-phenylenediamine solutions.[120]

The effect of Co^{2+}, Cu^{2+}, Ni^{2+}, Mn^{3+}, and Fe^{3+} tetrasulfo pcs on the autoxidation of hydrazine is studied, but only Co pc has a distinct catalytic effect. Differences in catalytic activity are explained by the differing abilities of the complexes to bind molecular oxygen reversibly. The ternary complex N_2H_4-Co tetrasulfo pc-O_2 is an active reaction intermediate[121]

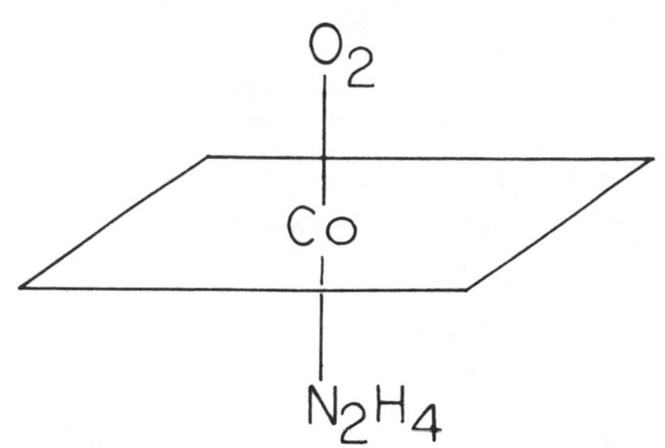

"Oxidation of hydroxylamine by molecular oxygen, similar to the oxidation of hydrazine, is catalyzed by Co^{2+} tetrasulfo pc. The oxidation products are nitrogen, dinitrogen oxide, and nitrite. Oxygen is reduced to water and hydrogen peroxide, whose reaction with hydroxylamine and/or hydrazine is also catalyzed by Co^{2+} tetrasulfo pc."* Co tetrasulfo pc catalyzes the autoxidation of cysteine as well.[123] However, "cysteine may be linked to Co tetrasulfo pc through the sulfur atom or through the nitrogen of the amino group. The former possibility seems much more likely, since Co tetrasulfo pc also catalyzes autooxidation of other thio compounds, with sulfur as the only donor atom."**

XXV. CATALYSTS IN EPOXIDATION

Metal pcs are studied as catalysts in the epoxidation of olefins. Several investigations have been published as patents by workers at the Institut Français du Pétrole, des Carburants et Lubrifiants,[124-128] and as articles appearing in the *Recueil des Travaux Chimiques des Pays Bas* from work undertaken at the Shell Laboratories in Amsterdam.[129,130]

Propylene is epoxidized in liquid phase with atmospheric oxygen with a selectivity of 42 to 76% of 1,2-epoxypropane and conversion of less than 10% by using molybdenum compounds as catalysts.[124,126] For example,[128] 60 g propylene and 80 g benzene are introduced into a reactor. The reactor is maintained at 190°C under a pressure of oxygen of 7 bars. Molybdenum pc, $MoOC_{32}H_{16}N_8$ is the catalyst (1×10^3g at Mo/kg of charge). For a conversion of 18% of propylene, the selectivity of epoxypropane is 45% with respect to the propylene consumed. Liquid phase epoxidation of 2,4,4-trimethyl-1-pentene with oxygen over molybdenum compounds yields 2,4-trimethyl-1,2-epoxypentane.[125] Conversions are probably less than 10%.

It is shown that after an initial period, during which the rate rapidly changes, the rate of the molybdenum catalyzed epoxidation of olefins with tert-butyl hydroperoxide become independent of the structure of the catalyst, including Mo pc catalysts.[129] The catalysts isolated from the reaction mixtures of the molybdenum catalyzed epoxidations of various olefins with tert-butyl hydroperoxide are all soluble Mo(VI)-1,2-diol complexes. For example, "when the heterogeneous catalysts MoO_2 pc and MoO_3 were

* From Wagnerova, D. M., Schwertnerova, E., and Veprek-Siska, J., *Collect. Czech. Chem. Commun.*, 39(11), 3036, 1974. With permission.

** From Dolansky, J., Wagnerova, D. M., and Veprek-Siska, J., *Collect. Czech. Chem. Commun.*, 41(8), 2326, 1976. With permission.

used as catalysts for the epoxidation of cyclohexene with tert-butyl hydroperoxide, virtually the same final rates were observed as with the homogeneous catalysts. Moreover, epoxidation continued, although at a lower rate, when the catalyst was filtered off and the filtrate allowed to react further. This suggested that soluble Mo(VI)-1,2-diol complexes may also be formed from heterogeneous molybdenum catalysts."*

Epoxycyclohexane is formed from cyclohexene and molecular oxygen in the presence of iron poly pc.[127]

XXVI. CATALYSTS IN AUTOXIDATION

"The term autoxidation is understood to mean the reaction of a substance with molecular oxygen without the intervention of a flame."[131]

It is found that "in autoxidation of tertiary, aryl or cycloalkanes the selectivity for organic hydroperoxides can be substantially increased by carrying out the reaction in the presence of metal free pc or chlorophyll. For example, an autoxidation of isopentane with 0.05 weight percent pc at 9.1 mole percent conversion gave selectivities of t-amyl hydroperoxides — 83.6 mole percent, acetone — 12.3 mole percent, and t-amyl alcohol — 1.7 mole percent. The same reaction without the pc at 10.0 mole percent conversion gave selectivities of t-amyl hydroperoxide — 56.8 mole percent, acetone — 31.2 mole percent, and t-amyl alcohol — 4.8 mole percent."[131]

The autoxidation of cumene is studied in the presence of Co pc catalyst[132] and in the presence of Mg pc catalyst.[133]

The autoxidation of *m*-diisopropylbenzene is accelerated by Cu pc; at high degrees of oxidation very high yields are obtained of predominantly *m*-isopropyl-7-cumyl-hydroperoxide and *m*-phenylene diisopropyl dihydroperoxide.[134] α to β conversion of the Cu pc catalyst also takes place.

The autoxidation of *m*-isopropyl-7-cumyl hydroperoxide in the presence of Cu pc gives considerable amounts of secondary products. The results are discussed with regard to the mechanism of catalysis.[134]

The catalytic effects of several metal acetylacetonates on the cleavage and autoxidation of polyvinyl alcohol are the subject of a research.[135] "The Co acetylacetonate and Co^{2+} pc were the most active catalysts in peroxide decomposition and autoxidation. These complexes also seem to be the most stable.[135]

"The autoxidation of 2,6-dialkyl substituted phenols has been studied using salcomines, the complex derived from Co^{2+} and schiff bases of salicylaldehyde and ethylene diamine, and metal pcs in *N,N'*-dimethylformamide. With salcomines and Co^{2+} pc, the predominant products are the corresponding 2,6-dialkylbenzoquinones along with some minor quantities of 3,3',5,5'-diphenoquinones. This is the first reported use of a pc complex in selective oxidation of disubstituted phenols."** Cu^{2+} pc, Mn^{2+} pc, and Fe^{2+} pc also exhibit catalytic behavior.

The autoxidation of 9,10-dihydroanthracene, by superperoxide ion, to anthraquinone and anthracene is catalyzed by Fe(II) or Co pc.[137,138] Intermediate formation of a superperoxide ion-metal pc complex is postulated.

XXVII. THE INFLUENCE OF EXTRAPLANAR LIGANDS ATTACHED TO THE CENTRAL METAL ATOM ON THE AUTOXIDATION OF Co^{2+} AND Fe^{2+} 4,4',4'',4'''-TETRASULFONATED PHTHALOCYANINES

The effect of the influence of extraplanar ligands attached to the central metal atom on the autoxidation of Co^{2+} and Fe^{2+} 4,4',4'',4'''-tetrasulfonated pcs is studied experi-

* From Sheldon, R. A., *Recl. Trav. Chim. Pays Bas*, 92(3), 367, 1973. With permission.

** From Kothari, V. M. and Tazuma, J. J., *J. Catal.*, 41(1), 180, 1976. With permission.

mentally in aqueous and 80% acetic acid solutions.[139] "Metal pcs and their derivatives are complexes which closely resemble metalloporphyrins constituting the active sites of many metalloenzymes. The biological function of the latter compounds is often determined by the presence of labile additional ligands above and below the porphyrin plane, or by the ability of their metal ions to undergo a reversible variation of their oxidation state. Studies on the influence of such ligands on properties of metal pc complexes could help explain the mechanism of action of natural metalloenzymes.

"One of the most important properties of metal pc complexes is their catalytic activity in aerial oxidation processes. That is connected with their ability to transport electrons or oxygen. Some of them show the capacity of a reversible oxygen binding resulting in labile oxygen adducts. Although a number of investigations have been made to clarify these processes, their detailed mechanisms are not known. Many complications in the studies arise from the existence of both monomer and dimer in aqueous solutions of these complexes."* A variety of influences are possible. The coordinated water molecule aids the transfer of metal from oxygen. The presence of KCN accelerates the autoxidation of the Co and Fe complexes and leads to their decomposition. Imidazole which forms a stable compound facilitates the oxidation of Co^{2+} to Co^{3+}. Pyridine forms stable complexes. LiBr strongly inhibits autoxidation by weakening the metal-oxygen bond through coordination with the metal on the reverse side of the pc ring.[139]

The catalytic effect of vanadyl tetrasulfo pc in complexing with molecular oxygen is also studied with ascorbic acid, hydroxylamine, and hydrazine adducts.[140]

XXVIII. CATALYSTS IN THE DECOMPOSITION OF HYDROGEN PEROXIDE

The catalytic decomposition of hydrogen peroxide,

$$2H_2O_2 \rightarrow O_2 + 2H_2O$$

by pcs has been studied by several workers.[141-148] Detailed reaction kinetic mechanisms are determined experimentally[141] for the decomposition of H_2O_2 in an aqueous solution catalyzed by the Fe^{3+} complex of 4,4′,4″,4‴-tetrasulfo pc at pH 5.5 to 10 by measuring the rate of increasing oxygen concentration. A similar study has been made for the Co^{2+} complex of 4,4′,4″,4‴-tetrasulfo pc at pH 3.8 to 10.[142] The authors state, "In the first transition series, iron and copper ions, and complexes thereof, are especially effective catalysts. From the catalase-like activity of Cu^{2+} complexes it is known that the reaction occurs within the coordination sphere. Therefore, the kind of ligand bound to the metal ion has a strong influence on the activity. This means, the activity is dependent (beside other parameters) on the saturation degree of the coordination sphere of the metal ion. This explains why Cu^{2+} complexes formed with tetradentate ligands, like 4,4′,4″,4‴-tetrasulfo pc are catalytically inactive. Of course, this does not mean that all metal ion tetrasulfo pc complexes are inactive; those formed with a metal ion having a coordination number greater than 4 may well be active."**

The kinetics of the catalytic decomposition of hydrogen peroxide with derivatives of ruthenium and osmium pc are also the subject of a study.[143]

In an investigation of hydrogen peroxide decomposition by pcs and poly pcs it is found that the catalytic activity of the poly pcs is in the sequence Fe > Co >> Ni > Cu.[144]

The catalytic activity of unsubstituted and tetrasubstituted pc complexes of Co in

* From Przywarska-Boniecka, H. and Fried, K., *Rocz. Chem.*, 50(1), 43, 1976. With permission.

** From Waldemeier, P., Prijs, B., and Sigel, H., *Z. Naturforsch. B.*, 27(2), 95, 1972. With permission.

hydrogen peroxide decreases in the order Co pc > Co 4,4′,4″,4‴-tetranitro pc > Co 4,4′,4″,4‴-tetrachloro pc. The reprecipitation of these catalysts from sulfuric acid decreases their surface area but does not affect their catalytic activity.[145] Another study of the catalytic properties of Co pc during hydrogen peroxide decomposition in an alkaline medium is reported.[146]

The fact that "metal chelates, cobalt chelates in particular, play an important role as oxygen carriers in biological systems" is the impetus for a spectral study "of the active centers on the Co pc surface and their activity in the hydrogen peroxide decomposition reaction"[147] before and after coordination of the Co pc with pyridine and triphenylphosphine. The catalytic activity is increased by coordination.

A platinum electrode is coated with a porous film of pc.[148] It is found that hydrogen peroxide decomposes on the illuminated surface coated with the pc film.

XXIX. CATALYSTS IN THE DECOMPOSITION OF HYDROPEROXIDES

The catalytic decomposition of ethylbenzene hydroperoxides, at 90 to 130°C, is studied using Cu, Co, Ni, Mn, and Ce oleates, acetylacetonates, and pcs. Rates of decomposition, frequency factors, and activation energies are determined. Oleates and acetylacetonates are equally efficient and are more efficient than pcs.[149] Reaction rate is determined and mechanism is discussed of the decomposition of 2-benzyl-2-propyl hydroperoxide in the presence of *N*-(2-naphthyl) aniline inhibitor with Co pc and Cu pc catalysts in various solvents. It is shown that the viscosity of the solvent influences the rate of reaction.[150] The decomposition is first order with respect to hydroperoxide. The same order of reaction holds for the decomposition of 7-cumyl hydroperoxide in the presence of Co pc and *N*-(2-naphthyl) aniline inhibitor.[151] α Cu pc is found to have little effect on the decomposition of 7-cumyl hydroperoxide; an $\alpha \rightarrow \beta$ conversion of the Cu pc is observed.[152]

The decomposition of 7-cumyl hydroperoxide is studied in the presence of Co pc in a variety of organic solvents. The rate of reaction increases with increasing polarity of the solvent.[153] Decay to free radicals or by molecular decomposition may be influenced by the solvent.

The thermal decomposition of azobisisobutyronitrile in an oxygen atmosphere is studied[154] and the interaction of metal ions with peroxyl radicals is observed. Mn pc is considered to stabilize the peroxyl radicals.

The vapor of hydrazine hydrate is decomposed in the temperature range 150 to 350°C with β Cu pc catalyst to ammonia and nitrogen with complete conversion at 300°C. "Though the mechanism of the reaction and the active site within the Cu pc molecule are not yet known, it is probable that hydrazine being able to migrate into the Cu pc lattice is decomposed not only at the surface but also in the interior of the Cu pc catalyst."[155]

The cleavage of poly(vinylalcohol) in the presence of tert-butyl hydroperoxide and Fe and Co pcs (as well as hemin, hematin, and cytochrome *C* and metal acetylacetonates) shows that the most effective promoters of the cleavage reactions are Fe pc and Co pc.[156] "The reactive pcs were more effective than the acetylacetonates" and "the order of reactivity is different for the pcs and the acetylacetonates, again demonstrating the importance of the ligand in attempting to assay the reactivity of metals in catalytic reactions."*

* From Jochsberger, T., Auerbach, A., Indictor, N., *J. Polym. Sci., Polym. Chem. Ed.*, 14(5), 1083, 1976. With permission.

XXX. CATALYSTS IN THE DECOMPOSITION OF FORMIC ACID

The activation energies of the dissociation of formic acid on Fe, Cu, Co, Ni, and Zn pcs are determined.[157] The α pc crystals give activation energies, with the exception of Zn, that are 3 to 4 kcal/mol lower than the β pc crystals. The activation energies for Mn, Fe, Co, Ni, Cu, and Zn pcs follow the electronegativity of Pauling.[158] A reexamination of this reaction over β Cu pc gives activation energies of 6.0 kcal/mol for the β crystal needles and 16.6 kcal/mol for the β crystal powder.[159] The chemisorption of formic acid over metal pcs, studied by X-ray photoelectron (XPS) and ultraviolet photoelectron (UPS) spectroscopy,[160] indicates that during the catalytic decomposition of formic acid over Fe pc, "formic acid is dissociatively adsorbed at the Fe atom of pc as

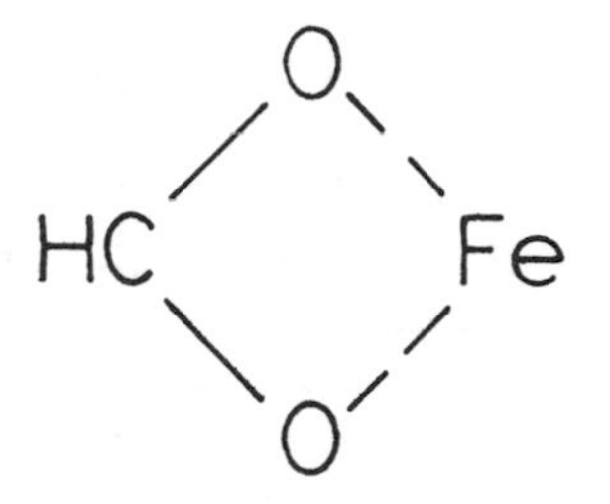

with a change of the oxidation state of the Fe atom from Fe^{2+} to Fe^{3+}. The shift of the N (1s) peak towards higher binding energy indicates the oxidation of an N atom or the adsorption of a positively charged substance. Therefore, the proton dissociated from formic acid would appear to be bound on an N atom in Fe pc, especially on one of the bridge N atoms which have higher electron density, forming an

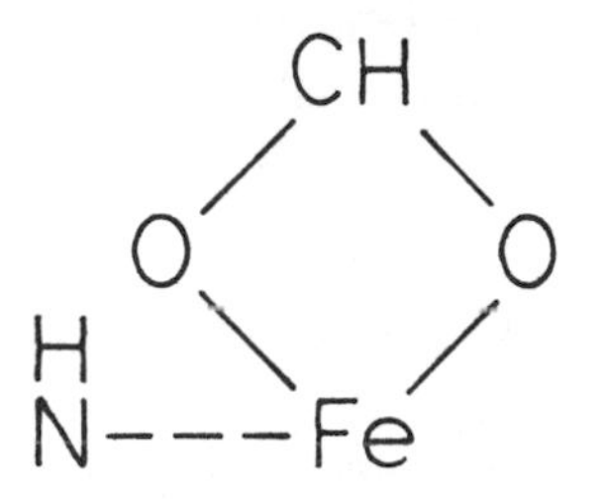

structure. . . This chemisorption bond is typically ionic."* Similar results are obtained for Co pc and Mg pc. However, formic acid is not adsorbed on H_2 pc.[160]

XXXI. POLYMERIZATION CATALYSTS

There is evidence that pcs can be used as polymerization catalysts.[161-164] Nitrile polymers can be made from perfluoroglutaronitrile and methyl cyanide that are thermally stable and have a low friction coefficient for use in antifriction drive train elements. They are prepared at high temperatures and pressures in the presence of basic catalysts such as copper pc.[161] High-molecular-weight polyoxymethylenes with increased thermal stability are made from gaseous formaldehyde in inert solvents in the presence of pc or copper pc.[162] Colored polyolefins are prepared by polymerizing or copolymerizing α olefins in the mass or in inert hydrocarbon solvents in the presence of metal pc catalysts.[163]

* From Kawai, T., Soma, M., Matsumoto, Y., Onishi, T., and Tamaru, K., *Chem. Phys. Lett.*, 37(2), 378, 1976. With permission.

The effectiveness of a pc catalyst in polymerization may be improved by including it in a system including alkyl peracids and acyl hydroperoxides.[164] Thus, the polymerization rate of monomers in the presence of copper pc, tert-butyl hydroperoxide, and peroxycapric acid increases 12-fold compared to the polymerization rate observed in the presence of copper pc.

XXXII. CATALYSTS IN AMMONIA SYNTHESIS

The Fe pc-graphite-potassium system is described for the catalytic synthesis of ammonia.[165] The catalyst is prepared in a vacuum at high temperature.

XXXIII. DECARBOXYLATION CATALYSTS

The activity of pcs in the decarboxylation and polymerization of pyruvic acid decreases in the order Fe > Co > Cu > metal free pc. Metal-free pc and $FeSO_4$ are essentially inactive. The apparent activation energy of decarboxylation at 40 to 80° is 14 kcal/mol for Fe pc and 9 kcal/mol for Co pc. Co pc loses its activity after about 50 min of reaction.[166] Oxidative decarboxylation of oxalic acid in the presence of Fe pc, FeCl pc, Co pc, and Co (SO_3H) pc is also studied; reaction rates and apparent activation energies are determined.[167] A semiqualitative correlation is obtained between the rate of catalytic oxidative decarboxylation of oxalic acid on Co pc and its derivatives and the theoretical calculated change of the electron density on the bridging nitrogen atoms.[168]

The catalytic activity of substituted Co pcs in the oxidative decarboxylation of oxalic acid is the object of enquiry.[169] Introduction of electron acceptor substituents into the pc ring either lowers the catalytic activity slightly (NO_2) or stops it altogether (Cl); electron donor substituents (NH_2, OH) increase the activity.

The effect of the nature of the metal and ligands on pc catalytic properties in oxalic acid decarboxylation by Fe, Co, Cu, and Al pcs is the subject of a review with 19 references.[170]

XXXIV. CATALYSTS IN THE FISCHER-TROPSCH SYNTHESIS

Heating an alkali metal (electron donor) with a metal or metal-free pc (electron acceptor) at above the melting point of the alkali metal yields catalysts that have a strong adsorption for hydrogen and carbon monoxide and that are useful as catalysts for the Fischer-Tropsch synthesis.[171]

XXXV. CATALYSTS IN THE PREPARATION OF ISOCYANATES

Organic isocyanate compounds can be prepared by treating aromatic nitro compounds with carbon monoxide in the presence of pcs complexed with Ru, Rh, Os, Ir, or Pt.[172] For example, 10 mℓ nitrobenzene, 100 mg Ru pc, and 60 kg/cm^2 carbon monoxide yield phenyl isocyanate at 88% conversion and 75% selectivity after 2 hr at 195°C.

XXXVI. CATALYSTS IN THE HYDROXYLATION OF AROMATIC COMPOUNDS

Aromatic compounds can be hydroxylated to phenols by treating them with aqueous hydrogen solutions containing 5 to 90% hydrogen peroxide at −10 to 100°C at one to 50 atm pressure "in the presence of a pc catalyst selected from the group consisting

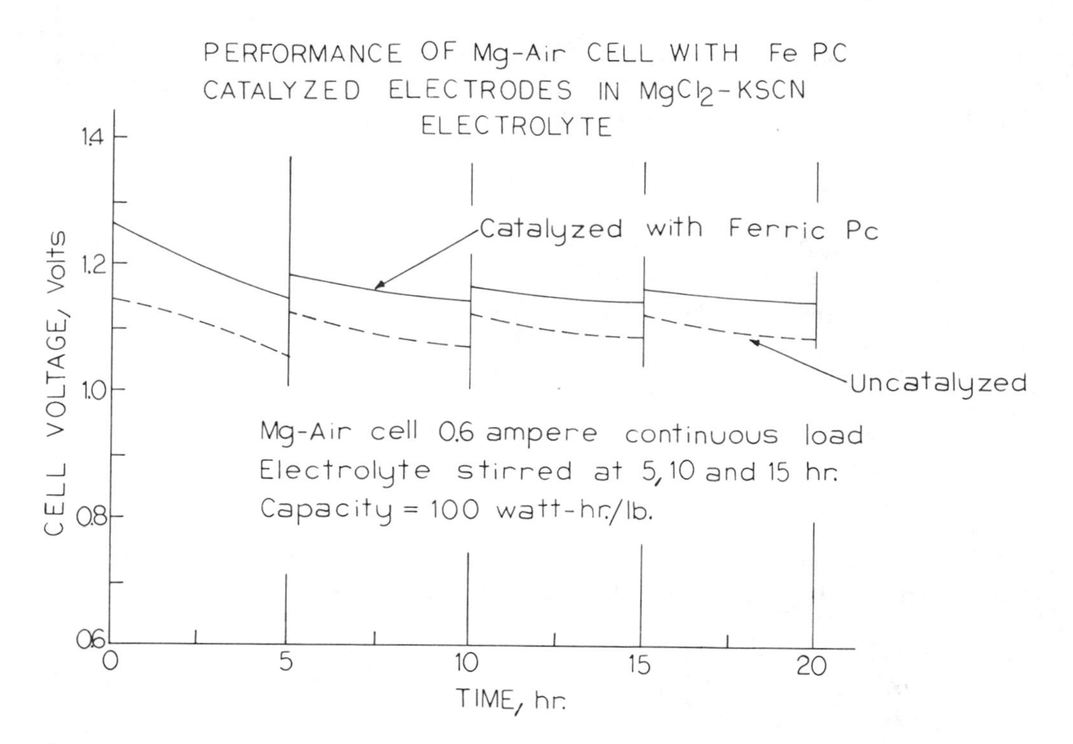

FIGURE 1. Performance of magnesium-air cell with ferric pc catalyzed electrodes in $MgCl_2$ KSCN electrolyte.

of cobalt, vanadium, manganese, iron, copper, nickel, molybdenum, chromium, and tungsten." A "particularly preferred catalyst is Co pc disulfonate."[173]

XXXVII. CATALYSTS IN A MAGNESIUM-AIR CELL WITH Fe^{3+} PHTHALOCYANINE CATALYZED ELECTRODES IN $MgCl_2$-KSCN ELECTROLYTE

Air depolarized cells are developed with an improved aqueous electrolyte.[174] The electrolyte may be neutral, slightly acid or slightly alkaline and contains a salt such as a halide of an alkali metal, alkaline earth metal, aluminum, zinc, or ammonia, and a cyanate or thiocyanate salt. These air depolarized cells "alleviate the problem of carbonation with attendant plugging of the pores in the cathode that is often encountered with air depolarized cells utilizing strong alkaline electrolytes." As an example, "a magnesium-air flat cell was constructed using a thin 'fixed zone' plastic bonded carbon electrode as cathode, a magnesium sheet anode and 75 ml of 1:1 (by volume) mixture of a 20 percent $MgCl_2$ solution and a 5M KSCN (40 percent by weight) solution (pH about 6). Apparent cathode area was 5 in^2. The cathode was catalyzed with 1mg/cm^2 of ferric pc catalyst. This cell, which had a capacity of 100 whr/lb, was placed on a 600 ma drain; its discharge performance is shown in Figure 1. The electrolyte was stirred at 5, 10, and 15 hour intervals. The increased cell voltage observed immediately after stirring indicates diffusion limitations arising from the precipitation of $Mg(OH)_2$ during cell discharge."

XXXVIII. CATALYTIC ACTIVITY OF COBALT PHTHALOCYANINE

Co pc deposited on silica is studied by reflection spectroscopy in UV and visible regions, to identify active centers in the Co pc structure.[175] It is demonstrated "that the processes taking place after adsorption on a solid surface of Co pc greatly resemble those in a liquid phase with cobalt complexes. The behavior of the cobalt ion on a pc surface is also the same as on an oxide surface."

XXXIX. CATALYSTS IN A VARIETY OF REACTIONS

Reactions of copper carbenoids with sulfoxides is the subject of a study.[176] The copper chelate catalyzed reaction of α-diazoacetophenone with substituted diphenyl sulfoxides gives two types of products, diaryl sulfides and oxosulfonium ylides. The oxosulfonium ylides exhibit a novel reaction with phenyl-glyoxal affording oxosulfonium glyoxalylphenacylides in the presence of copper acetylacetonate. When copper pc is used as a catalyst, the reaction gives a lactone besides diaryl sulfides.[176]

The specific action of Na_2 pc is studied as a base for aldol condensation.[177] It does so act for aldehydes but not for ketones. During the aldol condensation of isobutyl aldehyde, Na_2 pc is converted to H_2 pc by an intermediate, retarding the aldol condensation.

The reversible carbonylation of Fe pc provides a source of carbon monoxide, permitting carbonylated Fe pc to act as a catalyst in the carbonylation of alcohols:[178]

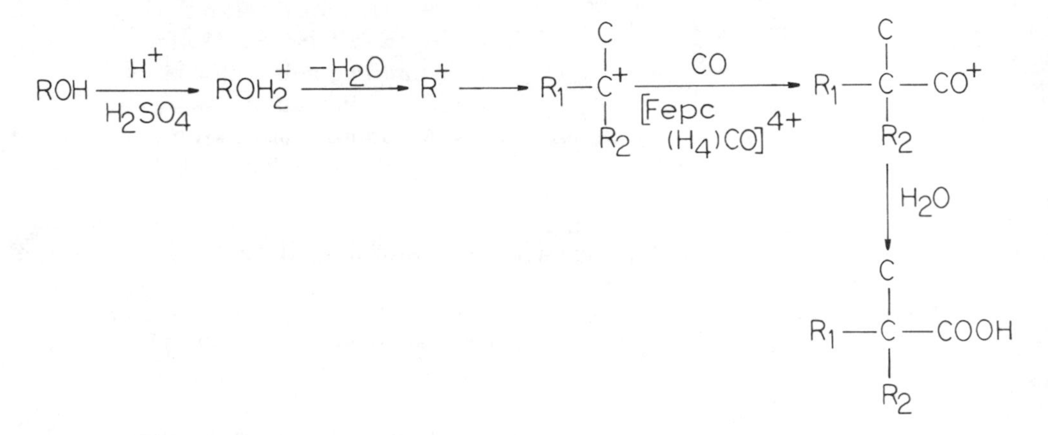

Magnesium air cells with a cyanate or thiocyanate containing electrolyte are developed with a plastic bonded carbon cathode with ferric pc catalyst.[179]

In a research entitled "Catalase Activity of Ferriheme", it is indicated that the monomeric and dimeric species of the Fe(III) complex of 4,4′,4″,4‴-tetrasulfo pc contribute to catalytic activity.[180]

XL. CATALYSTS RELATING TO SULFUR-BEARING COMPOUNDS

Pc catalysts are used in a patented process for the treatment of effluent aqueous streams.[181,182] The oxidizing catalyst must be capable of "effecting substantially complete conversion of the ammonium hydrosulfide salt contained in the waste stream. . . Particularly preferred metal pc compounds include those of cobalt and vanadium. Other metal pc compounds that may be used include those of iron, nickel, copper, molybdenum, manganese, or tungsten. Moreover, any suitable derivative of the metal pc may be employed, including the sulfonated derivatives and the carboxylated derivatives. . . the preferred carrier material is activated carbon." The inventors, in particular, derive "a process for the treatment of an aqueous waste stream containing NH_4HS to produce a treated water stream substantially free of $(NH_4)_2S_2O_3$ and, in preferred embodiments, to recover elemental sulfur and ammonia."

Another process[183] relates to "the production of sulfur which comprises the oxidation of a sulfide solution in the presence of a catalyst comprising a metal pc compound to produce sulfur and a polysulfide effluent. . ." Catalysts include Co pc and vanadium pc on such solid supports as charcoal or γ-alumina.

Sulfur-containing organic compounds are removed from wastewaters by treating the water with an oxygen-bearing gas in the presence of a catalyst such as Co and Cu pc tetracarbonates. Desulfurization and imidazole sulfur-organic derivatives and trithiocarbonic acid derivative destruction are carried out at pH 1 to 3.[184]

"H_2S is removed from gases by absorption in oxythioarsenate solutions with subsequent regeneration. To increase the sorption capacity of the solutions and to speed up their regeneration, metal pcs such as Co pc are introduced into the oxythioarsenate solutions."[185]

Mercaptans are removed from hydrocarbon raw materials also by oxidation with a pc catalyst, but in an alkaline medium and with a promoter, namely, a bromide of a Group I, VI, or VII metal or its complex with ethanolamine.[186]

The method of application of the pc catalyst to its carrier in the treatment of "sour hydrocarbon distillates" in order to oxidize the mercaptans in the distillates[187] is described.

Catalytic treatment of aqueous streams, principally industrial effluents that contain sulfide compounds, is directed at the recovery of elemental sulfur with a minimum of oxygen in a two-step cycle, using metallic pc sulfonates as catalysts.[188]

"An essential feature of the present invention is the utilization of a two step oxidation procedure wherein the first step is conducted at relatively mild conditions, thereby enabling the use of relatively low pressure oxygen or air, and the second step operates on the effluent stream from the first step to complete the oxidation of the residual sulfide present in the effluent stream from the first step in the form of a polysulfide. This method allows the first step to be operated at conditions which prevent the deposition on the solid catalyst employed therein by virtue of the fact that the sulfur formed therein is soluble in the excess sulfide charged to this step. The problem of sulfur deposition is avoided in the second step by operating at conditions resulting in the formation of liquid sulfur which separates easily from the solid catalyst." In the first and second steps the solid catalyst is advantageously cobalt pc monosulfonate combined with activated carbon.

A process is described for oxidizing 2-mercaptobenzothiazole (MBT) by heating an organic solution containing this compound with an oxygen-containing gas and a catalytic amount of a cobalt pc sulfonate at 25 to 200°C.[189] "More particularly, this invention relates to the production of 2,2′-thiobisbenzothiazole, 2,2′-dithiobisbenzothiazole (MBTS), and benzothiazylsulfenamides by oxidizing MBT. MBTS is presently used as a rubber vulcanization accelerator. In addition, it is an important intermediate for the production of various sulfenamides which themselves are very useful as rubber chemicals and especially as vulcanization accelerators."

Another process for the preparation of sulfenamides by catalytic oxidation[190] includes metal pcs as oxidation catalysts "in the reaction of primary or secondary amines with 2-mercaptobenzothiazole, an alkali metal salt of 2-mercaptobenzothiazole, a dithiocarbamate, a dithiocarbamic acid, a thiuram disulfide, or 2,2′-dithiobis (benzothiazole)." The metal of the pc can be Co, Mn, V, Cr, Ni, Fe, Cu, or Pt.

In the oxidation of cysteine and hydrogen sulfide with sulfo derivatives of Co pc,[191] it is found that tetrachloro and tetra-SO_3Na Co pc complexes increase the stability of the Co^{2+} oxidation state in the pc complex whereas other ligand arrangements are less active Co pc catalysts because they tend to stabilize the Co^{3+} oxidation state.

Co^{2+} pc 4,4′,4″,4‴-tetraamino pc attached to cross-linked polyacrylamide produces a stable catalyst with enhanced activity in the oxidation of thiols such as $HS(CH_2)_2OH$.[192]

The stability of β Cu pc as an oxidation catalyst is studied with 2-propanol and 2-propanethiol reactants. By a gas chromatographic technique, 2-propanol is oxidized to $(CH_3)_2CO$ with about 40% reaction. Under parallel conditions, 2-propanethiol reacts with about 100% reaction to disulfide and by H_2S loss to propene.[193]

REFERENCES

1. Kawai, T., Soma, M., Matsumoto, Y., Onishi, T., and Tamaru, K., *Chem. Phys. Lett.*, 37, 2, 378, 1976.
2. Kropf, H. and Steinbach, F., Eds., (Thieme: Stuttgart), *Katalyse an Phthalocyaninen, Symposium am 10. Mai 1972 in Hamburg*, Thieme, Stuttgart, 1973, 162.
3. Manassen, J., *Catal. Rev.*, 9(2), 223, 1974.
4. Steinbach, F., Schmidt, H. H., and Zobel, M., *Catal., Proc. Int. Symp.*, Elsevier, Amsterdam, 1975, 417.
5. Campadelli, F., Cariati, F., Carniti, P., Morazzoni, F., and Ragaini, V., *J. Catal., 44(1), 167, 1976.*
6. Schaldach, M. and Kirsch, U., *Trans. Am. Soc. Artif. Intern. Organs*, 16, 184, 1970.
7. Wood, J. H., Keenan, C. W., and Bull, W. E., *Fundamentals of College Chemistry*, 3rd ed., Harper & Row, New York, 1972.
8. Jasinski, R., *J. Electrochem. Soc.*, 112, 526, 1965; *Nature (London)*, 201, 1212, 1964.
9. Jahnke, H., *Ber. Bunsenges. Phys. Chem.*, 72, 1053, 1968.
10. Jahnke, H. and Schoenborn, M., paper presented at the 19th CITCE Meeting, Detroit, September 16, 1968; *Journées Int. Étude Piles Combust. C. R.*, 3rd, 60, 1969.
11. Jahnke, H., Schoenborn, M., and Zimmermann, G., *Proc. Symp. Electrocatal.*, Electrochem. Soc., Princeton, N.J., 1974, 303.
12. Alt, H., Binder, H., and Sandstede, G., *J. Catal.*, 28(1), 8, 1973.
13. Manassen, J., *J. Catal.*, 33(1), 133, 1974.
14. Brodd, R. J., Kalnoki-Kis, T., Kozawa, A., and Zilionis, V. E., U.S. NTIS PB Rep. No. 212018, National Technical Information Service, Springfield, Va., 1970.
15. Kozawa, A., Zilionis, V. E., and Brodd, R. J., *J. Electrochem. Soc.*, 118(10), 1705, 1971.
16. Contour, J. P., Lenfant, P., and Vijh, A. K., *J. Catal.*, 29(1), 8, 1973.
17. Savy, M., Andro, P., Bernard, C., and Magner, G., *Electrochim. Acta*, 18(2), 191, 1973.
18. Randin, J. P., *Electrochim. Acta*, 19(2), 83, 1974.
19. Jahnke, H. G., Schoenborn, M. F., and Zimmermann, G., *Proc. Symp. Electrocatal.*, Electrochem. Soc., Princeton, N.J., 1974, 303.
20. Alferov, G. A. and Sevast'yanov, V. I., *Elektrokhimiya*, 11(5), 827, 1975.
21. Alferov, G. A., *Tr. Mosk. Fiz. Tekhn. In-Ta. Ser. Obshch. Mol. Fiz.*, 6, 152, 1975.
22. Alferov, G. A., *Tr. Mosk. Fiz. Tekhn. In-Ta. Ser. Obshch. Mol. Fiz.*, 6, 161, 1975.
23. Shumov, Yu. S. and Heyrovsky, M., *J. Electroanal. Chem. Interfacial Electrochem.*, 65(1), 469, 1975.
24. Savy, M., Magner, G., and Peslerbe, G., *C. R. Acad. Sci. Ser. C*, 275(3), 163, 1972.
25. Savy, M. and Meyer, G., *C. R. Hebd. Seances Acad. Sci. Ser. C*, 282(16), 707, 1976.
26. Savy, M., Bernard, C., and Magner, G., *Electrochim. Acta*, 20(5), 383, 1975.
27. Beck, F., *Ber. Bunsenges. Phys. Chem.*, 77(5), 353, 1973.
28. Beck, F., Dammert, W., Heiss, J., Hiller, H., and Polster, R., *Z. Naturforsch. Teil A*, 28(6), 1009, 1973.
29. Tezuka, M., Ohkatsu, Y., and Osa, T., *Bull. Chem. Soc. Jpn.*, 48(5), 1471, 1975.
30. Osa, T., Ohkatsu, Y., and Tezuka, M., *J. Fac. Eng. Univ. Tokyo Ser. A*, 10, 66, 1972.
31. Beyer, W. and Von Sturm, F., *Angew. Chem. Int. Ed. Engl.*, 11(2), 140, 1972.
32. Savy, M., Andro, P., and Bernard, C., *Electrochim. Acta*, 19, 403, 1974.
33. Savy, M., Andro, P., and Bernard, C., *Croatica Chemica Acta*, 44(1), 107, 1972.
34. Johansson, L. Y., Mrha, J., and Larsson, R., *Electrochim. Acta*, 18, 255, 1973.
35. Larsson, R. and Mrha, J., *Electrochim. Acta*, 18(6), 391, 1973.
36. Musilová, M., Mrha, J., and Jindra, J., *J. Appl. Electrochem.*, 3(3), 213, 1973.
37. Meier, H., Albrecht, W., Tschirwitz, U., and Zimmerhackl, E., *Ber. Bunsenges. Phys. Chem.*, 77(10/11), 843, 1973.
38. Radyushkina, K. A., Burshtein, R. Kh., Berezin, B. D., Tarasevich, M. R., and Levina, S. D., *Elektrokhimiya*, 9(3), 410, 1973.

39. Pobedinskii, S. N., Bazanov, M. I., Trofimenko, A. A., Aleksandrova, A. A., Belonogov, K. N., and Al'yanov, M. I., *Vopr. Kinet. Katal.*, 2, 108, 1974.
40. Pobedinskii, S. N., Trofimenko, A. A., Aleksandrova, A. N., Belonogov, K. N., Al'yanov, M. I., and Bazanov, M. I., *Vopr. Kinet. Katal.*, 2, 105, 1974.
41. Pobedinskii, S. N., Trofimenko, A. A., Bazanov, M. I., Al'yanov, M. I., Aleksandrova, A. N., Erin, V. A., and Belonogov, K. N., in *Katalitich. reaktsii v zhidk. faze.*, 1974, 698.
42. Radyushkina, K. A., Levina, O. A., Tarasevich, M. R., Burshtein, R. Kh., Berezin, B. D., Shormanova, L. P., and Koifman, O. I., *Elektrokhimiya*, 11(6), 989, 1975.
43. Burshtein, R. Kh., Tarasevich, M. R., Radyushkina, K. A., and Zagudaeva, N. M., *Adsorbtsiya Adsorbenty*, 2, 34, 1974.
44. Pobedinskii, S. N., Trofimenko, A. A., Aleksandrova, A. N., Belonogov, K. N., and Al'yanov, M. I., *Tr. Ivanov. Khim. Tekhnol. In-Ta*, 16, 27, 1973.
45. Domnikov, A. A., Reznikov, G. L., and Yuppets, F. R., *Tezisy Dokl. Vses. Soveshch. Elektrokhim. 5th*, 2, 455, 1974, *Vses. Inst. Nauchm. Tekh. Int., Moscow.*
46. Dabrowski, R., Tomassi, W., and Witkiewicz, Z., *Biul. Wojsk. Akad. Tech.*, 23(2), 47, 1974.
47. Dabrowski, R., Twardowski, A., and Witkiewicz, Z., *Biul. Wojsk. Akad. Tech.*, 23(2), 57, 1974.
48. Kretzschmar, Chr. and Wiesener, K., *Z. Phys. Chemie, Leipzig*, 257(1), 39, 1976.
49. Jahnke, H., Schoenborn, M., and Zimmermann, G., *Katal. Phthalocyaninen. Symp.*, Thieme, Stuttgart, 1973, 71.
50. Tenygl, J., Musilová, M., and Mrha, J., Czechoslovakian Patent 156,983, January 15, 1975.
51. Meshitsuka, S., Ichikawa, M., and Tamaru, K., *J. Chem. Soc. Chem. Commun.*, 5, 158, 1974.
52. Zakharkin, G. I. and Tarasevich, M. P., *Elektrokhimiya*, 11(7), 1019, 1975.
53. Beck, F., Heiss, J., Hiller, H., and Polster, R., *Katal. Phthalocyaninen Symp.*, Thieme, Stuttgart, 1973, 53.
54. Pobedinskii, S. N., Aleksandrova, A. N., Trofimenko, A. A., Belonogov, K. N., and Al'yanov, M. I., *Tr. Ivanov. Khim. Tekhnol. In-Ta*, 16, 31, 1973.
55. Pobedinskii, S. N., Trofimenko, A. A., Aleksandrova, A. N., Belonogov, K. N., and Al'yanov, M. I., *Tr. Ivanov. Khim Tekhnol. In-Ta*, 17, 117, 1974.
56. Gerischer, H., Pettinger, B., and Luebke, M., Proc. Symp. Electrocatal. Electrochem. Soc., Princeton, N.J., 1974, 162-77.
57. Mochida, I., Takeyoshi, K., Fujitsu, H., and Takeshita, K., *Chem. Lett.*, 4, 327, 1976.
58. Ichikawa, M., Sonoda, R., and Meshitsuka, S., *Chem. Lett.*, 7, 709, 1973.
59. Ichikawa, M., and Meshitsuka, S., *J. Am. Chem. Soc.*, 95(10), 3411, 1973.
60. Inoue, H., Aoki, R., and Imoto, E., *Chem. Lett.*, 10, 1157, 1974.
61. Inoue, H., Hata, H., and Imoto, E., *Chem. Lett.*, 1241, 1975.
62. Ichikawa, M., Japanese Kokai 74 116,009, November 6, 1974.
63. Ichikawa, M., Japanese Kokai 74 116,010, November 6, 1974.
64. Kropf, H. and Mueller, J., *Justus Liebigs Ann. Chem.*, 7—8, 1236, 1976.
65. Stolfa, F., (to Universal Oil Products Company), U.S. Patent 3,773,655, November 20, 1973.
66. Ichikawa, M., Soma, M., Onishi, T., and Tamaru, K., *Trans. Faraday Soc.*, 63(5), 1215, 1967.
67. Naito, S., Ichikawa, M., and Tamaru, K., *J. Chem. Soc. Faraday Trans. 1*, 68 (Part 8), 1451, 1972.
68. Ichikawa, M., Naito, S., Saito, S., and Tamaru, K., *J. Chem. Soc. Faraday Trans. 1*, 69(4), 685, 1973.
69. Naito, S. and Tamaru, K., *Z. Phys. Chem. (Frankfurt am Main)*, 94(1—3), 63, 1975.
70. Naito, S. and Tamaru, K., *Z. Phys. Chem. (Frankfurt am Main)*, 94(1—3), 156, 1975.
71. Ichikawa, M., Soma, M., Onishi, T., and Tamaru, K., *Trans. Faraday Soc.*, 63(8), 2012, 1967.
72. Turkevich, J. and Sato, T., *Catal., Proc. Int. Congr., 5th*, Vol. 1, North-Holland, Amsterdam, 1973, 587.
73. Manassen, J. and Bar-Ilan, A., *J. Catal.*, 17(1), 86, 1970.
74. Bar-Ilan, A. and Manassen, J., *Catal. Proc. Int. Congr., 5th*, Vol. 2, North-Holland, Amsterdam, 1973, 1149.
75. Borisenkova, S. A., Rudenko, A. P., and Herrero-Palensueva, V. E., *Vestn. Mosk. Univ. Khim.*, 16(1), 122, 1975.
76. Borisenkova, S. A., Il'ina, L. M., Rudenko, A. P., and Savilov, A. P., *Vestn. Mosk. Univ. Khim.*, 14(4), 494, 1973.
77. Borisenkova, S. A., Il'ina, L. M., and Rudenko, A. P., *Zh. Fiz. Khim.*, 50(7), 1712, 1976.
78. Naito, S. and Tamaru, K., *Z. Phys. Chemie Neue Folge*, 94, 150, 1975.
79. Kropf, H. and Witt, D. J., *Katal. Phthalocyaninen, Symp.*, Georg Thieme, Stuttgart, 1973, 139.
80. Bobrovski, A. P. and Kholmogorov, V. E., *Khim. Vys. Energ.*, 6(2), 125, 1972.
81. Siemens Aktiengesellschaft, British Patent 1,213,364, November 25, 1970.
82. Hara, T., *Kagaku Kogyo*, 23(4), 490, 1972.
83. Inoue, H., Kida, Y., and Imoto, E., *Bull. Chem. Soc. Jpn.*, 40(1), 184, 1967.
84. Suzuki, Y., Okushima, H., and Nitta, I., Japanese Kokai 72 34,309, November 21, 1972.

85. Kothari, V. M. (to Goodyear Tire and Rubber Company), U.S. Patent 3,935,247, January 27, 1976.
86. Omura, Y., Nakamura, M., Oka, M., Fujiwara, Y., and Itoi, K., (to Kuraray Company Ltd.), Japanese Kokai 74 127,937, December 7, 1974.
87. Meshitsuka, S., Ichikawa, M., and Tamaru, K., *J. Chem. Soc. Chem. Commun.*, 9, 360, 1975.
88. Steinbach, F. and Hiltner, K., *Z. Phys. Chem. (Frankfurt am Main)*, 83(1—4), 126, 1973.
89. Steinbach, F. and Hiltner, K., *Katal. Phthalocyaninen. Symp.*, Georg Thieme, Stuttgart, 1973, 122.
90. Ionescu, N. I., Schuster, R. H., Trestianu, D., and Mihailescu, A., *Z. Chem.*, 15(12), 496, 1975.
91. Rutledge, T. F. (to ICI America Inc.), U.S. Patent 3,860,642, January 14, 1975.
92. Rutledge, T. F. (to ICI America Inc.), German Offenlegungsschrift 2,360,369, June 12, 1974.
93. Ragaini, V. and Saravalle, R., *React. Kinet. Catal. Lett.*, 1(3), 271, 1974.
94. Ragaini, V., Carniti, P., Rainaldi, U., and Morazzoni, F., *Chim. Ind. (Milan)*, 58(7), 473, 1976.
95. Allara, D. L. and Chan, M. G., *J. Polym. Sci. Polym. Chem. Ed.*, 14(8), 1857, 1976.
96. Allara, D. L. and Chan, M. G., *Am. Chem. Soc. Div. Org. Coat. Plast. Chem. Pap.*, 34(2), 114, 1974.
97. Berezin, B. D. and Shlyapova, L. N., *Izv. Vyssh. Ucheb. Zaved. Khim. Khim. Tekhnol.*, 14(11), 1665, 1971.
98. Osa, T., Ohkatsu, Y., and Tezuka, M., *Chem. Lett.*, 2, 99, 1973.
99. Kwaskowska-Chec, E., Fried, K., Przywarska-Boniecka, H., and Ziȯlkowski, J. J., *React. Kinet. Catal. Lett.*, 2(4), 425, 1975.
100. Tsanev, D., *God. Nauchoizsled. Inst. Khim. Prom.*, 1970 (Pub. 1971), 6(1), 147-54 (Bulgarian).
101. Kropf, H. and Knaack, K., *Justus Liebigs Ann. Chem.*, 757, 109, 1972.
102. Kropf, H. and Knaack, K., *Justus Liebigs Ann. Chem.*, 757, 121, 1972.
103. Kropf, H., *Angew. Chem. Int. Ed. Engl.*, 11(3), 239, 1972.
104. Kropf, H. and Guertler, D., *Justus Liebigs Ann. Chem.*, 3, 519, 1976.
105. Yurchak, S., Williams, R. H., Gorring, R. L., and Silvestri, A. J., *Catal., Proc. Int. Congr., 5th*, North-Holland, Amsterdam, 1973, 867.
106. Williams, R. H., Silvestri, A. J., and Gorring, R. L. (to Mobil Oil Corporation) U.S. Patent 3,816,548, January 11, 1974.
107. Kast, H. Baumann, H., Mayer, U., and Oberlinner, A. (to Badische Anilin- & Soda-Fabrik AG), U.S. Patent 3,828,071, August 6, 1974.
108. Kast, H., Baumann, H., Mayer, U., and Oberlinner, A. (to Badische Anilin- & Soda-Fabrik AG), German Offenlegungsschrift 2,138,931, February 15, 1973.
109. Allara, D. L. and Chan, M. G., *Am. Chem. Soc., Div. Org. Coat. Plast. Chem. Pap.*, 34(2), 114, 1974.
110. Komissarov, G. G., Asanov, A. N., and Shumov, Yu. S., *Dokl. Akad. Nauk S.S.S.R.*, 206(6), 1468, 1972.
111. Andrianova, T. I., Sherle, A. I., and Berlin, A. A., *Izv. Akad. Nauk S.S.R. Ser. Khim.*, 3, 531, 1973.
112. Dabrowski, R. and Witkiewicz, Z., *Zesz. Nauk. Inst. Ciezkiej Syn. Org. Blachowni Slask.*, 4(20), 87, 1972.
113. Dabrowski, R., Waclawek, W., and Witkiewicz, Z., *Biul. Wojsk. Akad. Tech.*, 25(5), 81, 1976.
114. Hara, T., Ohkatsu, Y., and Osa, T., *Chem. Lett.*, 2, 103, 1973.
115. Vofsi, D., Cohen, J. F., and Martan, M., (to Yeda Research and Development Company Ltd.), U.S. Patent 3,954,876, May 4, 1976.
116. Vofsi, D., Martan, M., and Cohen, J. F. (to Yeda Research and Development Company Ltd.), German Offenlegungsschrift 2,302,751, July 26, 1973.
117. Kropf, H. and Knaack, K., *Tetrahedron*, 28(4), 1143, 1972.
118. Kropf, H. and Kasper, B., *Justus Liebigs Ann. Chem.*, 12, 2232, 1975.
119. Kropf, H., Vogel, W., Kyburg, I., *Justus Liebigs Ann. Chem.*, 7 - 8, 1229, 1976.
120. Anan'eva, T. A., Al'yanov, M. I., Dolikhina, I. V., *Tr. Ivanov, Khim. Tekhnol. In-Ta*, 17, 72, 1974.
121. Wagnerova, D. M., Schwertnerova, E., and Veprek-Siska, J., *Coll. Czech. Chem. Commun.*, 38(3), 756, 1973.
122. Wagnerova, D. M., Schwertnerova, E., and Veprek-Siska, J., *Coll. Czech. Chem. Commun.*, 39(11), 3036, 1974.
123. Dolansky, J., Wagnerova, D. M., and Veprek-Siska, J., *Coll. Czech. Chem. Commun.*, 41(8), 2326, 1976.
124. Bocard, C., Gadelle, C., Mimoun, H., and Serée de Roch, I. (to Institut Français du Petrole, des Carburants et Lubrifiants), French Patent 2,044,007, February 19, 1971.
125. Busson, C., Gadelle, C., Serée de Roch, I. and Van Landeghem, H. (to Institut Français du Petrole, des Carburants et Lubrifiants), German Offenlegungsschrift 2,207,399, November 2, 1972.
126. Gadelle, C. and Serée de Roch, I. (to Institut Français du Petrole, des Carburants et Lubrifiants), French Patent 2,125,788, November 3, 1972.

127. Mimoun, H., Gadelle, C., Bocard, C., Serée de Roch, I., and Baumgartner, P. (to Institut Français du Petrole, des Carburants et Lubrifiants, French Demande 2,187,774, January 18, 1974.
128. Institut Français du Petrole, des Carburants et Lubrifiants, British Patent 1,330,602, September 19, 1973.
129. Sheldon, R. A., *Recl. Trav. Chim. Pays Bas,* 92(2), 253, 1973.
130. Sheldon, R. A., *Recl. Trav. Chim. Pays Bas,* 92(3), 367, 1973.
131. Barone, B. (to Petro-Tex Chemical Corporation), U.S. Patent 3,873,625, March 25, 1975.
132. Kropf, H., Ivanov, S. K., and Diercks, P., *Justus Liebigs Ann. Chem.,* 12, 2046, 1975.
133. Kropf, H. and Kasper, B., *Justus Liebigs Ann. Chem.,* 12, 2232, 1975.
134. Kropf, H. and Vogel, W., *Justus Liebigs Ann. Chem.,* 11, 2010, 1975.
135. Auerbach, A., Indictor, N., and Jochsberger, T., *Macromolecules,* 8(5), 632, 1975.
136. Kothari, V. M. and Tazuma, J. J., *J. Catal.,* 41(1), 180, 1976.
137. Osa, T., Ohkatsu, Y., and Tezuka, M., *J. Fac. Eng. Univ. Tokyo, Ser. A,* 10, 66, 1972.
138. Tezuka, M., Ohkatsu, Y., and Osa, T., *Bull. Chem. Soc. Jpn.,* 48(5), 1471, 1975.
139. Przywarska-Boniecka, H. and Fried, K., *Rocz. Chem.,* 50(1), 43, 1976.
140. Schwertnerová, E., Wagnerová, D. M., and Vepřek-Šiška, J., *Z. Chemie,* 14(8), 311, 1974.
141. Waldmeier, P. and Sigel, H., *Inorg. Chim. Acta,* 5(4), 659, 1971.
142. Waldmeier, P., Prijs, B., and Sigel, H., *Z. Naturforsch. B,* 27(2), 95, 1972.
143. Shpyalova, A. N. and Berezin, B. D., *Izv. Vyssh. Ucheb. Zaved. Khim. Khim. Tekhnol.,* 14(12), 1810, 1971.
144. Meier, H., Zimmerhackl, E., Albrecht, W., and Tschirwitz, U., Katalyse an Phthalocyaninen, Symp., Hamburg, 1973, 104.
145. Borisenkova, S. A., Erokhin, A. S., and Rudenko, A. P., *Vestn. Mosk. Univ. Khim.,* 16(4), 472, 1975.
146. Erin, V. A., Troshina, S. N., Pobedinskii, S. N., and Belonogov, K. N., *Vopr. Kinet. Katal.,* 2, 99, 1974.
147. Andreev, A. A., Prakhov, L. T., Stankova, M. G., and Shopov, D. M., *Catal. Proc. Int. Symp.,* Elsevier, Amsterdam, 1975, 407.
148. Shumov, Yu. S. and Komissarov, G. G., *Zh. Fiz. Khim.,* 47(1), 221, 1973.
149. Hrusovsky, M., Rattay, V., Fancovic, K., Mrazova, M., Rupcikova, L., and Zeriavova, M., *Zb. Pr. Chemickotechnol. Fak. SVST,* p.193, 1974.
150. Kropf, H., Ivanov, S. K., and Zink, I., *Justus Liebigs Ann. Chem.,* 12, 2055, 1975.
151. Kropf, H., Spangenberg, J., Hofer, A., and Wenck, H., *Justus Liebigs Ann. Chem.,* 7—8, 1242, 1976.
152. Kropf, H., Spangenberg, J., and Hofer, A., *Justus Liebigs Ann. Chem.,* 7—8, 1253, 1976.
153. Kropf, H., Ivanov, S. K., and Diercks, P., *Justus Liebigs Ann. Chem.,* 12, 2046, 1975.
154. Okuma, K., Niki, E., and Kamiya, Y., *Nippon Kagaku Kaishi,* 6, 962, 1976.
155. Steinbach, F. and Zobel, M., *Z. Phys. Chem. (Frankfurt am Main),* 87(1—3), 142, 1973.
156. Jochsberger, T., Auerbach, A., and Indictor, N., *J. Polym. Sci. Polym. Chem. Ed.,* 14(5), 1083, 1976.
157. Hanke, W., *Naturwissenschaften,* 52(16), 475, 1965.
158. Hanke, W. and Gutschick, D., *Z. Anorg. Allg. Chem.,* 366(3—4), 201, 1969.
159. Hankel, D., Von Freyberg, H. G., and Doiwa, A., *Z. Naturforsch. B,* 27(2), 204, 1972.
160. Kawai, T., Soma, M., Matsumoto, Y., Onishi, T., and Tamaru, K., *Chem. Phys. Lett.,* 37(2), 378, 1976.
161. Johns, I. B., U.S. Patent 3,609,128, September 28, 1971.
162. Militskova, E. A. and Parakhina, G. V., Russian Patent 394,396, August 22, 1973.
163. Pomogailo, A. D., Matkovskii, P. E., and Lisitskaya, A. P., Russian Patent 443,862, September 25, 1974.
164. Shchennikova, M. K., Artemova, E. A., Chuev, I. I., and Metelev, A. K., *Pr. Khim. Khim. Tekhnol.,* 1, 86, 1973.
165. Tsai, M.-K., Li, W.-H., Hsu, T.-L., Chiao, S.-W., Li, P.-C., and Chao, S.-T., *K'o Hsueh T'ung Pao,* 19(1), 20, 1974.
166. Il'ina, L. M., Borisenkova, S. A., Rudenko, A. P., and Lavrova, E. V., *Vestn. Mosk. Univ. Khim.,* 13(2), 249, 1972.
167. Borisenkova, S. A., Il'ina, L. M., Leonova, E. V., and Rudenko, A. P., *Zh. Org. Khim.,* 9(9), 1827, 1973.
168. Anikin, N. A. and Rudenko, A. P., *Vestn. Mosk. Univ. Khim.,* 17(1), 122, 1976.
169. Borisenkova, S. A., Erokhin, A. S., Novikov, V. A., and Rudenko, A. P., *Zh. Org. Khim.,* 11(9), 1977, 1975.
170. Borisenkova, S. A. and Rudenko, A. P., Vestn. Mosk. Univ. Khim., 17(1), 3, 1976.
171. Tamaru, K., Ohnishi, K., Soma, M., Ichikawa, K., and Suto, M., Japanese Kokai 72 08,284, March 9, 1972.

172. Hiraoka, M. and Ito, M. (to Nippon Soda Company Ltd.), Japanese Kokai 74 14,731, April 10, 1974.
173. Massie, S. N., U.S. Patent 3,692,842, September 19, 1972.
174. Kordesch, K. V. (to Union Carbide Corporation), U.S. Patent 3,783,026, January 1, 1974.
175. Andreev, A., Prahov, L., and Shopov, D., *C. R. Acad. Bulg. Sci.*, 26(12), 1637, 1973.
176. Takebayashi, M., Kashiwada, T., Hamaguchi, M., and Ibata, T., *Chem. Lett.*, 8, 809, 1973.
177. Inoue, H., Kunikawa, K., and Imoto, E., *Bull. Chem. Soc. Jpn.*, 46(2), 518, 1973.
178. Gaspard, S., Viovy, R., Bregeault, J. M., Jarjour, C., and Yolon, S., *C. R. Hebd. Séances Acad. Sci. Ser. C*, 281(22), 925, 1975.
179. Kordesch, K. V. (to Union Carbide Corporation), U.S. Patent 3,783,026, January 1, 1974.
180. Jones, P., Robson, T., and Brown, S. B., *Biochem. J.*, 135(2), 353, 1973.
181. Urban, P. and Rosenwald, R. H. (to Universal Oil Products Company), British Patent 1,318,794, May 31, 1973.
182. Urban, P. and Rosenwald, R. H. (to Universal Oil Products Company), British Patent 1,319,619, June 6, 1973.
183. Urban, P. (to Universal Oil Products Company), U.S. Patent 3,972,988, August 3, 1976.
184. Mazgarov, A. M., Bursova, S. N., Mel'nikova, N. S., and Novikova, G. V. (to All Union Scientific Research Institute of Chemicals for Plant Protection, Shchelkovo), Russian Patent 431,124, June 5, 1974.
185. Golyand, S. M., Kopteva, R. F., and Kundo, N. N., Russian Patent 385,605, June 14, 1973.
186. Mazgarov, A. M., Tukov, G. V., Fomin, V. A., and Akhmadullina, A. G. (to All Union Scientific Research Institute of Hydrocarbon Raw Materials), Russian Patent 513,069, May 5, 1976.
187. Gleim, W. K. T. and Urban, P. (to Universal Oil Products Company), U.S. Patent 3,108,081, October 22, 1963.
188. Brown, K. M. (to Universal Oil Products Company), British Patent 1,321,475, June 27, 1973.
189. Goulandris, G. C., (to American Cyanamid Company), U.S. Patent 3,654,297, April 4, 1972.
190. Campbell, R. H. and Wise, R. W. (to Monsanto Company), U.S. Patent 3,737,431, June 5, 1973.
191. Simonov, A. D., Keier, N. P., Kundo, N. N., Mamaeva, E. K., and Glazneva, G. V., *Kinet. Katal.*, 14(4), 988, 1973.
192. Maas, T. A. M. M., Kuijer, M., and Zwart, J., *J. Chem. Soc. Chem. Commun.*, 3, 86, 1976.
193. Steinbach, F. and Schmidt, H., *J. Catal.*, 29(3), 515, 1973.

Chapter 12

PHYSICAL PROPERTIES

I. INTRODUCTION

The physical properties of pcs have excited the curiosity of many investigators who are impressed by the variety of pc particles and crystals which are straightforward in preparation and which smoothly lend themselves to elegant crystalline systems and dispersions. One author has expressed this sentiment and logic in a discussion of the physics of organic solids when he wrote "The pcs are the model substances for the entire physics of organic solids."[1]

II. VAPOR PRESSURES AND HEATS OF SUBLIMATION

The vapor pressures of Co pc, Sn pc, Cl_2 Sn pc, diphenyl tin pc, Cl_{16}pc, and Br_{16}pc are determined by the effusion method, and the heats and entropies of sublimation are obtained.[2] Mass spectra show that Co pc, Sn pc, and diphenyl tin pc sublime without decomposition.

The Knudsen method is modified to measure the vapor pressures at 600 to 700°C of dichloro derivatives of Si pc and Ge pc.[3] The temperature dependence of vapor pressure is determined as well[4,5] for Ni pc, VO pc, pc, and Cl_2 Ge pc at 350 to 950°C, as well as heats of sublimation.

The heats of sublimation are determined for Sn, H_2, VO, Zn, Ni, Pd, Pt, Pb, Mo, Cu, Co, and Fe pcs.[6] "The theory is very simple, but many experimental precautions are required in order to obtain accurate results." For example, decomposition of Mn pc prevented a determination.

III. ADSORPTION BY PHTHALOCYANINES

A theoretical and experimental study is made of the adsorption of a variety of organic compounds on pc. The guiding principle is based on the additivity of the contributions of various groups of atoms of pc to the total adsorption effect.[7] "The results indicate that there are certain privileged sites for adsorption in pc, at the center and along certain axes."*

Analysis is also made of complex organic compounds by gas-solid chromatography on various pcs coated on graphitized carbon black.[8] "By gas-solid chromatography the selectivities of the various pcs were compared and their affinity towards different organic functions was determined."[9]

Gas-solid chromatography for analysis separation of condensed aromatic compounds and polar compounds is demonstrated with graphitized carbon black thin layers coated with another adsorbent, Co pc.[8]

The adsorption properties of soluble and insoluble pcs (Li, Na, Co, Zn, and Cu) deposited on silochrome S-80 are studied by gas-solid chromatography.[10] The columns with the soluble pcs are more selective and the chromatograms have more symmetric peaks.

Adsorption effects of silica supported pcs are studied by reflection spectroscopy.[11,12] The addition of pyridine and ammonia to Co pc adsorbent indicates that they stabilize the Co^{3+} by the formation of octahedral complexes. Ni pc and Cu pc are also studied

* From Kuznetov, A. V., Vidal-Madjar, C., and Guiochon, G., *Bull. Soc. Chim. Fr.*, 5, 1440, 1969. With permission.

and it is indicated here as well that ammonia is adsorbed on them in octahedral coordination.

The mechanism of adsorption of gelatin layers by dyes, presumably including Cu pc derivatives, is studied.[13] "The differences in the degree of adsorption are attributed to differences in the degree of ionization of the various dyes."

The specific surface area, s, of β Cu pc decreases as a result of heating *in vacuo* at 300°C, from about 90 to about 45 m^2/g and its colorizing power diminishes by 10%. α Cu pc *in vacuo* has an s of 30, 11, and 4 m^2/g at 20, 200, and 300°C, respectively. The measurements are complicated by the fact that the crystals exhibit $\alpha \rightarrow \beta$ growth at temperatures exceeding 200°C.[14]

Adsorption isotherms and heats of adsorption are measured for various molecules adsorbed on β Cu pc. The molecules include nitrogen, alcohols, hydrocarbons, ketones, esters, aromatic hydrocarbons, and water.[15-17]

Adsorption isotherms are also obtained for propyl alcohol, butyl alcohol, hexyl alcohol, and phenol on Cu pc blue and green.[18]

The adsorption of linear poly(dimethylsiloxane) at a pigment-liquid interface is determined by taking β radiation measurements with ^{31}Si.[19] "To study the effect of polarity on the adsorption, carbon black and pc blue were selected as nonpolar adsorbents and TiO_2 (anatase), chrome yellow, and red iron oxide as polar ones."*

An investigation is made to increase the wettability by water of Cu pc by surfactant adsorption, in order to improve the characteristics of the aqueous dispersions of Cu pc pigments.[20]

The adsorption of sodium dodecylsulfate and dodecyltrimethylammonium bromide are determined for Cu pc and chlorinated Cu pc. With both anionic and cationic surfactant adsorbents, the polar group is oriented toward the solution.[20]

IV. IMPRINTING

An effort is made to determine whether phenol-melamine-HCHO polycondensate will remember the presence of adsorbed Cu pc tetrasulfonic acid after it is removed, and whether it will react toward Cu pc tetrasulfonic acid in a selective manner analogous to silica gel and some organic substances.[21,22] The "recognition" of this acid by the resins is tested by taking adsorption isotherms. They indicate that only a small amount of nonspecific adsorption takes place and no imprinting can be proven.

V. ABSORPTION

A. By Phthalocyanines

A NO_x-containing gas is scrubbed with a solution containing a Co complex of pc sulfonic acid or its salts to absorb the NO_x.[23] "The method is useful for treating boiler flue gas and is not affected by oxygen or SO_2. In one example, a gas containing 320 ppm NO_x was purified to a gas containing 5 ppm NO_x."

B. Of Phthalocyanines

The absorption of pcs in thermoplastic media in the 0.5 to 0.8-μm region is studied.[24] From 75 to 100 g of polymer is mixed with a few milligrams of dye for several hours, subjected to pressure and heat and cooled. After this treatment, solution instead of crystalline pc spectra appeared (with only two exceptions) and no evidence of solid pc could be detected under a microscope.

* From Sato, T., Tanaka, T., and Yoshida, T., *J. Polym. Sci. Part B*, 5(10), 947, 1967. With permission.

VI. DISPERSIONS

A. Dispersions and Sedimentation Rates in Various Media

The sedimentation rate of pc blue is measured in methyl benzene, toluene, and ethyl acetate containing polymethacrylate polymers.[25] The sedimentation rate is of interest because it is a measure of the dispersibility of a pc pigment in a polymer host medium. The settling of hindered Cu pc is studied by a modified Steinhour hindered settling method in hydrocarbon liquids.[26]

The preparation and measurement of particles of easily crystallized organic pigments including Cu pc and polychloro Cu pc is studied, including the determination of sedimentation rate.[27,28]

Dispersions of pigments bonded to the finish are discussed in detail.[29]

B. Particle Size of Phthalocyanines in Dispersions

Ultrasonic treatment of an aqueous 1% pc blue solution in the presence of Triton® X-100 as a dispersing agent gives a stable dispersion, the particle size distributions of which are measured by the buffered-layer technique.[30]

The technique of using the Joyce-Loebl disk centrifuge is extended to determining the particle size of pigments, including Cu pc, in printing inks.[31]

VII. LIGHT AND COLOR EFFECTS OF PARTICLE SIZE AND PARTICLE SIZE DISTRIBUTIONS OF PHTHALOCYANINES

The intensity of visible light reflected from a pigment decreases when the size of the pigment particles, such as Cu pc, is decreased by grinding.[32] A method is given for determining the degree of pigment grinding from light intensity measurements.

The coloristic properties of organic pigments including Cl_{16} Cu pc are discussed in terms of the applications of the Kubelka-Munk-Mie theory to their particle size distribution.[33]

The coloristic properties of β Cu pc depend on the particle size and on the shape of the primary crystals. Measurements on oriented pigment crystals in films show that the transmission color is reddish blue in the direction of the acicular axis of β Cu pc and is greenish blue at right angles to the axis.[34]

VIII. BEER'S LAW BEHAVIOR OF DYE SOLUTIONS

An experimental study is made to show that Beer's Law — that the absorbance or optical density of a solution in monochromatic light is directly proportional to the concentration of the absorbing solute — is not always accurate, particularly when the solvent is water.[35] "The solutions are then said 'not to obey Beer's Law' and it is assumed that the cause is association of the solute. This assumption is often correct, but the converse, viz. that obedience of the law implies non-association, is not necessarily so. When the particle-size distribution in the concentration range under study remains constant, the law is obeyed, even though the solute is highly associated."* Copper pc pigment milled for different times to give different degrees of association is one of the systems studied.

Difficulties in preparing dye solutions for accurate Beer's Law strength measurements are discussed in terms of a variety of factors: additives, concentration, dissolution aids, pH, solvent, dye interaction, fluorescence, irradiation, plating, solution heating, temperature, hydrolysis, interfering ions, ionic strength, light fading, redox potential, and turbidity.[36]

* From Duff, D. G. and Giles, C. H., *J. Soc. Dyers Colour*, 88(5), 181, 1972. With permission.

IX. DIFFUSION

A. Of Substances in Phthalocyanines

Molecular iodine diffuses in single crystals of Zn pc and forms a paramagnetic complex with Zn pc. At 125 to 155°C, the diffusion coefficient D of iodine in Zn pc single crystals is $D = 5 \times 10^4 e(-24{,}000/RT) cm^2sec^{-1}$.[37]

B. Of Phthalocyanines in Substances

The importance of penetration or diffusion of dye molecules is indicated by a wood dyeing process.[38] "Wood, especially veneer, was treated with a uniform, deeply penetrating dye, which allows the repairing of wood parts without repeated dyeing and makes unnecessary the staining of furniture parts on dyeing. The wood was immersed in solutions of soluble or difficultly soluble dyes, for example, Cu pcs. . ."

X. DISSOLUTION OF GASES IN PHTHALOCYANINE SOLUTION

The solubility and thermodynamices of oxygen dissolution in aqueous solutions of Fe, Mn, Ni, and Co tetrasulfo pcs[39] and of argon dissolution in aqueous solutions of various transition metal tetrasulfo pc complexes[40] are the subject of two studies.

XI. EXPANSION COEFFICIENT

The expansion coefficient of α and β Cu pc is determined at −170 to +180°C and −170 to +285°C by X-ray diffraction.[41]

XII. EFFECT OF PHTHALOCYANINES ON MELT VISCOSITY

The effect of small amounts of pigments such as titania, carbon black, and pc on the melt viscosity of polypropylene is determined.[42] The interaction coefficient between polypropylene and the added pigment is in the order pc > carbon black > TiO_2. The melt viscosity of polypropylene relating to the adjuvant blend is expressed by

$$\eta_o = \eta_{pp}(1 + A\phi + B\phi^2)$$

where η_o = melt viscosity of polypropylene with adjuvant, η_{pp} = melt viscosity of polypropylene at zero shear rate, ϕ = volume fraction of impurity, and A,B = coefficient of interaction between polypropylene and the adjuvant.

XIII. EFFECT OF PHTHALOCYANINES ON ELASTIC MODULUS

The elastic modulus of polypropylene increases with increasing pigment concentration until a pigment content of greater than 5% is attained, following which the modulus decreases.[43] Organic pigments such as azo compounds, pcs, and carbon black, provide greater reinforcement than inorganic pigments, such as titania, calcium carbonate, and cadmium sulfide.

XIV. COMPRESSION OF PHTHALOCYANINE POWDERS

Cu pc and polychloro Cu pc derivative powders are compressed with an instrumented tabletting machine, and the strength of the resulting agglomerates is measured and is related to their tendency to form during handling, storage, and mixing with plastics.[44,45] The ease of pigment agglomeration and the strength of the agglomerates are the cause of difficulties encountered during dispersion in plastics.

The effect of pressure on the low energy peak is measured on pc and its salts with Cu, Zn, Fe, Ni, Co, Cr, Sn, and Pb.[46] In all cases the low energy absorption peak is shifted to lower energies with increase in pressure.

XV. ANALYSIS OF PHTHALOCYANINES BY CHROMATOGRAPHY

Elemental analysis is made of metal pcs by gas chromatography (GC).[47]

A gas chromatographic method, accurate to ± 0.2%, for the determination of the chlorine content of chlorinated Cu pc is described.[48] The sample is decomposed, the chlorine evolved reacts with ammonia, and the amount of unreacted ammonia is determined by GC.

A thesis concerns itself with an investigation of the pcs using GC.[49]

Pc pigments are identified by their R_f values in thin layer chromatography.[50]

XVI. EFFECT OF PHTHALOCYANINES ON THE STABILIZATION OF THERMOSTABLE POLYMERS

The effect of the addition of 0.5% Mg, Cu, Ni, and Co pcs toward inhibiting thermal degradation of polyesters, polyamides, and aromatic heterocyclic polymers at 350°C is studied. The most effective of the pcs tested is Mg pc.[51]

XVII. SLIDING BEHAVIOR AND COEFFICIENT OF FRICTION

The coefficient of friction, f, of pc sliding on pure copper in an ultrahigh vacuum is found to be 0.33 ± 0.03 and in air 0.35 ± 0.02.[52,53]

XVIII. UV IRRADIATION OF POLYMERS CONTAINING PHTHALOCYANINES

Young's modulus and the loss tangent δ are measured for polyvinylchloride and its derivatives with Cu pc after irradiation with UV radiation.[54]

XIX. CRYSTALLIZATION NUCLEI

The crystallization of high-density polyethylene from dilute solution is influenced by the nucleation with such substances as pigments including Cu pc and chlorinated Cu pc.[55] The nucleation effect cannot be correlated with the chemical structure, particle size, or morphology of the pigment. However, they cause smaller crystalline grain size and altered morphology, especially in spherulite structures.[56]

Analysis of the crystallization for high-density polyethylene indicates that crystallization takes place in three stages, and that this behavior is influenced by the presence of additives such as pc derivatives.[57]

Pc derivatives in the amount of ≤0.05 wt% act as crystallization nuclei to polymerizing caprolactam monomers, improving the mechanical properties and decreasing the heterogeneity and size of spherulites in polycaprolactam, as well as dyeing the polymers in uniform colors.[58]

XX. ELECTROPHORESIS

The zeta potentials of β Cu pc and monochloro α Cu pc in benzene or heptane in the presence of surfactants are determined as a function of the concentration of water or methanol polar additives and the resistance to flocculation is studied.[59-61]

The zeta potential, as well as heat of wetting and particle size distribution of pc blue are measured in 1% aqueous suspensions containing 0.1 to 1.0% polyethylene glycol monylphenyl ether suspended by 22 kHz ultrasound.[62]

The complexing effect of neutralizing agents and of dyes, including a Cu pc complex, as well as electrokinetic potential are measured.[63]

XXI. SURFACE EFFECTS

A variety of surface effects are studied. They include the effects of subdivision or milling, surface ionization energies, orientation overgrowth, the change of surface properties at crystal transformation, electric and photoelectric effects, interphase surface activity, and measurement of surface free energy and polarity. Electric and photoelectric surface effects are discussed in the section on electrical and photoelectrical properties.

Surface ionization energies are determined for pcs by introducing them as a fine powder into an argon-filled Millikan apparatus fitted with quartz windows.[64] The surface ionization energies for pc, Cu pc, and Mg pc are 5.20, 5.00, and 4.96 EV, respectively.

Ionization energies and their measurement are the subject of a discussion.[65]

The orientation overgrowth of metal pcs on the surface of single crystals, in terms of Cu pc vacuum condensed on muscovite[66] and of Cu pc vacuum condensed on alkali halides[67] are studied.

The changes in surface properties on transformation from α to β Cu pc and Ni pc are examined by electron microscope and X-ray diffraction analysis. Their heats of immersion are also measured calorimetrically.[68]

The interfacial tension at the interphase between aqueous metallic salts such as $MgCl_2$ and $ZnCl_2$ and 0.5 *M* pc in α-chloronaphthalene passes through a minimum at 1:1 molar ratios; this effect is attributed to interactions between pc and the metallic salts.[69]

Hydrophilic properties of pigments including Cu pc are measured including the heat of wetting which is determined calorimetrically.[70]

The surface free energies and polarities of H_2 pc, Cu pc, and chlorinated Cu pc are measured.[71]

The effect of roughness of surface on the wetting of Cu pc with water, glycerol, thiodiglycerol, or ethylene glycol is studied by scanning electron microscopy.[72]

The specific surface areas of mixtures of α Cu pc and alumina are measured against grinding time.[73]

XXII. ELECTRONIC STRUCTURE WITH EMPHASIS ON LINEAR COMBINATION OF ATOMIC ORBITALS — MOLECULAR ORBITAL (LCAO-MO) CALCULATIONS

The C–C and C–N bond orders in pyrrole and pyridine are calculated by the LCAO-MO method in the Hueckel approximation.[74] The difference of electronegativities of nitrogen and carbon atoms is introduced in the form: $\alpha_N = \alpha_C + \beta\delta$, where α_N and α_C are Coulomb integrals for N and C atoms, respectively; β is the standard resonance integral, and δ is a dimensionless empirical parameter, called the "electronegativity parameter". δ is $\sim$ 1 and $\sim$ 3 for pyridine and pyrrole molecules, respectively. Similar calculations for porphine and pc give $\delta\sim$ 1.0 to 1.1 and $\delta\sim$ 0 to 0.7, respectively. The results agree with diamagnetic susceptibilities and NMR spectra.

Hueckel MO for H_2 pc are calculated.[75] "An interesting consequence of using the correct symmetry of the molecule is that charge density at atomic site *2* is not minimum as reported by Basu but is minimum at sites *1* and *3.* This explains the oxidation reactions of pcs with acyl peroxide, organic hypochlorite, etc., and the formation of various dyes from metal pcs.''*

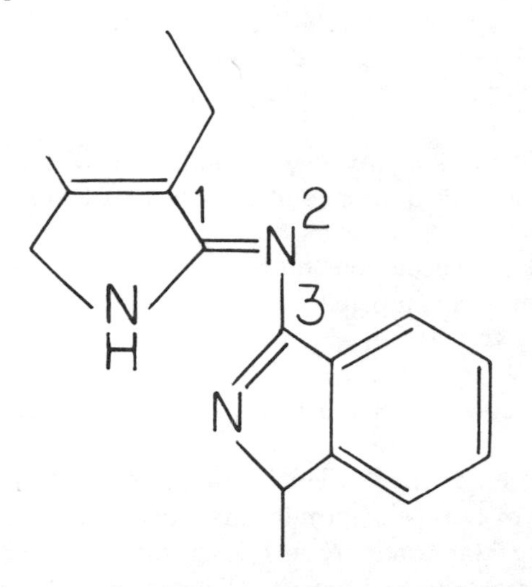

The minimum π-electron densities at positions *1* and *3* is discussed.[76] "Mathur's assumption of a 40-π-electron system is open to discussion. Instead, a complete delocalization of two π-electrons of the imino nitrogen atoms is believed to be more adequate."[76]

Calculation of the diamagnetic susceptibility of H_2 pc by the simple LCAO-MO method is made with the assumptions that the π-electron system of H_2 pc reduces to

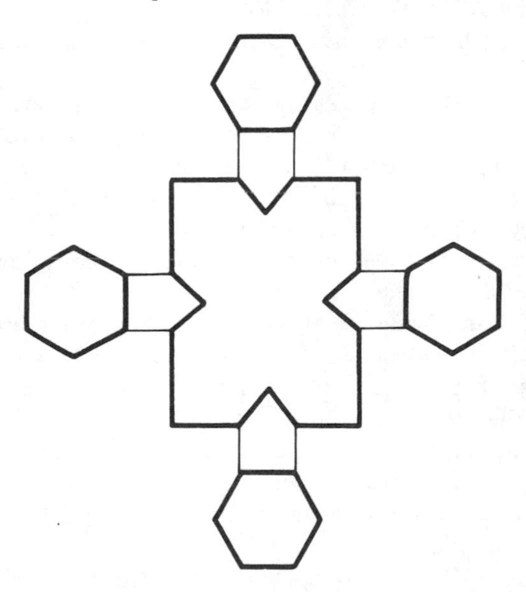

and carbon and nitrogen atoms are equivalent.[77]

The ω (omega) technique is used to make LCAO-MO calculations for H_2 pc.[78] "These corrections improve the eigen valued and so, hopefully, one gets better eigen vectors."[78]

Excess electron and excess hole energy bands for β H_2 pc are calculated by using the tight-binding approximation ω technique LCAO-MOs as basis functions for construct-

* From Mathur, S. C., *J. Chem. Phys.,* 45(9), 3470, 1966. With permission.

ing the Bloch functions and a modified molecular potential.[79] The energy bands are highly anisotropic and the mobilities of electrons and holes are independent of temperature at 200 to 500 K.

A new average procedure ameliorates the convergence problem in the ω technique.[80] "The chemical reactivities of H_2 pc give excellent correlation with the calculated charge densities which indicates the reliability of the latter as well as the suitability of the ω technique to large molecules."

Extended Hueckel calculations for porphin and pc show that $n - \pi^*$ transitions introduced by the bridge aza nitrogen may be the cause of Soret band broadening of pc and tetrazaporphin.[81]

Extended Hueckel (EH) calculations on Mg, Zn, Cu, Ni, Fe, Mn, and VO pc and tetrazaporphin are made and compared with similar calculations on porphyrins.[82] The bridge nitrogen atoms give rise to $n - \pi^*$ transitions which are probably in the region of the Soret band.

Extended Hueckel calculations on porphyrins, pcs, and related rings are the subject of a thesis.[83]

The π MO of H_2 pc are obtained by Hueckel-type MO calculations.[84] "By comparison with the MO levels of tetrabenzporphin, the intense absorption of pc in the visible range is attributed to the large Coulomb energies of the four bridge N."

Wave functions of pcs are constructed from Hueckel LCAO-MO orbitals.[85]

"The band structures for an excess electron and an excess hole in the H_2 pc crystal were calculated in the tight-binding approximation. The LCAO wave functions for the cation and anion radicals were calculated by the unrestricted self-consistent-field method."[86]

The excess electron and hole band structures of β H_2 pc are calculated with the tight-binding approximation, using the Hueckel MO.[87]

The energy band structure of H_2 pc is calculated using the tight-binding approximation.[88]

Excess electron and excess hole energy bands for γ-Pt pc are calculated using the tight-binding approximation, extended Hueckel technique LCAO-MOs as basis functions for contructing the Bloch functions and a modified molecular potential.[89] The energy bands obtained are highly anisotropic. It is argued that the holes play a dominant role in the charge carrier transport in Pt pc.

A review with 20 references describes the chemical constitution of transition metal pcs by a simple LCAO-MO model.[90]

The molecular electronic structure of metal pcs (Fe, Co, Ni, and Cu) is determined by a MO treatment.[91] The calculated π-electron charge densities are higher on the nitrogen atoms compared to the other atoms in the molecule. This is in agreement with the ESR studies of the metal pcs. The results of MO calculations for the remaining 3d-transition metal (Sc, Ti, V, and Mn) pcs are also given.[92] π-electron charge densities over the atomic sites and the optical properties of these metal pcs are calculated.

The energy levels and spectral transitions in Co and VO pcs are observed and compared with Hueckel MO calculations.[93]

A simplified LCAO-MO calculation according to the Hueckel method, and X-ray analyses, are carried out for the pc complexes of 3d elements Mn through Ni.[94] In this manner, a model is derived which represents the electronic structures of the transition metal complexes and makes possible the further understanding of the structures based on the nuclear charges of the central atoms and the total electronic charges of the complexes.

The π and σ lone pair electron system of the pc molecule is studied by a semiempirical self-consistent field molecular (SCF-MO) method.[95] Electronic transitions of both $\pi - \pi^*$ and $n - \pi^*$ types are considered. The excited states are calculated by means of the method of superposition of configurations where all singly excited states are included. Assignments for the electronic spectrum of pc are made in good agreement with experiment. The position of the lowest electronically allowed n $-\pi^*$ transition is predicted to be found in the region of the strong Soret band.

A semiempirical method, the PEEL method, is applied in a study of the electronic structure and excited states of Cu pc.[96] "In the PEEL method the metal ion is explicitly taken into account. This method permits a study of the metal ligand bond and an analysis of the different types of electronic transitions (d − d, CT, and $\pi-\pi^*$) found in the electronic spectrum."*

The electronic eigenvalue problem of the Cu pc molecule is approximated using MO formalism.[97] The symmetry of the molecule allows a group theoretical reduction of the secular equations and a classification of the eigenvalues and eigen function coefficients. The model statements are compared with the experimental molecular spectra in the vapor phase and in solution from the near UV to the near IR.

In a study entitled "Chemical Reactivities of Tetrapyrrole Pigments: A Comparison of Experimental Behavior with the Results of SCF-π-MO Calculations", Pariser-Parr-Pople π calculations are used to compare reactivities and physical properties of a number of tetrapyrrole pigments including the pcs.[98]

A theoretical consideration is made of the reactivities of the porphin and pc nucleus by the simple Hueckel method.[99]

A method is proposed for calculating electronic structure of large molecules within the Hueckel approximation by applying perturbation theory to the electron dense matrix.[100] The calculations are demonstrated on Co pc derivatives and a semiqualitative correlation is obtained between the rate of catalytic oxidative decarboxylation of $H_2C_2O_4$ on Co pc derivatives and its derivatives and the theoretical calculated change of the electron density on the bridging nitrogen atoms.

Using wave functions given by the projected electron dense method the energies of excited π-electron states are calculated.[101] The electronic spectra are calculated and agree with experiment.

A quantitative attempt is made to apply a four-orbital model to porphyrin molecules. The model is a combination of LCAO-MO and a simplified treatment of configuration interaction.[102] Spectra calculations include pc; the calculations are in fair agreement with observed spectra.

Self-consistent molecular orbital (SCMO) calculations on the porphin ring by the method of Pariser, Parr, and Pople confirm earlier conclusions that the visible and Soret bands of the porphin absorption spectrum can be explained by a four-orbital model. The reassignment of the UV bands of tetrazaporphin and pc are made.[103]

SCMO-CI (self-consistent molecular orbital configuration interaction) calculations are performed on several molecules including porphin and pc.[104] Q state angular momentum is calculated to be $4.35\hbar$ for porphin and $3.13\hbar$ for pc.

Photoemission techniques show an electron energy distribution in good agreement with free electron MO calculations for several pcs, chlorinated pcs, and protoporphyrin.[105] "The spectrum is due to excitation from the orbitals of the conjugated central ring: below 4 eV, $\pi-\pi^*$ transitions; from 4 to 11.8 eV, excitonic transitions from bonding orbitals to higher continuum states."

In a research "Electron Capture by Some Molecules of Biological Significance and the Determination of Absolute Electron Affinities", the electron capture method is applied to pc and chlorophyll, among other molecules.[106] The energy quantities obtained correlate with calculated energies of the lowest empty molecular orbitals of these molecules.

From Henriksson, A., Roos, B., and Sundbom, M., *Theor. Chim. Acta*, 27(4), 303, 1972. With permission.

In a study "On the Simultaneous Calculation of Perturbation Effects and Lower Excited States of Molecules",[107] an "iterative process is suggested for the evaluation of eigenvectors X_1 and X_2 which correspond to the two lowest eigenvalues of the common equation A(X) = W, where A is the supermatrix of the given operator representation, and describes the unperturbed Hamiltonian and W stands for the perturbation correction matrix . . . The semiempirical methods on either the SCF (self consistent field) or the LCI (limited configuration interaction) level were used for the construction of A. The results for pc are given."*

In a comparison of the structures of Cu pc and Pt pc, it is calculated that the width of the 3d electrons zone is about two orders higher in Pt pc than in Cu pc, but it is negligible in comparison with the π-electron zone width. [108] The density of π-electrons is 10^5 in the direction of the b axis and 1.6×10^{-2} in the directions of the a and c axes for Cu pc and is 0.7×10^2 and 10^{-3}, respectively, for Pt pc.

A study of the "electronic structure and properties of cumulated systems" indicates that qualitative information can be obtained from $(pcSiO)_\infty$, $(pcGeO)_\infty$, and $(pc SnO)_\infty$.[109] "The following conclusions are drawn: In these compounds, with the transition from Si to Ge to Sn, the ionization potential decreases and the affinity to the electron increases. In the same direction, the energy gap in the π-electron spectrum narrows. The valence and the conductivity zone also narrow."

Pc forms stable 18-electron conjugated ring systems which obey Hueckel's rule.[110] The most important outline of conjugation in pc which determine its optical properties in the visible absorption spectrum and the near UV is the porphyrazine macro ring. Calculation of π-electron charges of bonds indicate a high degree of participation in the conjugation by noncovalent pairs of nitrogen atoms.

Experimental studies of additional hyperfine structure of nitrogen and phosphorus atoms in the VO and Cu complexes of pc, porphyrins, thiophosphates, and thiophosphinates are reviewed with 29 references.[111] The unpaired electron is delocalized to the remote phosphorus atoms owing to a direct overlap of the metal $3d_{xy}$ orbital with the 3s or 4s orbitals of the phosphorus atoms.

A gilsonite asphaltene from Cowboy Vein, Utah petroleum, doped with 2000 ppm of vanadyl pc, shows distinct 9-line superhyperfine (SHF) patterns superimposed on the ^{51}V anisotropic hyperfine spectrum of 4 $\perp$ region due to the coupling of unpaired electron spins to ^{14}N nuclei. The SHF splittings observed are discussed in terms of recent findings on aromatic systems, orbital coefficients, and the association-dissociation model for the structure of asphaltenes.[112]

XXIII. LIGHT ABSORPTION AND ABSORPTION SPECTRA

A. Phthalocyanine

The optical absorption spectrum of metal-free pc single crystals is investigated at 3500 to 25,000 Å. "The spectrum of the pc molecule retains much of its identity in the crystal, as evidenced by the similarity in the region of main absorption. This is a consequence of the weak intermolecular binding forces."[113]

The IR spectrum of deuterated pc, D_2pc, sublimed on a NaCl plate is recorded in the region 4000 to 600 cm^{-1} (25,000 to 167,000 Å).[114] The spectrum differs from that of H_2pc in that new bands appear at 2480, 1143, 1076, 977, and 964 cm^{-1}. These bands are considered to be the ND absorption bands.

Disks made of silica are used for obtaining IR spectra of vapor deposited pc films.[115] "Very useful spectra can be obtained by placing a thin layer of gold between the pc sample and glass."

* From Luzanov, A. V. and Umanskii, V. E., *Opt. Spektrosk.*, 40(1), 201, 1976. With permission.

IR spectra of the β forms of H_2pc and of D_2pc are recorded.[116] Absorptions due to N-H vibrations are assigned to bands at 3273, 1539, and 735 cm^{-1}. Differences between the spectra of the parent compound and the D_2pc in the 760 to 710 cm^{-1} region are explained in terms of a lattice vibration modified by a second order isotope effect.

Pc dissolved in 1-chloronaphthalene (0.1 to 5 μM) is irradiated by flash photolysis, the light being passed through a red filter. Absorption peaks near 700 mμ (7000 Å) disappear completely and a new absorption maximum is observed at 484 mμ (4840 Å) due to pc in the triplet state.[117]

In a research entitled "Singlet Exciton States of Crystalline Metal-Free Pc: a Calculation of the Visible Transitions", the lower singlet exciton states of crystalline metal-free pc are calculated and compared with the known experimental spectra.[118] "The interactions between molecules are computed directly by the use of π-electron theory including configuration interaction. Good agreement is obtained when interactions with vibronic components are included. The experimental Davydov splittings obtained by Lyons, Walsh, and White (550 cm^{-1} and 200 cm^{-1}) are practically reproduced, together with the observed polarizations (*ac, b, ac, b*) and positions of the maxima. Calculated relative intensities of the various peaks are only in approximate agreement with the experimental results. A small additional peak is predicted near 19000 cm^{-1}."*

B. Phthalocyanine and Metal Phthalocyanines

IR spectra of pc, chlorinated and unchlorinated Cu pcs, Mo pc, Na_2pc, and Al(OH) pc are recorded.[119]

The IR spectra of α forms of pcs at 500 to 90 cm^{-1} sublimed in vacuum with and without the divalent metals Be, Mg, Fe, Ni, Co, and Cu are studied.[120] In the case of the Mg, Fe, Co, and Cu pcs the spectra of β forms are also recorded.

The absorption spectrum of pc in sulfuric acid is characterized by a band which is displaced as follows for various π-ligand metal ions (displacement in cm^{-1}): Zn 700, Co 650, Ni 850, Cu 600, Pd 980, Pt 960, Ga 280, Ir 1130, Al 330, Rh 1010, Ru 680, Sn 620, V 330, and Os 1010.[121] The ions are classed in three groups, each of which shows a linear relation between band displacement and variation in free energy of solution: (1) Sn, Al, V, Ga, (2) Zn, Co, Cu and (3) Pt group metals and Ni.

IR spectra of 4,4′,4″,4‴ tetranitro pc salts of Co, Ni, and Cu and Fe and of the 4,4′,4″,4‴ tetraamino pc salts of Co, Ni, and Cu are determined at 3800 to 700 cm^{-1}.[122] "The character of the IR spectra of the Cu nitro and amino salts is much like that of chloro and hydroxy substituted salts . . . Bands at 1536—1534 cm^{-1} characterize the presence of nitro groups, while bands at 1572—1565 cm^{-1} characterize deformed oscillations of amino groups."

The IR spectra of Cu, Ni, and Co tetrachloro, and of Ca hexa and octa chloro pcs are compared with that of Cu pc over the range 400 to 700 nm.[123] The IR spectra of pc and several divalent metal derivatives are recorded at 400 to 4000 cm^{-1},[124] and the far IR spectra (400 to 30 cm^{-1}) of pc and its metal derivatives are studied with assignments proposed for the individual absorption bands observed.[125]

Major band positions and intensities are tabulated for 96% sulfuric acid of the Cu, Ni, and Co derivatives of 4,4′,4″,4‴ tetrachloro pc and 3,3′,3″,3‴,5,5′,5″,5‴ octachloro pc, and for the Cu and Ni derivatives of hexadecachloro pc.[126]

The shift from an absorption maximum at 390 to 400 mμ to 410 to 420 mμ (Soret region) of a porphyrin that forms an octahedral complex with a nitrogenous base is ascribed to steric interference between the ligand and the π electron system of the porphyrin ring.[127]

Vapor absorption spectra in the range 800 to 200 mμ are given for pc, and Mg, TiO,

* From Devaux, P., *Mol. Phys.*, 23(2), 265, 1972. With permission.

VO, Fe, Co, Ni, Cu, Zn, $SnCl_2$, and Pb pcs.[128] Vacuum UV spectra of pc and Zn pc are also given, in the region 200 to 145 mμ.

Near IR to vacuum UV absorption spectra and the optical constants of Cu and Zn pcs and porphyrin films are measured.[129]

Spectroscopic studies on porphyrins and pcs are the subject of a thesis.[130]

π-electron absorption spectra in the 14,000 to 60,000 cm^{-1} region, in pc and porphyrins, are calculated using the MO method.[131] It is also attempted to explain the new transitions beyond the Soret band of pc and metal pcs observed previously[128,129] "on the basis of molecular orbital calculations under D_{4h} symmetry, with the help of basically a kind of Hueckel technique."

Single crystal, film, and selected vapor and solution near IR absorption spectra are obtained for pc, and Cu, Ni, Zn, Pd, Fe, Mn, and Cr pcs.[132]

IR spectra of sulfonated metal pcs are studied.[133]

The IR absorption spectra at 670 to 3800 cm^{-1} are studied for sublimed pc and its Mg, Zn, Fe(II), and Cu derivatives in their interaction with gaseous acetic acid, HCl, DCl, HBr, and DBr.[134]

An investigation is made of the IR absorption spectra of sublimed metal pcs of Be, Mg, Zn, Cu, Fe, Co, and Ni and of metal-free pc in the presence of water and D_2O vapors.[135]

The IR absorption spectra of solid layers of monoanions, dianions, trianions, and tetraanions of metal pcs containing Fe, Co, Ni, and VO are measured.[136] The anions are obtained by treating layers of metal pcs with sodium vapors.

If the spectral shifts arising from temperature dependence and broadening due to crystal field effects are accounted for, there is a 1:1 correspondence between peaks observed in vapor, sublimed film, and single crystal absorption spectra.[137] "The simple band model offers a reasonably good description of crystalline Cu, Pt, and metal free pcs."

The "effect of mechanical and anharmonicity on the temperature dependence of integrated apparent intensities in the IR region of metal pcs and inorganic ions isolated in alkali-halide matrices" is the subject of a thesis.[138]

A detailed study is made of the IR spectra 4000 to 650 cm^{-1} of the NO derivatives from the β forms of Cr, Mn, Fe, and Co pcs.[139]

The absorption spectra of polyamide films dyed with fiber reactive pc dyes are measured and studied as a function of the bond between the dye and fiber.[140]

The frequencies of the IR bands of the NH group vibrations are studied in porphyrins with pyrrole derivatives as model compounds.[141] In comparison with the ν(NH) band at 3495 cm^{-1}, the largest frequency shift is found with pc with the ν(NH) wave number of 3290 cm^{-1}. "The smooth course of the frequency shifts from pyrrole to pc and within the porphine compounds series, relates to the complication of the molecular structure, indicating that the shifts are not due to the intramolecular hydrogen bond."

The "quasi continuous spectrum of local levels in layers in pcs" is the title of a review with seven references that may relate to IR absorption.[142]

A study of the electronic absorption spectra of anhydrous solutions of metal complexes (Cu, $AlCl_3$, VO, Zn, Co, Ni, Pd) of pc and its derivatives hexadecachloro-, dodecabromo-, trichloro-, tetra-3-phenyl-, and 2,3-naphthalocyanine complexes with Cu, VO, Zn, $AlCl_3$, Co, Ni, Pd, with Lewis acids ($GaBr_3$, $AlCl_3$, $AlBr_3$ and halides of B, Si, and Sn) in organic solvents (ketones, ethers, alkyl halides, amides, acetyl, phosphoryl, and sulfonyl chlorides, and nitro derivatives) reveals a strong long wavelength shift of all the absorption bands compared with the position of these bands measured in organic solvent solutions.[143]

The electronic absorption spectra are obtained for the anions of pc and its complexes with Mg, Fe, Co, and Ni formed during reduction of pigments by sodium in an ether-THF mixture.[144] The localization of electrons in the anions is elucidated.

IR spectroscopy shows that in the Fe pc monoanion, formed by reaction of Fe pc with sodium in THF-$(C_2H_5)_2O$, the negative charge is distributed between the ring and the Fe atom at the center of the ring.[145] In the Ni pc monoanion, the charge is localized on the ring.

Electronic absorption spectroscopic observations of Zn, Ni, Fe(II), Mn(II), Mg, Al(III)Cl, and Co(II) pc, indicate that all the complexes, with the exception of Co(II) pc and Fe(II) pc, give reduced species in which the additional electrons are confined essentially to the ring orbital.[146] Reduction of Co(II) pc and Fe(II) pc and possibly also Mn(II) pc appears to involve electron addition to both ring and metal orbitals. Reduction of the metal pcs is accomplished with sodium in THF or electrochemically at a mercury pool or Pt cathode.

The translucency-wave phenomenon is analyzed. This phenomenon arises when a pulse of powerful monochromatic electromagnetic radiation passes through an optically dense medium containing resonance-absorbing centers, including pc-type organic compounds.[147]

C. Copper Phthalocyanines

An IR study is made of chloro substituted Cu pcs with various chlorine contents.[148] Absorption spectra are observed of hydroxy substituted Cu pc complexes.[149]

The visible absorption spectra of 10^{-6} *M* Cu pc in anhydrous nitrobenzene containing 0.001 to 1.06 *M* $AlCl_3$ show successive appearances of Cu pc · $nAlCl_3$ with $n = 0$ to 4.[150]

IR spectroscopic studies are made of the formation of the adsorption solvate layer on Cu pc particles from fatty acid solutions in a nonpolar liquid.[151]

A comparison of the spectra of characteristic losses of slow electrons in anthracene and Cu pc crystals is made with optical absorption spectra.[152]

Analogy considerations and an analog computer for the quantum mechanical treatment of the light absorption of dyes are the subject of a study that includes sulfonated Cu pc and pc.[153]

The forms $P(H^+)$, $P(H^+)_2$, $P(H^+)_3$, and $P(H^+)_4$ can be differentiated spectroscopically in solutions of Cu, 4,4′,4″,4‴ pc tetrasulfonate (P) in the acids H_2O-H_2SO_4, H_2SO_4-oleum, HAc-H_2SO_4, and dioxane-H_2SO_4.[154] The absorption spectra of Cu 4,4′,4″,4‴ pc tetrasulfuric acid are the subject of a thesis.[155]

The 5000 to 8000 Å spectra of pc and Cu pc embedded in an argon matrix at 5 K at molar ratios of 4000:1 are very similar to their spectra in liquid solutions.[156]

Absorption spectra of pc and of Cu pc are measured in 1-chloronaphthalene at 4000 to 8500 Å.[157]

For porous films of Cu pc, in 2 *M* KCl, the absorption maximum, at 6350 Å, coincides with the maximum of photovoltage and photocurrent.[158]

Spectra are observed of Cu pc, Cu 1,2-naphthalocyanine, and Cu 1,2-anthracyanine.[159] The principal IR bands shift toward longer wavelengths, along with lowering of activation energy, with increased number of polyconjugated rings in the ligand.

An $\alpha \rightarrow \beta$ rearrangement of pc and Cu pc is observed by mixing α pc or αCu pc with KBr in a vibrator and characterizing the mixture by IR spectra.[160] Complete conversion is obtained after 5 or 10 min.

A spectroscopic study shows that for systems containing fine colloidal particles, the spectral dependence of the attenuation coefficient, K, has a shape analogous to the absorption spectrum.[161] The experimental results are given for the colloidal solutions of Cu pc complex, methyl chlorophyllide a + b, and tetraphenylporphin, containing particles with particle diameter $\leqslant$ 10 mμ.

Cu pc protective dyes (I) and (II) are added to red sensitive Ag(Cl,Br) emulsions to eliminate red diffuse light.[162] The amount of I or II used is 1 gm/kg emulsion. They absorb light in the 660 to 710 nm range.

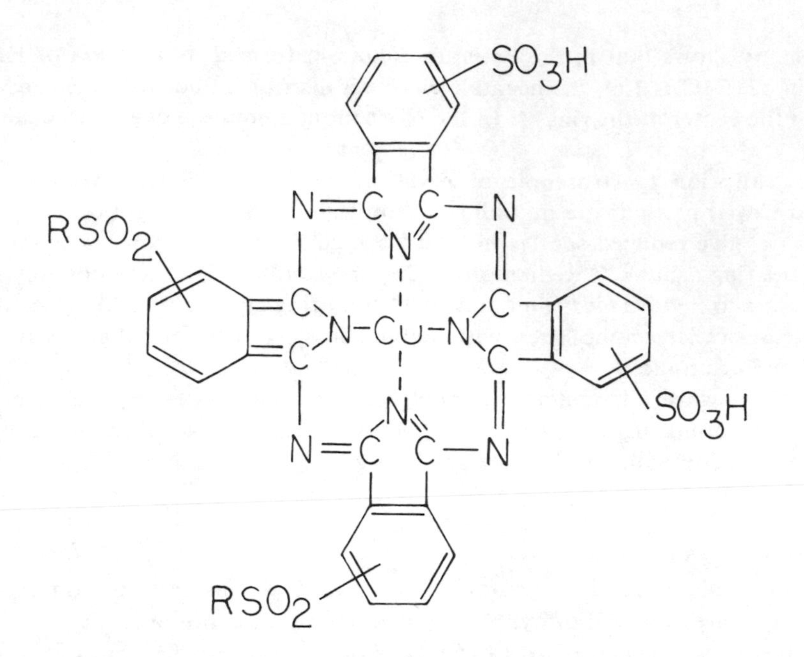

Cu pcs of the formula (I), when used as the bluish green or green filter dyes in the red sensitive Ag halide emulsion layer of color film increases the color sharpness of the images obtained.[163]

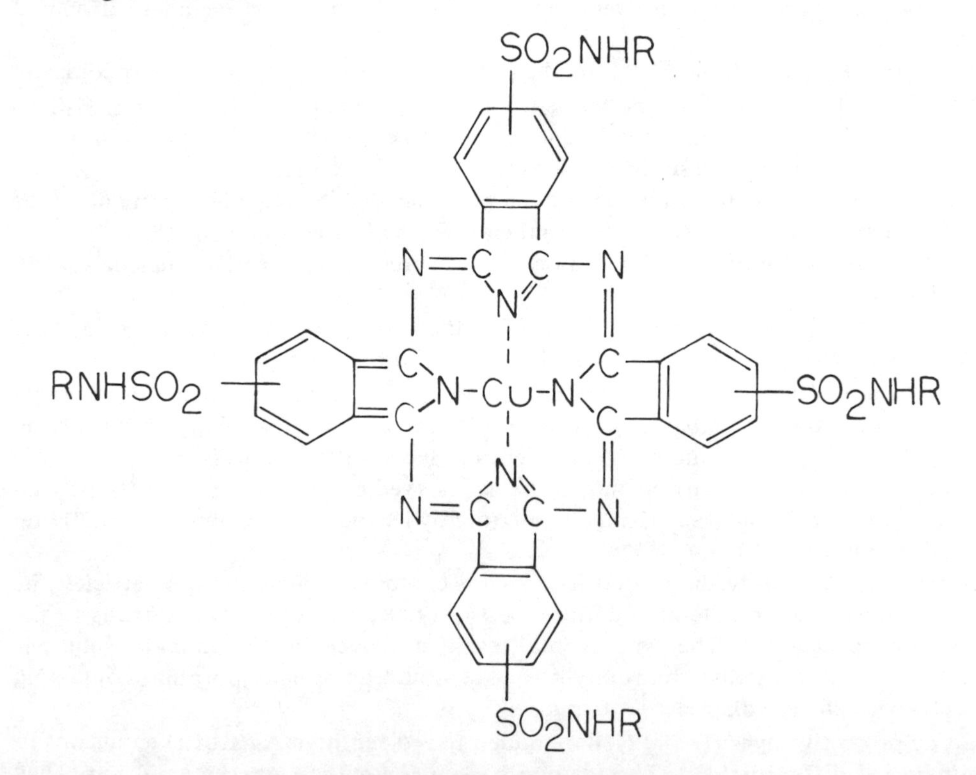

where R = H, *m*- $C_6H_4SO_3H$ or 2-hydroxy-6-sulfo-1-naphthyl.

The solution variables that affect the accuracy and precision of the spectrophotometric measurement of dyes including the Cu complex of pc disulfonic acid are discussed.[164]

UV spectra of Cu pc and its azo derivatives in pyridine are measured with a Beckman DU-2 spectrophotometer employing 1 cm. quartz cells.[165] These spectra are "examined on the basis of symmetry operation, ligand field theory and molecular orbital considerations. The above treatment was also employed to determine the structure of the synthesized complexes."

D. Lanthanide Phthalocyanines

Lanthanide pcs form, with acids and nonsolvate solvents such as *o*-dichlorobenzene, adducts with spectra that differ sharply from the usual spectra of pcs and contain three strong and broad bands.[166]

Absorption spectra of DMF solutions of mono and di pc complexes of Nd and Lu are given.[167] A sandwich model is proposed for the di pc complexes on the basis of the spectra.

E. Titanium Phthalocyanine

A spectrophotometric study of the complexing between Ti(III) with pc tetrasulfonic acid is made.[168] The complex has a maximum at 520 nm.

F. Vanadyl Phthalocyanine

The IR spectrum is observed for the vanadyl ion.[169] Absorption spectra are measured of excited singlet-singlet absorption molecules of the pc vanadium complex and of cryptocyanine in toluene.[170]

G. Cobalt and Iron Phthalocyanines

A spectroscopic study is made of the reduction of cobalt and iron pcs.[171] Reducing agents used are N_2H_4 or H_2S in solution.

The absorption of NO on solid Fe(II) pc is studied by IR spectroscopy [172] in the region 700 to 5000 cm^{-1}. Fe(II) pc absorbs NO in the molar ratio 1:1, causing the appearance of a maximum absorption relative to NO in the investigated region.

The effect of axial ligands on the absorption spectrum of Fe(II) pc is measured at 12000 to 35000 cm^{-1}.[173] In general, the spectra consist of four major bands and several minor ones.

The absorption spectra and magnetic circular dichroism of Fe(II), Co(I), Co(II), Co(III), and Zn(II) pcs are observed and charge transfer bands are examined.[174] "The band in Co(III) pc dicyanide at 22.5 kK is not a charge transfer band but must be a transition of the pc ring."[174]

Electronic spectra of Fe^{2+} complexes with 4-*tert*-butyl pc and nitrogen bases are determined in organic solvents.[175]

The quantum yield of radiationless conversion from the first excited singlet to the lowest triplet state is determined from the measurement of the saturation of absorption of a light pulse in solutions of organic dyes including tetra-4-*tert*-butyl pc and several of its metallic complexes in benzene and polymethyl methacrylate.[176]

H. Platinum Group Phthalocyanines

The IR spectra of pc derivatives of Pd, Pt, Rh, Ir, Ru, and Os in the region 220 to 5000 cm^{-1} are observed.[177] Bands relatively sensitive to and characteristic of the Pt group metals in their pos are observed. A possible tentative assignment of the weak band varying between 910 to 918 cm^{-1} is as a rocking vibration of the metal pc nucleus.

I. Zinc Phthalocyanine

Far IR spectra (30 to 400 cm^{-1}) are recorded for charge-transfer complexes formed by Zn pc with various amines and show new absorption bands in the 120 to 220 cm^{-1} region which are due to the Zn pc-amine intermolecular bond.[178,179] The absorptions by fundamental vibrations of aromatic amines shift to higher frequencies, whereas those of aliphatic amines tend to decrease in frequency.[179]

Polarization spectra are observed for metal porphyrins presumably including Zn pcs in the visible region at 77 K in $(CH_3)_2CHOH$-$(C_2H_5)_2O$-petroleum ether mixture (2:5:5) and at room temperature in castor oil.[180]

J. Aluminum Phthalocyanine

The absorption spectra of a solution of chlorinated Al pc in *o*-dichlorobenzene at 20 and 100°C are reproduced, showing a strong temperature dependence.[181]

K. Silicon and Germanium Phthalocyanines

IR absorption spectra, 300 to 1620 cm^{-1}, are obtained for $pcSiCl_2$ and $pcGeCl_2$ films deposited on KBr crystals at 50°C from their solutions in α chloronaphthalene.[182]

IR spectra at 400 to 4000 cm^{-1} are determined for $pcSiCl_2$, $pcSi(OH)_2$, and the product obtained by precipitating $pcSiCl_2$ from concentrated sulfuric acid over ice, and the product obtained by dehydrating $pcSi(OH)_2$ in a vacuum at 430 to 50 °C for 10 hr.[183]

The IR absorption spectra of $pcSiCl_2$ are recorded at 100 to 4000 cm^{-1}.[184] Twenty-six bands are identified. The spectra are quite similar to those of uranyl pc. The two chlorine atoms remain attached to Si and the 180 cm^{-1} band is most likely due to the Cl-Si-Cl bending motion.

IR and NMR studies are carried out on Ge hemiporphyrazines and pcs.[185] The NMR studies show that the hemiporphyrazine ring, in contrast to the pc ring, is not aromatic. The IR studies yield a set of Ge-X and Ge-O frequencies when X = OH, OD, F, Cl, and I.

The UV and visible spectra of Cl_2Ge pc and $(HO)_2Ge$ pc are nearly identical and are more complex than those of the corresponding Si derivatives.[186] The pseudo-first-order kinetics of hydrolytic degradation of the X_2Ge pc are studied in 16 to 18 *M* sulfuric acid at 100 to 130°C.

L. Effects of Dimerization

The effects of dimerization in the electronic spectra of pcs are studied over the range 5000 to 45,000 cm^{-1} in THF solution, using Si pc dimers where the Si atoms are connected by an oxygen atom bridge and the outer side of the Si atoms of the dimer are connected to siloxy methyl side chains.[187] The spectral differences between monomer and dimer "are interpreted in terms of exciton coupling of neutral-excitation transitions of the two rings of the dimer."*

Effects of dimerization may also be observed in other pc complexes. "Of particular note are the tetrasulfonated pc dimers which exist in aqueous solution. These are not chemically bonded, but exist as loosely associated species, which can be dissociated by changes of solvent. However, they exhibit completely analogous spectra to those of the Si pc dimer, and it may be concluded that they exist in a similar sandwich type configuration."*

M. Lead and Tin Phthalocyanines

A review of metal-like compounds with 55 references discusses their physicochemical properties including IR spectral and magnetoresistance.[188] Compounds discussed are

* From Hush, N. S. and Woolsey, I. S., *Mol. Phys.*, 21(3), 465, 1971. With permission.

K_2 [Pt(CN)$_4$] $Br_{0.3} \cdot 3H_2O$, tetrathiofulvalene-tetracyanoquinodimethan (TTF—TCNQ) *N*-methylphenazinium tetracyanoquinodimethan (NMP-TCNQ), and Pb pc.

The IR spectra of SnF_2pc, $SnCl_2$pc, $SnBr_2$pc, SnI_2pc, Sn(OH)$_2$pc, Sn pc, α Sn pc$_2$, β Sn pc$_2$, and Pb pc are observed in the 286 to 5000 cm^{-1} region.[189] An interpretation of features of their spectra is offered.

IR and NMR studies are made of Sn pcs and hemiporphyrazines.[190] "The [IR] assignments for $SnBr_2$pc and SnI_2pc are new, previous assignments not having been made. Those for Sn(OH)$_2$pc, SnF_2pc, and $SnCl_2$pc are in essential agreement with earlier assignments except that the O-H bending assignment for Sn(OH)$_2$pc has been modified."

N. Tetrabenzoporphin Molecules and Their Suggested Existence in Interstellar Space

There is intriguing evidence that certain tetrabenzoporphin molecules may be present in interstellar space.[191] This evidence is provided by IR spectroscopy. Indeed, if tetrabenzoporphins are present in interstellar space, may not pcs be there also?

"The diffuse interstellar (d.i.) lines have proved an enigma for many decades. This paper will summarize the results of laboratory simulation studies designed to duplicate these diffuse interstellar lines.

"Following an 18 year constant search and analysis, one was led to a group of molecules, namely the porphyrins, whose spectra were suggestive of the diffuse interstellar lines. Finally, a single porphyrin, namely Mg tetrabenzporphin (χ) was isolated and studied in great detail. This molecule placed in an appropriate paraffin matrix provides the spectra upon which the laboratory simulation and identification is based.

"A search was made for some of the stronger and sharper lines shown in Table 1, using interferometric techniques and the largest available telescopes.

"A comparision of the laboratory spectra of Mg TBP (χ) with corresponding d.i. astronomical data is given in Table 2. In order to match the d.i. lines, however, it was necessary to attach to an axial ligand, the pyridine molecule, on either side of the central atom." (See Figure 1).

"This work has considerably solidified the spectral matching with the diffuse interstellar lines: using photographic techniques, there are 6 matched lines to ± 2 Å, leaving only 4 (or possibly 6) *weak* laboratory lines unaccounted for. In addition, the line widths match to within a factor of 2; the intensities are not inconsistent, to first order, with the d.i. lines. If the molecule χ is indeed present in the interstellar medium, then the present picture of the interstellar dust and its environment need to be altered; some of the dust apparently is in local environments, having high intensity IR pump sources in its vicinity. Note that the broad spectral feature of χ centered at 6350 Å in absorption at room temperature, and observed 77 K in fluorescence has been recently discovered in the d.i. spectra by York."[343]*

XXIV. STARK EFFECT

The effect of an electric field on the visible absorption spectrum is known as the Stark effect.

The Stark effect and ferroelectricity are examined in polycrystalline films of the semiconductor Cu pc by electroabsorption.[192] Measurements are carried out at the frequency of the feeding voltage, 600 Hz. The spectrum corresponds to broadening of absorption bands without shift, which indicates a polar character of the centers which absorb light.

* From Johnson, F. M., *Mem. Soc. R. Sci. Liege, 6th ser.*, 3, 391, 1972. With permission.

Table 1
IR ABSORPTION MEASUREMENTS (cm^{-1})

ZnTBP	TBP	MgTBP	H_2pc	Porphyrin	FeTBP	Mg pc[a]	H_2 pc[a]
					2985		
						1609	1610
1555		1590m		3300w	2930	1586	1600
1490m							
1470w		1475w	1485m	1530w		1481	1503
		1465m					
		1455m				1454	1461
1422s	1435m	1430m	1420m	1410m		1408	1439
		1400m					
1335	1310m	1322m	1320s			1333	1336
		1287m	1305s				
1300s			1285s			1283	1277
1245s	1225m	1232s (doublet)	1260s		1260		
1205s	1200w		1170m	1222m	1245		
1160w	1160m	1212m	1140m	1135m	1220	1163	1160
1115vs	1110m	*1110vs*	1100vs		1100	1116	1119
	1090s	*1115s*	1080s		1070	1086	1094
1060	1070w				1060	1060	
	1040m	*1051* (triplet)		1050m		1060	
		1032					1007
1016s	1010m	1014			1010		
972w	968w	1000m	900	970			
940w	935w						
897vs	870s	887m-s	860vs	950vs			
870m	860s					890	874
840s	840s	823s		850vs		872	870
835	830vs			840vs			
755vs	*732vs*	*753vs*	730vs	770vs	755	752	736
740vs	*718vs*	737vs	718vs	740w	735	728	714
			712vs	725s	700		
702vs		703	700s	715s			
	690m sharp	*693s.v.*[b] sharp	665w	690vs			
630s	660w	620w	600m			642	620
						575	557
	610m	610w (doublet)	475w				492
	430vs		415m			435	434

Note: w, weak, m, medium, and s, strong.

[a] Crystal modification of phthalocyanines, data from A. N. Sidorov and I. P. Kotlyar, *Opt. Spectrosc.*, 10, 92, 1961.

[b] $693 = [(231 \pm 1) \times 3]\ cm^{-1}$.

By using a differential method the Stark effect is investigated on films of organic semiconductors, in particular pc and its metal complexes. All Stark effects observed have a characteristic form with a sharp peak at the long wavelength absorption edge.[193]

The polarizability of the first excited singlet state in tetra-*tert*-butyl pc and its copper complex is studied by measuring the Stark effect.[194] The visible absorption spectra are determined on solid solutions (about 5×10^{-6} *M*) in polymethylmethacrylate. A plot of the electric field effect resembles closely the first derivative of the optical density with respect to wavelength. This implies that there is a change in the polarizability upon going from the ground to the excited state, but that the dipole moment remains zero.

The Stark effect of electronic levels in the semiconductor H_2 pc is measured, "and

Table 2
LABORATORY SPECTRA OF Mg TBP COMPARED WITH CORRESPONDING D.I. ASTRONOMICAL DATA

Laboratory data			Astronomical data[a]	
Mg TBP (F. M. Johnson)	Width (Å)	Mg TBP (Sevchenko)	Wavelength (Å)	Width (Å)
f 6663 Å	1—2		6661	1
f 6633vw				
f 6628w		6623 Å		
f 6614	1—2		6614	1
f 6610vw		6570		
f 6334vw				
		6404		
		6377	6376	2
6289s				
6284s	1—2	6282	6284	4
abs 6174	14 Å (total)		6175	30
abs 4428vs	40		4428vs	20

Uncertain assignment (H_2 TBP?)
6274
6278

[a] G. H. Herbig, private communication.

several features, i.e., the field dependency, the wave length dependency and the dependence to dielectricity are examined.[195]

"The electro-optical effect is observed as the change in transmittance under an application of sinusoidal voltage (f = 60 Hz) to the samples, which are sandwiched between glass plates coated with SnO_2 and semitransparent aluminum electrodes."

The Stark effect depends mainly on the aligned molecule, that is, aligned dipole. This observation agrees with previous findings that the electroabsorption correlates with the dielectric hysteresis loop observed on evaporated films of Cu pc." [344]*

XXV. LUMINESCENCE

A. Fluorescence and Phosphorescence

The phenomenon of luminescence is revealed by fluorescence and phosphorescence.

Since some references discuss both fluorescence and phosphorescence, and still other references refer to luminescence rather than to fluorescence and phosphorescence, the subject of luminescence is narrated without attempting to discuss fluorescence and phosphorescence as separate topics.

The luminescence of rare earth pcs is studied.[196] Absorption and luminescence spectra in the uv and visible range of Eu pc are given. The shape of the spectra for Gd and Yb pcs is similar. No phosphorescence is found in the luminescence spectra in the range 600 to 1000 mμ at low temperature in alcoholic solution. Because the triplet state in the molecules of metal pcs has a very low energy level (<10,000 cm^{-1}), the phosphorescence is beyond the spectral sensitivity of the photomultiplier used.

Eu pc has the strongest fluorescence, among the rare earth elements. Gd pc, which has a very high magnetic moment, exhibits a weaker fluorescence, Y pc does not flu-

* Masahiko M. and Yatabe, K., *J. Phys. Soc. Jpn.*, 33(4), 1176, 1972. With permission.

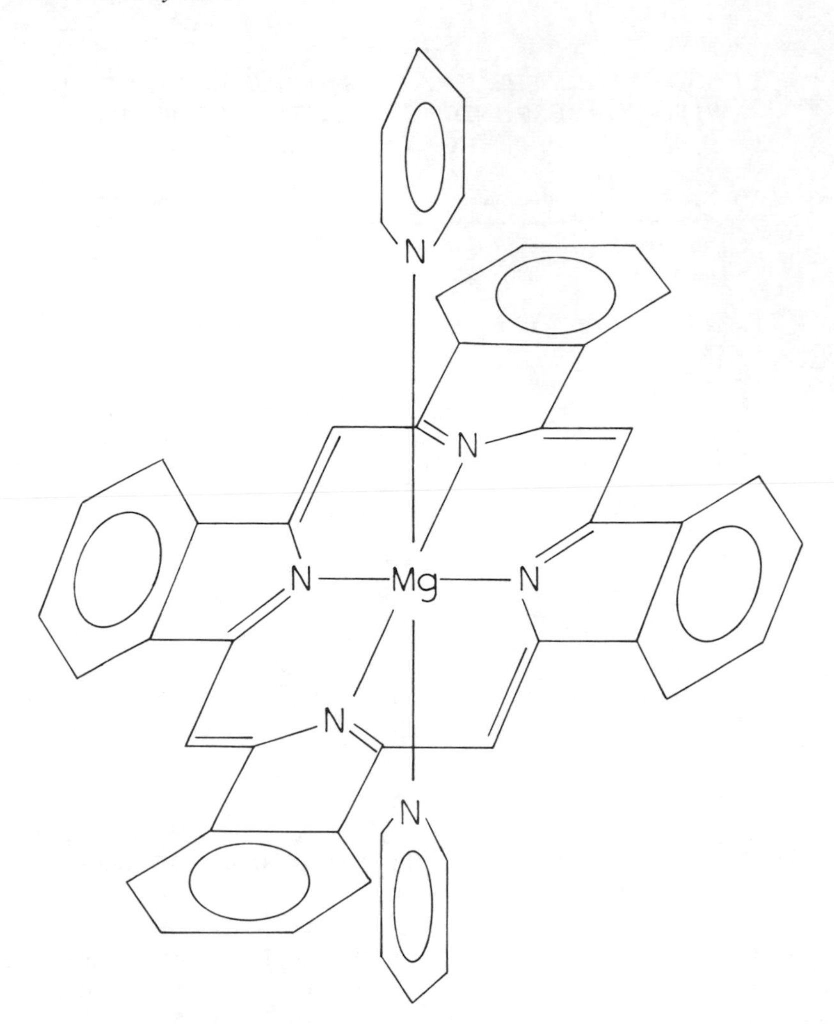

FIGURE 1. Porphyrin molecule (x) $MgC_{46}H_{30}N_6$. Note the four benzene rings and the two pyridine moieties attached to the central atom.

oresce at all (the excited level of Y is lower than that of the pc ring, while that of Eu is higher).[197]

Spectral luminescence properties of europium, gadolinium, and ytterbium pcs are observed.[198]

The effect of 10^{-3} to 10^{-4} *M* uranyl nitrate or acetate on the luminescence spectra of 10^{-4} to 10^{-5} *M* solutions of H_2pc in dioxane and of Mg pc in ethanol is investigated at 290 and 77 K.[199]

Luminescence spectra of uranyl pc complexes, complexes of divalent metal cations, such as Mg, with pc, and H_2 pc are compared in such solvents as dioxane, octane, and nonane.[200]

The luminescence spectrum of dioxane and nonane solutions of uranyl pc reveal that the UO_2^{++} ion only slightly perturbs the energy levels of the π-conjugated system.[201]

The absorption and fluorescence spectra of H_2 pc are investigated in octane, nonane, and decane at 77K with an ISP-67 spectrograph and excitation from a Xe lamp with a green-blue filter (transmission region 3600 to 6200 Å).[202]

Fluorescence measurements are made of pc.[203] Normal and flash excitation are used. Intersystem crossing yield is studied in ethanol with the aid of heavy atom quenchers such as CH_3I. The sum of the quantum yields of fluorescence and intersystem crossing are approximately unity, indicating that internal conversion plays a negligible role.

The excitation time effect on the imino-hydrogen shift on the fluorescence polarization of pc is determined at 77 K.[204]

The optical absorption and fluorescence spectra of H_2 pc and Cu pc are measured.[205]

The fluorescent spectral emitted band at 1555 cm^{-1} of H_2pc is probably due to a symmetrical vibration.[206]

Luminescence spectra of certain porphyrins are changed by reduction owing to illumination.[207] However, no photochemical changes are observed on pc.

A blue or short-wave (λ_{max} between 425 and 445 nm) derivatives presumably including pc is observed.[208] Heavy atoms in the solvent (ethanol-methyl iodide) repress the ordinary red fluorescence but not the blue.

The effect of intramolecular spin-orbital perturbations on luminescence is presumably considered for pc among other porphyrin complexes.[209]

The energy levels of low lying electronic derivatives including pcs are analyzed in terms of Wigner chains containing 5 or 18 electrons in the porphyrin ring system, and position of spectral bands in the flash fluorescence spectra is predicted.[210]

An investigation of the absorption and luminescence spectra and polarization measurements of pc solutions in acid media reveal that all the ionic forms of pcs in these media are nonsymmetrical.[211]

A survey of available and more recent data on the polarization spectra of protoporphyrin in caster oil shows that the two fluorescence bands in porphyrins are due to two different oscillators.[212]

A vibration analysis of the fluorescence and absorption spectra of Zn and Mg pcs is carried out.[213]

Lifetimes of the excited states of pcs are studied by the irradiation of solutions with a Q-switched ruby laser.[214] By detailed measurements the saturation of fluorescence and absorption with increasing laser power, the radiative and nonradiative lifetimes as well as the time constants for the $S_1 \rightarrow T_1$ and $T_1 \rightarrow S_0$ transitions are determined.

The spectrum of the luminescent radiation induced in Pd pc is examined at 77 K to determine the triplet level position of Pd pc.[215] The sample dissolved in quinoline is excited by a beam from the LG-55 Ne laser to ensure a high spectral purity of the exciting light, necessary for the intensification of the expected weak luminescence. The measurement discloses an evident triplet-singlet phosphorescence with a maximum at 996 mμ. No phosphorescence is detected in the Cu, Zn, V, and Yb complexes.

Fluorescence of H_2pc monomer and dimer in dioxane solution is observed.[216]

Excited singlet lifetimes of about 2 nsec for cryptocyanine and about 6 nsec for chloroaluminum pc are measured directly from fluorescence.[217] The fast excitation source is a ruby laser. A model is proposed to explain these relaxation measurements.

By increasing both the excitation intensity (a He-Ne laser at 6328 Å) and detection sensitivity, radiative emission, with msecond lifetimes, from the lowest excited states of the pcs is observed.[218] Emissions at 8000 to 9500 Å are associated with $S_0 \leftarrow T_1$ phosphorescence. The luminescence spectra of H_2, Li, Na, Mg, Al, VO, vanadyl, Co, Cu, Zn, and Pt pcs in benzene at 77 or 293 K, in CCl_4 at 77K and in hydrocarbon glass (methylcyclohexane:isopentane 4:1) at 77 K are tabulated.

In a study of the theory for the luminescent state in vanadyl, cobalt, and copper porphyrin complexes, existing experimental evidence shows that luminescence comes from both the tripdoublet and the quartet states, their relative importance varying with metal, porphyrin structure, and temperature.[219]

Spectra of O-O transitions are examined at about 4K with H_2pc and Zn and Mg pcs. The spectral traces shown for solutions in decane are also accompanied by spectral data on fluorescence of these pcs at 4K.[220]

The fine structure of the fluorescence spectra of some pc complexes are studied.[221] The quasi-line electronic spectra of Mg-Zn and Be pcs are studied at 77 K.

The line shape in quasilinear luminescence spectra of organic molecules includes H_2pc and Zn pc in frozen crystalline *n*-paraffin solutions at 4.2 K.[222]

Absorption and fluorescence spectra of pc and Zn pc are studied in matrices of Ar, Kr, Xe, CH_4, N, and SF_6 at liquid hydrogen temperature.[223] Zn pc is also studied in CO. The spectra show considerable fine structure whose resolution decreases in the sequence Ar > $CH_4 \geqslant$ Kr $\geqslant$ Xe > N $\geqslant$ SF_6 > CO. Only part of the fine structure seen in absorption appears in emission.

The effect of concentration, temperature, and composition of binary solvent on absorption and emission spectra, lifetime of excited states, and on the quantum yield of luminescence of porphin dyes are studied.[224] The solvent mixtures are ethanol-water $(CH_3)_2CO$-water, ethanol-glycerin, and cyclohexanol-glycerin.

The effect of aza substitution on the spectral luminescent properties of benzoporphyrins is studied.[225] The position of the lowest triplet level of the pc ring is determined.

The dependence of the fluorescence spectra of some aromatic hydrocarbons and pcs in solid solutions at 4.2 K on the frequency of laser excitation is studied.[226] It is found that "multiplets" exist in the fluorescence spectra in the case when laser excitation occurs in the vibronic transition region.

The fluorescence spectrum of single crystal pc has bands at 763, 808, 860, and 919 nm. The 763-nm peak may correspond to the emission edge; the remaining peaks may be vibrational in origin.[227] Zn pc single crystals exhibit fluorescence maxima at 795, 888, and 948 nm.

"For the first time, phosphorescence has been observed from and clearly assigned to the molecular triplet of several metallo-pc molecules. The pcs of Pt, Pd, Cu, Zn, and Cd have O-O bands in solution (77K) at the extraordinarily low energies of 10,590, 10,100, 9390, 9150, and 9120 cm^{-1}, while those of Mg and VO have possible transitions at about 9000 and 8500 to 8800 cm^{-1}."[228]

Luminescence is observed for Sn pc at 120 K.[229]

The fluorescence decay of solutions of H_2pc in 1-chloronaphthalene is investigated in the concentration range of 1×10^{-7} *M* to 1×10^{-4} *M*.[230] "The prompt molecular fluorescence decay time of H_2pc was found to be 6.3 ± 0.3 nsec. From the measured intensity ratios of the delayed and prompt fluorescence components compared with the dye concentration it was deduced that the delayed fluorescence of the H_2pc solutions is mainly caused by triplet-triplet interactions."

"Fluorescence from upper vibrational levels of the first excited molecular singlet is observed in solutions of selected pcs at 77 K. Vibrational relaxation, internal conversion, and intersystem crossing rates are reported for Pd, Rh, Mg, and H_2pcs for the first time."[231]

The dynamics of the triplet state of pc complexes of the platinum metals in zero field are studied.[232] "Surprisingly, Pd and Rh pc were found to fluoresce as well as phosphoresce from essentially ligand-centered, excited states, an unusual occurrence in open *d* shell transition metal complexes."[232]

The luminescence of metal porphyrin molecules, presumably including Pd pc, is observed.[233]

Results of an investigation of chlorophyll-like molecules, presumably including Pd pc, due to transitions from high excited levels, are given.[234]

Bimolecular luminescence quenching constants are given for systems containing phenanthrene, anthracene, decafluoroanthracene, naphthacene, or Mg pc and salts of nitrogen heterocycles.[235]

Emission spectra of H_2pc single crystals and evaporated films are studied in the near IR region.[236] In H_2pc single crystals, several fluorescence peaks are observed at 760 to 1000 nm. In a Cu pc single crystal, one phosphorescence band with fine structure is observed at 760 to 1000 nm. In Zn pc single crystal, both the fluorescence peaks and the phosphorescence peak at 1.15 μm are observed. In Co pc and Ni pc, no emission is observed. These emission spectra are discussed in connection with singlet-triplet interaction caused by the central metals.

The temperature dependence of the phosphorescence spectra in Cu pc single crystals is studied.[237,238] The emission peaks near 1.12 μm at 4.2 K decrease markedly with increasing temperature and new emission peaks centered at 1.09 μm appear above 77 K. The decay time of phosphorescence emission in Cu pc at 1.12 μm is also investigated with ruby laser excitation and is observed to be about 2 to 3 μsec.

"The evaporated thin film of H_2pc has a very broad emission band centered at nearly 880 mμ which is different markedly from the emission spectra of the H_2pc single crystal."[239]

Phosphorescence is observed with Zn, Pd, and V pcs.[240]

Luminescence is observed in the near IR region of 9800 to 11,000 cm^{-1} for the Yb^{+3} pc.[241]

Luminescence spectra show that metalloporphyrins belong to three distinct classes: (1) those exhibiting both fluorescence and phosphorescence (complexes containing Mg, Al, Zn, Cd, or Sn), (2) those exhibiting only phosphorescence (containing Pd, Pt, Cu or VO), and (3) those without any luminescence (Ni, Mn, Co, Ag, Hg, or Fe).[242]

The fluorescence and absorption spectra of pc and of its complexes with Mg and Zn are measured in *n*-decane.[243] The dependence of the line widths in the O-O multiplet in the fluorescence spectrum of Zn pc on the rate of the crystallization and on the type of excitation is studied.

Relative fluorescence of vapors of complex organic compounds including Zn pc (molecular concentrations of 10^{16} to $10^{17}/cm^3$) are measured with an error of ± 25%.[244]

Polarization of fluorescence of Zn pc in solution (2×10^{-6} *M*, as an example) is studied.[245]

The luminescence of H_2 pc in an octane matrix at 4.2 K is observed.[246-248] Fluorescence of H_2pc is measured at 77 K.[249]

Phosphorescence and its decay time in Pt pc single crystals are investigated in the near IR at about 971 nm by a Q-switched ruby laser.[250] "We can conclude that the triplet exciton-triplet exciton collison was observed in Pt pc single crystals directly for the first time." The phosphorescence of Pt pc single crystals at about 971 mμ is also studied as a function of the magnetic field.[251,252]

A detailed temperature study is made of the phosphorescence lifetime of and quantum yield of Pt pc in α-chloronaphthalene in the range from 1.3 K to room temperature "with particular attention to the important but previously not covered range of 15°K. to 77°K. From our data we calcuate the activation energies for intersystem crossing for the molecule and discuss the implications of our findings."[253]

B. Shpol'skii Effect

Shpol'skii spectra are studied in frozen paraffin mixtures with respect to the polarization of multiplet components.[254] For spectra of impurities at low concentrations, it is necessary to select substances with intense and narrow absorption bands (3,4-benzpyrene and pc are suitable). Examination of the multiplet spectra in fluorescence (pc in octane matrix) show that the polarization effects in this case are the same as found in the absorption spectrum.

The fluorescence spectra (600 to 770 nm) and the fluorescence excitation spectra of pc and etioporphyrin in *n*-octane are measured at 77K.[255,256] The rates of photoinduced interconversion of luminescing centers in Shpol'skii matrixes agree with the changes of polarization degree of luminescence of porphyrins in low temperature glasses.

In a journal article entitled "Burnout of the Dip in the Contour of a Purely Electron

Line in Shpol'skogo Systems",[257] deformation of the contour of the purely electronic line in the absorption and luminescence spectra of frozen pc solutions in octane at 5 K as a result of resonance excitation by ruby laser radiation is studied.

The homogeneous, pure electronic linewidth in the spectrum of a H_2pc solution in *n*-octane at 5K is measured.[258]

"In [Gorokhovskii, A. A.; Kaarli, R. K.; Rebane, L. A., JETP Lett., 20, 216, 1974] a hole was observed in one of the Shpol'skii spectra, namely in a 6940 Å component of the O-O multiplet of H_2pc in *n*-octane. With this an inhomogeneous broadening of comparatively narrow lines (in this case 3—4 cm^{-1}) in the Shpol'skii spectra was confirmed. However, only an upper limit was obtained, namely $\Gamma_{00} \leqslant 0.2$ cm^{-1}."* A more precise measurement of Γ_{00} is obtained:[258] "Measurement of the homogeneous width of the pure electronic transition by the hole-burning method in the contour of a narrow (3—4 cm^{-1}) inhomogeneously broadened component of the O-O multiplet of H_2 pc in *n*-octane at 5°K. gives $\Gamma_{00} \approx 0.03$ cm^{-1}."*

XXVI. RAMAN SPECTRA

The Raman spectra of a Co pc excited with a He-Ne laser at 6328 Å are recorded.[259] Some of the lines have anomalously high values of the depolarization ratio (>0.75). This indicates an asymmetry of the tensor of the spontaneous Raman effect.

Laser excited resonance Raman spectra of pc and Mg, Fe, Co, and Ni pcs and their mono-, di-, tri-, and tetra-negative anions are investigated.[260] The spectra show in addition to lines corresponding to totally symmetric vibration, those corresponding to nontotally symmetric IR-active vibrations.

Resonance Raman spectra of pc and Mg, Zn, and Cu pcs are recorded and a detailed spectral analysis is given.[261]

By precision measurements, it is found that as a rule the large number of weak lines in the Raman scattering spectra of transparent substances possess degrees of polarization which exceed the extreme value, 0.75, characteristic for the symmetric shape of the scattering tensor and, in the majority of cases, low (<6) values of the reversal coefficients.[262]

The possibility of excitation of vibrations, forbidden by the usual selection rules, in resonant spontaneous Raman spectra is shown for pc and Mg, Fe, and Co pcs.[263]

XXVII. X-RAY ANALYSIS

In an address presented before "The Robert A. Welch Foundation Conferences on Chemical Research" on November 7 to 9, 1960, J. Monteath Robertson paid tribute to X-ray analysis as a method of structure determination in the address "The Structure of Complex Organic Molecules by Direct X-Ray Analysis". Professor Robertson told his audience: "Well, now, coming to the subject before us tonight, I think it is generally recognized that Laue's discovery of the diffraction of X-rays by crystals in 1912 was one of the greatest scientific events of this century. Not only did it immediately throw new light on the nature of X-radiation and on the structure of the atom; it provided a tool of immense power for exploring the structure of crystals on an atomic scale. As nearly all matter, whatever its molecular complexity, can be made to crystallize under some conditions, the field which was thus opened out for exploration and detailed study in this new way was enormous, embodying the whole of chemistry and quite a lot of biology . . .

* From Gorokhovskii, A. A., Kaarli, R., and Rebane, L., *Opt. Commun.*, 16(2), 282, 1976. With permission.

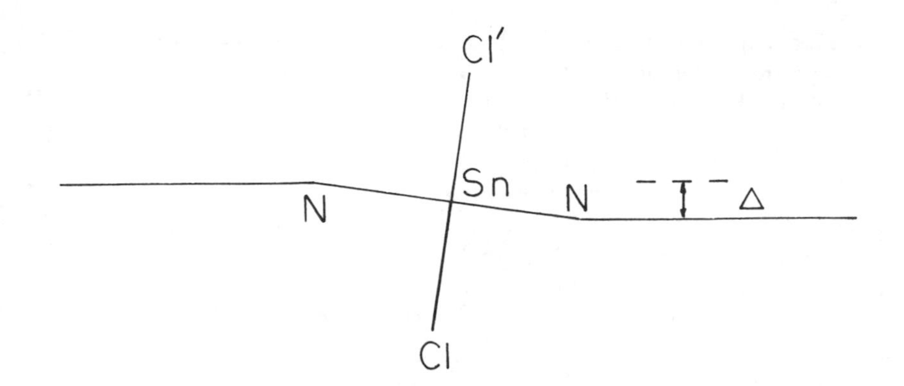

FIGURE 2. Lateral view of the dichloro pc tin (IV) molecule showing deformation of the molecule.

"I believe that we are now achieving a spectacular breakthrough in this field. I think we know how to solve the structures of molecules containing up to 100 atoms or more, even if the chemical formula is quite obscure. Further, I believe this can be done, or it will shortly be possible to do it, with little more trouble than we devoted twenty years ago to defining the positions of the atoms in a simple benzene derivative. The excitement attaching to this progress, and the importance of the results following from it, can, I think, only be compared with the events following the initial discovery of X-ray diffraction in 1912. Completely new fields then opened out, and this is now happening again today, over the whole domain of organic chemistry."[264]*

A study has been made in which "bis pc uranium (IV) ($C_{64}H_{32}H_{16}U$) is the first complex of an actinide element with the pc ion whose structure has been solved by X-ray diffraction."[265]

"The pc ring of dichloro pc tin (IV) is shown to be substantively 'crumpled' — a stepped deformation — apparently owing to an oversize tin atom," as shown in Figure 2.[266]

"Extensive studies of pc derivatives showed that metals such as zinc, cobalt, and iron, form crystalline stoichiometric molecular complexes with various amines. Fe (II) pc was found to form complexes with molecular ratios 1:2, 1:4 and 1:6, according to the temperature and the kind of amine . . . A single crystal x-ray diffraction study of the molecular complex between Fe (II) pc and 4-methylpyridine [$C_{32}N_8H_{16}Fe(C_6NH_7)_4$] shows that two 4-methylpyridine molecules are coordinated at an octahedral Fe (II) site with an N-Fe bond distance of 2 A and that the other two are free molecules occluded in the structure."[267]**

An X-ray diffraction study is made of thorium and uranium di pc complexes.[268]

An X-ray diffraction analysis if made of gallium trisodium trimetaphosphinate dodecahydrate [$Na_3Ga(P_3O_6N_3H_3)_2 \cdot 12H_2O$] · "The structure consists of complex anions of $\{Ga[(PO_2NH)_3]_2\}^{3-}$, Na cations, and water molecules."[269]

"Iron (II) and manganese (II) pc have been recognized as examples of a rare type of coordination compound in which the metal ion has an intermediate spin ground state (Fe, S = 1; Mn, S = 3/2). The basic stereochemistry of four-coordinate pcs has been known for some time, but surprisingly the quantitative stereochemistry of the much studied iron (II) and manganese (II) derivatives has not been determined."***

* From Robertson, J. M., *Proc. Robert A. Welch Foundation Conf. Chem. Res., IV., Molecular Structure and Organic Reactions,* Houston, Tex., November 7 to 9, 1960. With permission.

** From Kobayashi, T., Kurokawa, F., Ashida, T., Uyeda, N., and Suito, E., *J. Chem. Soc. Chem. Commun.,* 24, 1631, 1971. With permission.

*** From Kirner, J. F., Dow, W., and Scheidt, W. R., *Inorg. Chem.,* 15(7), 1685, 1976. With permission.

The molecular structure, therefore, of the two intermediate spin complexes, Fe (II) pc and Mn (II) pc are determined by X-ray diffraction.[270]

The structure of pc $Si[OSi(CH_3)_3]_2$ is determined by X-ray crystallography.[271] The compound, as crystallized from toluene, is monoclinic; the coordination arrangement of the central Si atom is approximately octahedral.

The octahedral silicon-oxygen, germanium-oxygen, and tin-oxygen bond length are determined by X-ray diffraction analysis from interplanar spacings in the pc polymers $(pcSiO)_x$, $(pcGeO)_x$, and $(pcSnO)_x$. The bond lengths are 1.7 to 1.9 Å.[272]

X-ray diffuse scattering from Pb pc at room temperature indicates linear lattice distortion parallel to the molecular column, which are interpreted in terms of a Kohn anomaly.[273]

"The low-energy structure of the x-ray K-absorption edge of Ni and Ni dimethylglyoxime, Ni pc, and K Ni cyanide is reexamined for evidence of the second order transition, the excitation of a 1s electron to a bound orbital, plus the simultaneous excitation of a plasmon. A perturbation calculation indicates that such a process may be observable."[274]*

The K absorption edges of X-ray spectra of various cobalt complexes [presumably including the pcs] and the significance of their fine structures regarding the electronic states of the complexes are discussed.[275]

The effect of the metal structure on the work function of systems composed of metals coated with a thin pc film including Co pc and Ni pc is studied by X-ray methods and electron microscopy.[276]

The equilibrium state attained between the piperidine complex of Zn pc and β form of Zn pc in dimethyl ketone is determined with the assistance of quantitative X-ray analysis.[277]

XXVIII. ELECTRON MICROSCOPE STUDIES

A. Crystal Lattices

The lattice image in electron micrographs is studied with relevance to Pt pc.[278] Calculations of the crystal spacing *d* are made and crystal models are compared with Pt pc.

Direct electron microscopic observations of Cu pc lattice spacings show that the β form consists of well-ordered crystals while the α form shows many distorted spacings.[279] The hexadecachloro form shows extremely regular lattice spacings. In aggregates, lattice planes frequently pass from one crystallite to another without interruption.

Crystals of Cu pc are studied with the electron microscope EM-7 and photomicrographs are presented.[280]

Electron microscope observation of the molecular disarray produced in the Pt crystal lattice is shown when a single fission fragment passes through it.[281,282] The extent of the damage can be measured to within one molecular plane.

Electron microscope studies are made of pc crystals under conditions which permit extinction contours to be photographed.[283] The images show striations which result from multiple elastic scattering. They are presumably caused by the crystal axis not being quite parallel to the incident electron beam over some areas.

The principles and operation of the scanning electron microscope and its application to color materials such as Cu pc are discussed.[284]

B. Molecular Images

Workers at Kyoto University have achieved molecular images of hexadecachloro Cu pc by means of electron microscopy.[285-287]

* From Mitchell, G. R., *J. Chem. Phys.*, 37, 216, 1962. With permission.

"Few or no attempts have been made to determine the structure of usual organic molecules (less than 20 A) through the direct observation of the image, since it is obvious that many practical difficulties will result, although the problem of resolution as well as that of image contrast has been treated theoretically. One of the anticipated troubles is that since the component atoms are mostly light elements, the amplitude of the electron waves which scatter from the atoms to form a molecular image is small, and the spherical aberration of the objective lens is too large to obtain a satisfactory phase contrast.

"Another major trouble is that a high magnification which is essential for observing the molecular image requires a bright electron source with increased current density. However, the molecules of the organic specimens are known to be so sensitive to radiation damage that they may become decomposed or deformed under the irradiation of an electron beam as dense as 1 Å/cm^2 at the object plane.

"Though it is also essential to keep the molecules in a proper orientation for observation, it has been considered almost unfeasible in view of the fact that the molecules make a fierce thermal vibration even at room temperature and also that the specimen-supporting membrane as well as the specimen molecule is composed of numerous similar atoms. We have sought a breakthrough in practical difficulties, in the belief that the solution of problems, mainly associated with the specimens, will be the prerequisite to the attainment of the molecular resolution. We used pc compounds for the test specimens because they were known to be thermostable and more resistant to radiation damage than other aliphatic compounds . . .

"It was also found that the chlorinated Cu pc is forty times more resistant to the electron radiation damage than the ordinary pcs.

"In order to determine the ultimate molecular resolution attainable with a conventional electron microscope, the direct observation of hexadeachloro Cu pc molecules was attempted. Since pc derivatives are known to form crystalline films with columns of parallel stacks of planar molecules, the specimens were prepared by epitaxial growth on KCl cleavage face through vacuum evaporation so that the column axis was directed almost normal to the thin-film surface holding an orientation suitable for the observation. The molecular orientation was determined by Patterson synthesis based on the laser optical transform of the electron diffraction pattern obtained from the individual crystallites places on the microgrid mesh. The direct observation was carried out with the 100-kV electron beam incident on the specimen along the column axis. The cross-like images arrayed in a centered rectangular net were clearly resolved, well representing the molecular shape of pc with the configuration like a four leaf clover."*

The effects of spherical aberration and accelerating voltage on atomic resolution in molecular images are studied by synthesizing by computer simulation for four cases: 500 kV, Cs (coefficient of spherical aberration for the objective lens) = 1.0 mm and 1.8 mm; and 100 kV, Cs = 1.4 mm and 0.35 mm.[286] "It was demonstrated that for 100 kV electrons even heavy atoms may not be recognizable unless Cs is less than 0.35 mm. Elevation of the accelerating voltage to 500 kV seems to be effective in obtaining atomic resolution, especially for the heavy atoms."

In a study of the effect of through-focusing on the bright and dark field molecular images in high resolution electron microscopy, the anticipated effect of through-focusing on the molecular images is investigated with computer simulated high resolution micrographs of a thin crystal of chlorinated Cu Pc.[287] "It was demonstrated that the molecular image reflecting the specimen structure will appear only at one focus position of through-focusing in the bright field mode for 500 kV electrons with the spher-

* From Uyeda, N., Kobayashi, T., Suito, E., Harada, Y., and Watanabe, M., *J. Appl. Phys.*, 43(12), 5181, 1972. With permission.

ical aberration coefficient Cs of 1.0 and 1.8 mm." A result of a through focusing calculation of the bright field image for 500-kV electrons with Cs = 1.0 is shown below in Figure 3.

It is the best image, obtained at a defocusing value of $\Delta f = 400$ Å, where the shape of isoindole rings as well as the heavy atoms are clearly observed.

XXIX. ULTRAVIOLET PHOTOELECTRON SPECTROSCOPY

"The photoelectric effect has the potential of providing new information on the structure and topography of cell surfaces. When a properly prepared specimen is subjected to ultraviolet light it will usually fluoresce. The emitted light is then focused and imaged with conventional glass optics in the powerful technique of fluorescence microscopy. If, however, the energy of the incident light is sufficiently high the sample can emit electrons (photoionize) as well as fluoresce. By evacuating the sample chamber, accelerating the electrons, and replacing the glass optics with electron optics, it is possible, in principle, to form an image of any biological surface based on its photoelectric properties. We refer to this technique as *photoelectron microscopy*. Photoelectron microscopy bears little resemblance to conventional or scanning electron microscopy but it may be viewed as a natural extension of fluorescence microscopy. The two essential differences are that photoelectron microscopy is a surface technique and the ultimate resolution is related to the wavelength of electrons rather than the wavelength of visible light."*

A photoelectron micrograph is obtained for solid Cu pc.[288] The photoionization threshold of solid Cu pc is on the order of 5.0 ± 0.1 eV (248 nm).

A photoelectron micrograph was taken of a 7-Å-thick sample of Cu pc at 90 K. The Cu pc was evaporated through a 200 mesh electron microscope grid onto the Butvar-coated end of a metal sample rod. "Since 7 Å is approximately the thickness of the pc molecule, *we conclude that it is possible to observe a monolayer of organic molecules in photoelectron microscopy.*[289]

A photoelectron study is made of pc placed on metal surfaces to study novel effects that sometimes arise that may be caused by the metal substrate.[290] "We propose a mechanism for the observed thickness dependence, based on the effect of ultraviolet light for studying organic and biological surfaces that minimize this effect."[290]

XPS and UPS are used to study the chemisorption of molecules such as pyridine and formic acid over metal pcs and with these measurements a discussion is possible relating to the catalytic decomposition of formic acid and adsorption of pyridine on Fe pc and other pcs.[341]

UPS is used as an aid to characterize surfaces of pcs used as pigments. "The uv photoelectron spectrum of Cu pc in the vapour phase showed characteristic peak positions corresponding to the uppermost filled molecular orbital levels in the Cu pc molecule. It was also shown that Cu pc pigments that had undergone different finishing treatments gave characteristically different spectra."[342]

XXX. X-RAY PHOTOELECTRON SPECTROSCOPY (XPS) OR ELECTRON SPECTROSCOPY FOR CHEMICAL ANALYSIS (ESCA)

In an article entitled "How ESCA Pays Its Way," the authors state "Our prejudice: Based on our dozen man-years using ESCA in two major chemical companies, we feel

* From Birrell, G. B., Burke, C., Dehlinger, P., and Griffith, O. H., *Biophys. J.*, 13(5), 462, 1973. With permission.

FIGURE 3. Artist's rendition of a bright field molecular image simulated in the vicinity of the best focus with C_s = 1.0 mm for 500-kV electrons. The molecular configuration is clearly observed in the image at the defocus values at 400 Å.

that it is the surface technique with the most answers for the chemical technologist. Though a dozen man-years may not sound like much, consider that this technique was discovered in the 50's with the first practical applications in the late 60's, the first commercially available instrumentation in the 70's. The beauty of ESCA is that it produces information about composition and bonding within the first few atomic layers of the surface of almost any solid or solidifiable material. And that's where a lot of the action is in such fields as plastics and heterogeneous catalysts. ESCA can examine such materials without damaging them, a capability unique among measurement techniques.''[291]*

''XPS (or ESCA) is rapidly developing as a powerful technique for the investigation of bonding in transition metal complexes. At present, however, the XPS or ESCA spectra of transition elements have only been systematically investigated in a few cases.''[292]**

In an article entitled ''Devaluation of the Gold Standard in X-Ray Photoelectron Spectroscopy'',[293] the study relates to vapor deposited gold which ''has been widely used as a calibrant for XPS.'' The authors state that ''it provides a good conducting surface which is free from contaminants, often introduced with internal standards. We have observed, however, that gold may react chemically with the sample, producing chemical shifts of the gold 4f peaks. The consequent nature and position of the gold peaks varies with the sample, time of deposition and probe temperature; but in each instance the gold no longer serves as a reliable calibrant.''[293]*** Similar effects may also arise in Cu pc, other pcs, and in porphyrins.[294]

An XPS measurement is made of VO pc.[295] The interest for this study resides in the observation that ''the possibility of establishing the effective charge on the atoms in a complex is especially of great importance. Our interest in finding a way of measuring such charges stems from the obvious importance of knowing the charge distribution within a molecule when considering its reactivity. For complexes active as homogeneous or heterogeneous catalysts, such a description is of paramount value.''

Although the XPS or ESCA technique as a tool for the study of structure is relatively of recent origin and application, it has generated a number of research articles focusing on pcs.

XPS is used to measure the electron-binding energies for a number of Fe compounds including Fe^{3+} Cl pc.[296]

''The correlation between the electrochemical properties of Fe pc and the variation of valence and spin state of Fe studied by the ESCA technique shows that the most active Fe pc for the reduction of oxygen is that whose central Fe atom has an oxidation state of + 3 and high spin.''[297]

The binding energy for the Fe 1s electron in Fe pc is measured and is found to be 7113.6 eV.[298]

The core electron-binding energies are determined by XPS for Cu pc[299] as well as for a number of other Cu complexes. These determinations ''were carried out to permit the conclusion whether or not copper is present in the Cu (I) or Cu (II) state.''

Binding energies are measured for paramagnetic Co (II) pc by XPS[292] and a study is made of the attendant satellite lines. A ''correlation between the intensity of the Co $2p_{3/2}$ satellite and the magnitude of its magnetic moment is shown.''[300] ''Intense satellite lines were observed for the 2p, 3s and 3p peaks in the case of the high spin cobalt (II) compounds, but not for low spin cobalt (III) complexes.''[292]**

The nature of the bonding in U pc_2 and Th pc_2 is also studied by XPS. The XPS

* From Riggs, W. M. and Beimer, R. G., *CHEMTECH*, November, 652, 1975. With permission.

** From Frost D. C., McDowell, C. A., and Woolsey, I. S., *Mol. Phys.*, 27(6), 1473, 1974. With permission.

***From, Betteridge, D., Carver, J. C., and Hercules, D. M., *J. Electron. Spectrosc. Relat. Phenomena*, 2(4), 327, 1973. With permission.

spectrum of each of the U (4f) levels reveals two electron "shake-up" satellites characteristic of 5f orbital participation in bonding.[301]

"XPS valence band spectra of H_2 pc, Cu pc, Ni pc, Co pc, Fe pc, and Pt pc [are] reported. The contribution of the d electrons in the metal pcs is already visible and can be interpreted in terms of final state structure. The π orbital structure can be interpreted surprisingly successfully by a combination of spectra of benzene and pyrrole, the ring system making up the pc structure. We find that in Fe pc, less in Co pc, and little in Cu pc, Ni pc, and Pt pc do the d electrons extend to the top of the valence band. This is consistent with the degree of catalytic activity of those compounds."*

The nitrogen 1s binding energies and some metal-binding energies are determined for porphyrins and pc compounds by XPS.[303] Both kinds of nitrogen in metals pcs are identical as determined by XPS.

XPS spectra of more than 600 compounds including Cu pc and Mg pc are described.[304] The XPS spectra of pc and Cu pc are measured.[294,305]

A review, with 26 references, discusses the XPS of porphyrins and pcs.[306]

XXXI. MOESSBAUER SPECTROSCOPY

Moessbauer spectra have been obtained for Fe pc.[307] The Moessbauer spectrum of ^{57}Fe(II) dipyridine pc at 295 K consists of a doublet with a quadrupole splitting of 2.04 mm/sec and an isomer shift of +0.06 mm/sec with respect to a ^{57}Co in Pd source . . . "The observed splitting is thus the largest yet recorded for a spin-paired 6-coordinated Fe(II) ion."[308] The ^{57}Fe Moessbauer spectra of Fe(II) pc and several of its derivatives are measured at 77 to 423°K.[309] In each case we obtain values of the isomer shift and of the quadrupole splitting. The former is related to the s electron density at the nucleus. The latter is proportional to the electric field gradient and is sensitive to the symmetry of the nuclear environment."** "Diamagnetic adducts of Fe(II) pc with several organic bases show quadrupole split spectra, and the peak separation varies with base strength. The electric field gradient arises from anisotropic covalent bonding."**

Moessbauer spectra of ^{57}Fe (II) pc(Z_2), where Z = ligand at the axial position, are measured at 77 K. "The spectra consists of quadrupole split doublets" from which the isomer shift and the quadrupole splitting are obtained.[310]

"The ^{57}Fe Moessbauer spectrum of a powdered sample of pc iron (II) in an applied magnetic field of 3.0 teslas has been measured as a function of temperature in the range 4.2°K. to 100°K. Measurements have also been made at 4.2°K. with 6.0 teslas applied, and on a single crystal specimen at 4.2°K with 3.0 teslas applied. Independent computer fits to the three measurements taken at 4.2°K. were found to be consistent with one another, and showed that detailed information concerning magnetic anisotropy can be obtained even from powdered samples of paramagnets by Moessbauer spectroscopy."[311]***

"Measurements of the ^{57}Fe Moessbauer spectrum of pc Fe (II) made at 4.2 and 100 °K., with the sample in applied magnetic fields at 30 koe., show that the electric field gradient at the iron nucleus is of positive sign and that there is no large asymmetry in the electric field gradient tensor."†

* From Hoechst, H., Goldmann, A., Huefner, S., and Malter, H., *Phys. Status Solidi B*, 76(2), 559, 1976. With permission.

** From Hudson, A. and Whitfield, H. J., *Inorg. Chem.*, 6(6), 1120, 1967. With permission.

*** From Dale, B. W., *Mol. Phys.*, 28(2), 503, 1974. With permission.

† From Dale, B. W., Williams, R. J. P., Edwards, P. R., and Johnson, C. E., *J. Chem. Phys.*, 49 (8), 3445, 1968. With permission.

"Moessbauer measurements made in the presence of large magnetic fields show that in pc bis (pyridine) iron (II) the electric field gradient at the iron nucleus is of positive sign. This indicates that the metal atom is bonded more strongly to the planar ligand than it is to the axial ligands."[313]

In a Moessbauer study of Fe (II) pc and bispyridine Fe (II) pc, at 295, 77, and 4.6 K., it is indicated from the wide quadrupole splitting and unusual temperature dependence of bispyridine Fe (II) pc that "the very extensive π system of the pc ring allows for even greater delocalization of electrons from pyrrole N π-orbitals; this character of the molecule, rather than the axial symmetry of the pyridine ligands, may cause the larger quadrupole splitting."[314]

Moessbauer spectra are measured on Fe (II) chloropc as well as Fe (II) pc down to 4.8 K.[315]

A Moessbauer spectrum of ^{57}Fe (II) pc of the β modification is also made. Moessbauer spectral studies as well as studies of magnetic properties of ^{57}Fe (II) pc "confirm that the ground state of iron (II) is a triplet."[316]

"Fe^{2+} pc has been the subject of a large number of measurements by different techniques and at different temperatures. The considerable interest in this complex is mainly due to its catalytic properties in redox processes. The stabilization of a triplet ground state is probably an essential function in these processes. The structure of Fe pc is very similar to that of iron porphin, which has a related behavior in redox processes in biological systems. Most experiments have been carried out at room temperature and below. The spectra recorded by the NGR/Nuclear γ-ray Resonance (Moessbauer) technique shows a quadrupole-split double which is almost independent of temperature . . ." Thus, Fe pc is studied by the NGR technique between room temperature and 557 K. "We can see a small change in the quadrupole splitting, and probably also in the chemical shift, with temperature."[317]

Iron polyazaporphines are also studied for the nature of their coordination and value of Fe by γ-resonance spectroscopy. They are prepared by reacting $FeCl_3$ with 1,2,4,5-tetracyanobenzene, tetracyanoethylene, or in H_2pc polymer.[318]

The observed difference in the Moessbauer parameters of α and β-crystal polymorphs[319] "permit us an insight into the nature of the interaction between the central metal atom and the axially situated nitrogens of the neighboring molecules."

The effect of pressure on the Moessbauer effect is a subject of interest. In one study, "the electronic structures of pc, ferrous pc, and the adducts with pyridine, 3-picoline, 4-picoline, and piperidin have been studied to 175 kbar pressure by using optical absorption, and Moessbauer resonance.[320] In another pressure study of Fe (II) pc, "the isomer shift decreases with increasing pressure while the quadrupole splitting increases from 2.52 to 3.30 at 165 kbar."*

Ferrous octaethyl porphyrin complexes are prepared and subjected to Moessbauer measurements between 4.2 and 300 K. "The quadrupole coupling constant is positive and the electric field gradient at iron is axially symmetric. The nearly temperature independent quadrupole splitting in Fe octaethylporphyrin suggests there are no low-lying excited orbital states in this complex. Comparison of the Moessbauer parameters reported here with data for related ferrous porphyrin and pc complexes provide insight into the σ and π bonding characteristics of the tetradentate ligands."[322]**

The Moessbauer spectroscopic study of some polyaminocarboxylates, presumably including Fe pc, is the subject of a thesis.[323]

* From Champion, A. R. and Drickamer, H. G., *Proc. Natl. Acad. Sci.*, 58(3), 876, 1967. With permission.

** From Dolphin, D., Sams, J. R., Tsin, T. B., and Wong, K. L., *J. Am. Chem. Soc.*, 98(22), 6970, 1976. With permission.

Isomer shifts and quadrupole splittings are determined from the Moessbauer spectra of ^{57}Co pc, ^{57}Fe pc, and ^{57}Fe pc · 2 pyridine ^{57}Co pc in pyridine, and Co pc · 2 pyridine at room and/or liquid nitrogen temperatures.[324]

In a study entitled "Emission Moessbauer Spectroscopy for Biologically Important Molecules. Vitamin B_{12}, its Analogs, and Cobalt Pc",[325] it is surmised that "Moessbauer spectra are characteristic of the entire molecules and offer interesting possibilities for the study of the chemistry of vitamin B_{12} systems, including coenzymes, and other biologically important compounds with conjugated ring temperatures."

Several studies have been directed at the Moessbauer spectra of ^{119}Sn pcs. "Quadrupole splitting and isomer shift data for some pc tin complexes indicate that simple correlations between chemical parameters, e.g., electronegativities, and Moessbauer parameters are not generally valid."[326] Intermolecular bonding effects in $^{119}Sn^m$ Moessbauer spectroscopy are studied[327] as are the Moessbauer recoilless fraction in tin compounds.[328] Moessbauer spectra of a series of pc tin (IV) complexes (pc SnX_2, where X = F, OH, Cl, Br, I) reveal isomer shifts "that vary linearly with the electronegativity of the X ligand."[329] The Moessbauer spectrum of crystalline triclinic $^{119}Sn(11)$pc is recorded at 82 K using a Sn-Pd source.[330] "The isomer shift relative to the source was $\delta = 1.44 \pm 0.01$ mm/sec, and the quadrupole splitting was $\Delta E_9 = 1.43 \pm 0.01$ mm/sec."

The effect of anharmonicity on the intensity of Moessbauer lines in the presence of local oscillations is considered.[331] "Anharmonicity changes the dependence of the Moessbauer line intensity on temperature and wave vector and particularly brings about an anisotropy of the Moessbauer effect in cubic crystals."

XXXII. MOESSBAUER AND ESCA OR XPS SPECTROSCOPY

A combination of Moessbauer and ESCA or XPS spectroscopy is used to study Fe pcs.[332-334]

In one study, "the ^{57}Fe Moessbauer and X-ray photoelectron spectra were determined for Fe pc Cl, Fe pc · $pO_3SC_6H_4CH_3$, Fe pc, Li[Fe pc] · 4.5 (THF), Li_2[Fe pc] · 5.5 (THF), Li_3[Fe pc] · 8(THF), and Li_4[Fe pc] · 9(THF). The magnetic moments, isomer shifts, quadrupole splittings, and electron distributions were determined."[332]

A combination of X-ray photoelectron spectroscopy and Moessbauer spectroscopy was used to study the chemical shifts of a series of Fe complexes including Fe pc.[333]

"The decrease in the electrocatalytic activity of Fe pc polymers in porous active carbon-Teflon battery electrodes was studied by ESCA and Moessbauer spectroscopy. The effects of oxygen and H_2SO_4 on these electrodes show that the activity decrease is not caused by a change in the Fe pc bond but rather by partial oxidation of the active carbon surface."[334]* We feel that the use of ESCA as a semiquantitative tool for surface studies, exemplified in this work, will yield valuable information also on other electrochemical problems."*

XXXIII. FIELD ION MICROSCOPE (FIM) OR FIELD ELECTRON MICROSCOPE (FEM)

"The field ion microscope, invented by Mueller, makes possible the visualization of individual atoms in the surface of a metal tip, with a resolution approaching 2 Å." Shadow outline of Cu pc are "formed by surrounding the molecules with an ordered layer of evaporated platinum. The image is the edge of the cavity in the Pt formed by the molecule being visualized. The key to the success of the work presented here is the

* From Larsson, R., Mrha, J., and Blomquist, J., *Acta Chem. Scand.*, 26(8), 3386, 1972. With permission.

relatively high degree of atomic order in Pt layers deposited by vacuum evaporation on iridium tips at temperatures less than 130°C . . . Without some degree of order, the resolution possible would be greatly reduced."*

"Total energy distribution (TED) curves of field emitted electrons were measured from various crystallographic directions from tungsten and molybdenum substrates with adsorbed pc and pentacene. Measurements which were confined to emission emanating from the well-known 'molecular patterns' associated with these adsorbates yielded considerable structure in the form of new peaks in the TED curves."[336]**

Results of the investigation of field emission through organic molecules (Cu pc and anthraquinone) are reported.[337] "Special attention was paid to the total energy distribution (TED) of the field emitted electrons.[11]

Since most adsorbates on the surface of the metal tip show a mobility or surface migration, further information may be deduced from emission current fluctuations and noise properties. Thus, "by means of field electron emission spectroscopy with Faraday collector and magnetic deflection the emission current fluctuations between two or more levels caused by single Cu pc adsorbed molecules have been recorded."[338]

XXXIV. INELASTIC TUNNELING SPECTROSCOPY

"Inelastic electron tunneling processes, involving the excitation of vibrational modes of molecular impurities in the insulating barrier of a metal-insulator-metal junction, were discovered by Lambe and Jaklevic in 1966 [Lambe, J. and Jaklevic, R. C., *Phys. Rev.*, 165, 821, 1968.]. Since that date, there has been interest in the possibility of observing transitions by inelastic tunneling."[339]*** "The study of inelastic electron tunneling has been extended to voltages above one volt. Excitation of electronic transitions is observed for several pc molecules, inserted as the insulator of a metal-insulator-metal junction (Al—Al oxide-pc-Pb). A new conductance channel opens at 1.15 V, corresponding to the first singlet-triplet transition. The results are compared to the selection rules of optical and electron impact spectroscopy . . . In conclusion, we believe the results show that excitation of electronic transitions can assist inelastic tunneling as vibrational transitions do."***

"A study of high bias tunneling to observe molecular electronic transitions is carried out resulting in the observation of both singlet-singlet and singlet-triplet transitions, including a singlet-triplet transition of β-carotene which had not been observed optically before. In tunneling spectrosocpy of electronic transitions, optical selection rules do not hold and a new theory of the phenomenon is needed. In addition to β-carotene, results are shown for Cu pc, xenocyanine, and tetracyanine."[340]

* From Graham, W. R., Hutchinson, F., and Reed, D. A., *J. Appl. Phys.*, 44(11), 5155, 1973. With permission.

** From Swanson, L. W., and Crouser, L. C., *Surface Sci.*, 23(1), 1, 1970. With permission.

*** From Leger, A., Klein, J., Belin, M., and Defourneau, D., *Solid State Commun.*, 11(10), 1331, 1972. With permission.

REFERENCES

1. Hamann, C., *Wiss. Z. Techn. Hochschule, Karl-Marx-Stadt,* 15(1), 107, 1973.
2. Bonderman, D. P., Cater, E. D., and Bennett, W. E., *J. Chem. Eng. Data,* 15(3), 396, 1970.
3. Markova, I. Ya., Lopratkina, I. L., Shaulov, Yu. Kh., and Priselkov, Yu. A., *Tr. Khim. Khim. Tekhnol.,* 2, 61, 1972.
4. Shaulov, Yu. Kh., Preselkov, Yu. A., Lopatkina, I. L., and Markova, I. Ya., *Zh. Fiz. Khim.,* 46(4), 857, 1972.
5. Shaulov, Yu. Kh., Lopatkina, I. L., Kiryukhin, I. A., and Krausulin, G. A., *Zh. Fiz. Khim.,* 49(1), 252, 1975.
6. MacKay, A. G., *Aust. J. Chem.,* 26(11), 2425, 1973.
7. Kuznetov, A. V., Vidal-Madjar, C., and Guiochon, G., *Bull. Soc. Chim. Fr.,* 5, 1440, 1969.
8. Vidal-Madjar, C., Ganansia, J., and Guiochon, G., *Gas Chromatography, Proc. Int. Symp. (Europe), 1970,* Vol. 8, 1971, 20.
9. Vidal-Madjar, C. and Guiochon, G., *J. Chromatogr. Sci.,* 9(11), 664, 1971.
10. Vetrova, Z. P., Karabanov, N. T., and Yashin, Ya. I., *Kolloidn. Zh.,* 37(5), 946, 1975.
11. Andreev, A. A., Prahov, L. T., and Shopov, D., *Dokl. Bolg. Akad. Nauk.,* 26(12), 1637, 1973.
12. Prahov, L. and Andreev, A., *React. Kinet. Catal. Lett.,* 3(3), 315, 1975.
13. Velichko, G. V. and Spasokukotskii, N. S., *Zh. Nauch. Prikl. Fotogr. Kinematogr.,* 17(2), 88, 1972.
14. Davydov, V. Ya., Kiselev, A. V., and Silina, T. V., *Kolloidn, Zh.,* 34(1), 30, 1972.
15. Davydov, V. Ya., Kiselev, A. V., and Silina, T. V., *Kolloidn. Zh.,* 36(2), 359, 1974.
16. Davydov, V. Ya., Kiselev, A. V., and Silina, T. V., *Kolloidn. Zh.,* 36(4), 762, 1974.
17. Davydov, V. Ya., Kiselev, A. V., and Silina, T. V., *Kolloidn. Zh.,* 36(5), 945, 1974.
18. Aristov, B. G., Vasina, A. F., and Feizulova, R. K., *Kolloidn. Zh.,* 37(1), 115, 1975.
19. Sato, T., Tanaka, T., and Yoshida, T., *J. Polym Sci. Part B.,* 5(10), 947, 1967.
20. Lörinc, A. and Steryopoulos, K., *Proc. Int. Conf. Colloid Surf. Sci.,* 1, 169, 1975; Acad. Kiadg Budapest.
21. Rackow, B., *Z. Chem.,* 7(10), 398, 1967.
22. Rackow, B., *Ber. Bunsenges. Phys. Chem.,* 72(1), 110, 1968.
23. Emoto, T., Saeki, K., and Kiso, Y., Japanese Kokai 76 25, 488, March 2, 1976.
24. Lucia, E. A., Marino, C. P., and Verderame, F. D., *J. Mol. Spectrosc.,* 26(1), 133, 1968.
25. Terada, Y. and Nemoto, Y., *Nagoya-Shi Kogyo Kenkyusho Kenkyu Hokoku,* 54, 17, 1975.
26. McKay, R. B., *J. Appl. Chem. Biotechnol.,* 26(2), 55, 1976.
27. Honigmann, B., *Farbe Lack,* 82(9), 815, 1976.
28. Herrmann, M. and Honigmann, B., *Farbe Lack,* 75(4), 337, 1969.
29. Bille, H. E., Krueger, R. P., Gulbins, K. E., and Faulhaber, G., *Textilia,* 47(12), 29, 1971.
30. Toyoshima, Y., Ono, M., and Yauchi, O., *Shikizai Kyokaishi,* 43(7), 325, 1970.
31. Carr, W., *J. Oil Colour Chem. Assoc.,* 55(9), 794, 1972.
32. Mai, L. S., Baranov, B. A., Gulyaev, N. N., and Gorenko, V. N., *Lakokrasoch. Mater. Ikh Primen.,* 6, 57, 1972.
33. Hauser, P. and Honigmann, B., *DEFAZET — Deutsch. Farben Z.,* 29(6), 238, 1975.
34. Hauser, P., Horn, D., and Sappok, R., *Fatipec Congr.,* 12, 191, 1974.
35. Duff, D. G. and Giles, C. H., *J. Soc. Dyers Colour.,* 88(5), 181, 1972.
36. Commerford, T. R., *Textile Chem. Colour.,* 6(1), 39, 1974.
37. Markosyan, E. A., Samuelyan, A. A., and Sharoyan, E. G., *Zh. Fiz. Khim.,* 47(1), 184, 1973.
38. Kambeitz, W., Riedel, G., and Senninger, R., German Offenlegungsschrift 1,901,255, August 6, 1970.
39. Al'yanov, M. I., Nedel'ko, B. E., Maizlish, V. E., and Zakharova, O. V., *Tr. Ivanov. Khim. Tekhnol. In-ta,* 17, 126, 1974.
40. Nedel'ko, B. E., Krestov, G.A., Al'yanov, M. I., Borodkin, V. F., Golyand, S. M., and Maizlish, V. E., *Tr. Ivanov. Khim. Tekhnol. Inst.,* 15, 75, 1974.
41. Hamann, C. and Schenk, M., *Krist. Tech.,* 6(6), K103, 1971.
42. Nagano, M., Okuda, M., and Ohashi, W., *Nagoya Kogyo Diagaku Gakuho,* 24, 417, 1972.
43. Murakami, K., Tamura, S., and Yamada, K., *Tohoku Daigaku Hisuiyoeki Kagaku Kenkyusho Hokoku,* 22(1), 25, 1972.
44. Smith, M. J., *J. Oil Colour Chem. Assoc.,* 56(2), 126, 1973.
45. Smith, M. J., *J. Oil Colour Chem. Assoc.,* 56(4), 155, 1973.
46. Riggleman, B. M. and Drickamer, H. G., *J. Chem. Phys.,* 35, 1343, 1961.
47. Larionov, V. R., Al'yanov, M. I., Khlyupin, Yu. M., Borodkin, V. F., Yashin, Ya. I., and Smirnov, R. P., *Tr. Ivanov. Khim. Tekhnol. Inst.,* 15, 71, 1973.

48. Al'yanov, M. I., Borodkin, V. F., Larionov, V. R., Yashin, Ya. I., and Smirnov, R. P., *Izv. Vyssh. Ucheb. Zaved. Khim. Khim. Tekhnol.*, 17(8), 1170, 1974.
49. Vary, E. M., Investigations of Pcs Using Gas Chromatography, thesis, avail. Univ. Microfilms, Order No. 66-11, 965, Ann Arbor, Mich., 1966; *Diss. Abstr. B*, 27(6), 1763, 1966.
50. Schlegelmilch, F. and Kuss, W., *DEFAZET-Deut. Farben*, 27(10), 484, 1973.
51. Rafikov, S. R., Rode, V. V., Zhuravleva, I. V., Bondarenko, E. M., and Gribkova, P. N., *Vysokomol. Soedin. Ser. A*, 11(9), 2043, 1969.
52. Flom, D. G., Haltner, A. J., and Gaulin, C. A., *ASLE (Am. Soc. Lubrication Eng.) Trans.*, 8(2), 133, 1965.
53. Haltner, A. J., *ASLE (Am. Soc. Lubrication Eng.) Trans.*, 9(2), 136, 1966.
54. Kwei, K.-P.S. and Kwei, T. K., *J. Appl. Polym. Sci.*, 12(7), 1551, 1968.
55. Alfonso, G. C., Olivero, L., Turturro, A., and Pedemonte, E., *Br. Polym. J.*, 5(2), 141, 1973.
56. Turturro, A., Olivero, L., Pedemonte, E., and Alfonso, G. C., *Br. Polym. J.*, 5(2), 129, 1973.
57. Stafford, J. W., *J. Polym. Sci. Polym. Symp.*, 42(2), 837, 1973.
58. Frunze, T. M., Shleifman, R. B., Syagaeva, S. I., and Sorokina, A. G., *Plast. Massy*, 3, 13, 1975.
59. Cooper, W. D. and Wright, P., *J. Chem. Soc. Faraday Trans. 1*, 70(5), 858, 1974.
60. Cooper, W. D. and Wright, P., *J. Colloid Interface Sci.*, 54(1), 28, 1976.
61. Cooper, W. D. and Wright, P., *Am. Chem. Soc. Div. Org. Coat. Plast. Chem. Pap.*, 34(2), 566, 1974.
62. Nukina, K., Toyoshima, Y., and Ueno, S., *Kolloid, Z. Z. Polym.*, 250(2), 116, 1972.
63. Oniciu, L., Baldea, I., Zoldi, V., and Judeu, E., *Ind. Usoara*, 23(2), 61, 1976.
64. Pope, M., *J. Chem. Phys.*, 36, 2810, 1962.
65. Kronick, P. L., *Conf. Surface Effects Detection*, Washington, D.C., 1965, 123.
66. Ashida, M., *Bull. Chem. Soc. Jpn.*, 39(12), 2625, 1966.
67. Ashida, M., *Bull. Chem. Soc. Jpn.*, 39(12), 2632, 1966.
68. Meguro, K., Koishi, M., Aizawa, M., Uchino, N., Matsumoto, K., Kimoto, H., Kumaki, Y., and Sawai, H., *Kogyo Kagaku Zasshi*, 69(6), 1724, 1966.
69. Garbielli, G. and Puggelli, M., *J. Colloid Interface Sci.*, 33(3), 484, 1970.
70. Schmitz, O. J., Sell, P. J., and Hamann, K., *Farbe Lack*, 79(11), 1049, 1973.
71. Wu, S. and Brzozowski, K. J., *J. Colloid Interface Sci.*, 37(4), 686, 1971.
72. Neumann, A. W., Renzow, D., Reumuth, H., and Richter, I. E., *Fortschr. Kolloide Polym.*, 55, 49, 1971.
73. Kawashima, N. and Meguro, K., *Shikizai Kyokaishi*, 46(12), 670, 1973.
74. Nowakowski, J., *Rocz. Chem.*, 39(12), 1877, 1965.
75. Mathur, S. C., *J. Chem. Phys.*, 45(9), 3470, 1966.
76. Klasinc, L. and Nowakowski, J., *J. Chem. Phys.*, 49(7), 3326, 1968.
77. Nowakowski, J., *Rocz. Chem.*, 40(1), 141, 1966.
78. Mathur, S. C. and McKannan, E. C., *Int. J. Quantum Chem. Symp.*, 1, 247, 1967.
79. Mathur, S. C., Singh, J., and Singh, D. C., *J. Phys. C*, 4(18), 3122, 1971.
80. Mathur, S. C. and Singh, D. C., *Indian J. Pure Appl. Phys.*, 9(12), 1031, 1971.
81. Schaffer, A. M. and Gouterman, M., *Theor. Chim. Acta*, 25, 62, 1972.
82. Schaffer, A. M., Gouterman, M., and Davidson, E. R., *Theor. Chim. Acta*, 30(1), 9, 1973.
83. Schaffer, A. M., Extended Hueckel Calculations on Porphyrins, Pcs, and Related Rings, thesis, Avail. Univ. Microfilms, Ann Arbor, Mich., Order No. 73-3782; *Diss. Abstr. Int. B*, 33(8), 3585, 1973.
84. Chen, I., *J. Mol. Spectrosc.*, 23(2), 131, 1967.
85. Ponomarev, O. A. and Kubarev, S. I., *Teor. Eksp. Khim.*, 4(4), 492, 1968.
86. Sukigara, M. and Nelson, R. C., *Mol. Phys.*, 17(4), 387, 1969.
87. Chen, I., *J. Chem. Phys.*, 51(8), 3241, 1969.
88. Mathur, S. C. and Ramesh, N., *Proc. Nucl. Phys. Solid State Phys. Symp.*, 16C, 64, 1973.
89. Mathur, S. C. and Singh, J., *Chem. Phys.*, 1, 476, 1973.
90. Taube, R., *Pure Appl. Chem.*, 38(3), 427, 1974.
91. Mathur, S. C. and Singh, J., *Int. J. Quantum Chem.*, 6(1), 57, 1972.
92. Mathur, S. C. and Singh, J., *Int. J. Quantum Chem.*, 8(1), 79, 1974.
93. Gal'pern, E. G., Luk'yanets, E. A., and Gal'pern, M. G., *Izv. Akad. Nauk, S.S.S.R. Ser. Khim.*, 9, 1976, 1973.
94. Taube, R., *Z. Chem.*, 6(1), 8, 1966.
95. Henriksson, A. and Sundbom, M., *Theor. Chim. Acta*, 27(3), 213, 1972.
96. Henriksson, A., Roos, B., and Sundbom, M., *Theor. Chim. Acta*, 27(4), 303, 1972.
97. Malter, H., *Phys. Status Solidi B*, 74(2), 627, 1976.
98. Fuhrhop, J. H. and Subramanian, J., *Phil. Trans. R. Soc. London, Ser. B*, 273(924), 335, 1976.
99. Pack, B. K. and Hong, Y. S., *Daehan Hwahak Hwoejee*, 12(3), 89, 1968.

100. Anikin, N. A. and Rudenko, A. P., *Vestn. Mosk. Univ. Khim.*, 17(1), 122, 1976.
101. Foersterling, H. D. and Kuhn, H., *Int. J. Quantum Chem.*, 2(3), 413, 1968.
102. Gouterman, M., Wagnière, G., and Snyder, L. C., *J. Mol. Spectrosc.*, 11(2), 108, 1963.
103. Weiss, C., Jr., Kobayashi, H., and Gouterman, M., *J. Mol. Spectrosc.*, 16(2), 415, 1965.
104. McHugh, A. J., Gouterman, M., and Weiss, C., Jr., *Theor. Chim. Acta*, 24(4), 346, 1972.
105. Schechtman, B. H. and Spicer, W. E., *Chem. Phys. Lett.*, 2(4), 207, 1968.
106. Fulton, A., Lyons, L. E., and Morris, G. C., *Aust. J. Chem.*, 21(12), 2853, 1968.
107. Luzanov, A. V. and Umanskii, V. E., *Opt. Spektrosk.*, 40(1), 201, 1976.
108. Kubarev, S. I. and Ponomarev, O. A., *Teor. Eksp. Khim.*, 4(4), 498, 1968.
109. Popov, N. A. and Shustorovich, E. M., *Zh. Strukt. Khim.*, 7(2), 258, 1966.
110. Berezin, B. D., *Zh. Fiz. Khim.*, 39(2), 321, 1965.
111. Larin, G. M., *Zh. Neorg. Khim.*, 17(10), 2662, 1972.
112. Yen, T. F., *Naturwissenschaften*, 58(5), 267, 1971.
113. Heilmeier, G. H. and Warfield, G., *J. Chem. Phys.*, 38, 893, 1963.
114. Onishi, T., Uyematsu, T., Watanabe, H., and Tamaru, K., *Spectrochim. Acta Part A*, 23(3), 730, 1967.
115. Wagner, H. and Hamann, C., *Spectrochim. Acta Part A*, 25(2), 335, 1969.
116. Sammes, M. P., *J. Chem. Soc. Perkin Trans. 2*, 2, 160, 1972.
117. Villar, J. G. and Lindqvist, L., *C. R. Acad. Sci. Paris, Ser, A, B*, 264B(26), 1807, 1967.
118. Devaux, P., *Mol. Phys.*, 23(2), 265, 1972.
119. Shurvell, H. F. and Pinzuti, L., *Can. J. Chem.*, 44(2), 125, 1966.
120. Aleksandrov, A. N., Sidorov, A. N., and Yaroslavskii, N. G., *Opt. Spektrosk.*, 22(4), 560, 1967.
121. Berezin, B. D., *Izv. Vyssh. Ucheb. Zaved. Khim. Khim. Tekhnol.*, 11(5), 537, 1968.
122. Klyuev, V. N., Al'yanov, M. I., and Shiryaeva, L. S., *Izv. Vyssh. Ucheb. Zaved. Khim. Khim. Tekhnol.*, 12(12), 1738, 1969.
123. Klyuev, V. N., Al'yanov, M. I., and Shiryaeva, L. S., *Izv. Vyssh. Ucheb. Zaved. Khim. Khim. Tekhnol.*, 12(8), 1106, 1969.
124. Kobayashi, T., Kurokawa F., Uyeda, N., and Suito, E., *Spectrochim. Acta, Part A*, 26(6), 1305, 1970.
125. Kobayashi, T., *Spectrochim. Acta Part A*, 26(6), 1313, 1970.
126. Klyuev, V. N., Berezin, B. D., ahd Shiryaeva, L. S., *Izv. Vyssh. Ucheb. Zaved. Khim. Khim. Tekhnol.*, 13(7), 997, 1970. (Russian).
127. Corwin, A. H., Whitten, D. G., Baker, E. W., and Kleinspehn, G. G., *J. Am. Chem. Soc.*, 85(22), 3621, 1963.
128. Edwards, L. and Gouterman, M., *J. Mol. Spectrosc.*, 33(7) 292, 1970.
129. Schechtman, B. H. and Spicer, W. E., *J. Mol. Spectrosc.*, 33(1), 28, 1970.
130. Bajema, L. L., Spectroscopic Studies of Porphyrins, Pcs, and Carbon Disulfide in Inert Matrices, thesis, avail. Univ. Microfilms, Order No. 71-934, Ann Arbor, Mich., *Diss. Abstr. Int. B*, 31(7), 3945, 1971.
131. Mathur, S. C., Singh, J., and Kumar, B., *Phys. Status Solidi B*, 48(2), 843, 1971.
132. Fielding, P. E. and MacKay, A. G., *Aust. J. Chem.*, 28(7), 1445, 1975.
133. Shorin, V. A., Maizlish, V. E., Borodkin, V. F., and Al'yanov, M. I., *Dokl. Nauch. Tekhn. Konf. Ivanov Khim. Tekhnol. In-ta*, p. 83, 1973.
134. Sidorov, A. N. and Terenin, A. N., *Opt. Spektrosk.*, 11, 325, 1961.
135. Sidorov, A. N., *Opt. Spektrosk.*, 13, 668, 1962.
136. Sidorov, A. N., *Opt. Spektrosk.*, 40(3), 492, 1976.
137. Chadderton, L. T., *J. Phys. Chem. Solids*, 24(6), 751, 1963.
138. Leles, B. K., Effect of Mechanical Anharmonicity on the Temperature Dependence of Integrated Apparent Intensities in the Infrared Region of Metallic Pcs and Inorganic Ions Isolated in Alkali Halide Matrices, thesis, avail. Univ. Microfilms, Order No. 70-11,350, Ann Arbor, Mich., *Diss. Abstr. Int. B*, 31(1), 150, 1970.
139. Ercolani, C. Neri, C., and Sartori, G., *J. Chem. Soc. A*, 9, 2123, 1968.
140. Zabashta, V. N., Ershov, A. P., and Kharkharov, A. A. *Izv. Vyssh. Ucheb. Zaved. Tekhnol. Tekstil'n. Prom.*, 6, 98, 1964.
141. Gurinovich, I. F., *Zh. Prikl. Spektrosk.*, 6(5), 657, 1967.
142. Usov, N. N. and Benderskii, V. A., *Fiz. Tekh. Poluprov.*, 4(2), 405, 1970.
143. Lebedev, O. L., Luk'yanets, E. A., and Puchnova, V. A., *Opt. Spektrosk.*, 30(4), 640, 1971.
144. Sidorov, A. N., *Zh. Strukt. Khim.*, 14(2), 255, 1973.
145. Sidorov, A. N. and Maslov, V. G., *Teor. Eksp. Khim.*, 8(6), 828, 1972.
146. Clack, D. W. and Yandle, J. R., *Inorg. Chem.*, 11(8), 1738, 1972.
147. Khartsiev, V. E., *Physics*, 3(3), 129, 1967.
148. Starke, M. and Wagner, H., *Z. Chem.*, 9(5), 193, 1969.

149. Al'yanov, M. I., Borodkin, V. F., and Benderskii, V. A., *Izv. Vyssh. Ucheb. Zaved. Khim. Khim. Tekhnol.*, 13(6), 857, 1970.
150. Gaspard, S., Verdaguer, M., and Viovy, R., *C. R. Acad. Sci. Ser. C*, 277(18), 821, 1973.
151. Tikhonov, V. P., Todosienko, S. S., and Fuks, G. I., *Kolloid Zh.*, 36(1), 76, 1974.
152. Bubnov, L. Ya. and Frankevich, E. L., *Fiz. Tverd. Tela (Leningrad)*, 16(5), 1533, 1974.
153. Kuhn, H., *Chimia (Aarau)*, 15, 53, 1961.
154. Foersterling, H. D. and Kuhn, H., *Chimia (Aarau)*, 19(5), 322, 1965.
155. Ahrens, U., The Light Absorption of Sodium 4,4',4",4''' Pc Tetrasulfonic Acid in Solution, thesis, Philipps University at Marburg (German), 1962.
156. Lucia, E. A., Verderame, F. D., and Taddei, G., *J. Chem. Phys.*, 52(5), 2307, 1970.
157. Assour, J. M. and Harrison, S. E., *J. Am. Chem. Soc.*, 87(3), 651, 1965.
158. Shumov, Yu. S., *Zh. Fiz. Khim.*, 47(3), 718, 1973.
159. Yurlova, G. A., *Zh. Obshch. Khim.*, 41(6), 1325, 1971.
160. Hanke, W., Gutschick, D., and Malewski, G., *Monatsber. Dtsch. Akad. Wiss. Berlin*, 11(8—9), 656, 1969.
161. Pribytkova, N. N. and Savost'yanova, M. V., *Spektrosk. Tr. Sib. Soveshch, 6th*, 1973, 250.
162. VEB Filmfabrik Wolfen, French Patent 1,558,999, March 7, 1969.
163. Schindler, W., Ruschitzky, E., Pietrzok, H., German Offenlegungsschrift 1,917,589, April 23, 1970.
164. Commerford, T. R., *Text. Chem. Color.*, 6(1), 39, 1974.
165. Cho, N.-S., Kim, K.-H., and Hahn, C.-S., *Daehan Hwahak Hwoejee*, 16(6), 378, 1971.
166. Lebedev, O. L. and Kolesnikova, V. K., *Opt. Spektrosk.*, 22(6), 998, 1967.
167. Moskalev, P. N. and Kirin, I. S., *Opt. Spektrosk.*, 29(6), 1149, 1970.
168. Tserkovnitskaya, I. A. and Perevoshchikova, V. V., *Zh. Anal. Khim.*, 26(5), 913, 1971.
169. Selbin, J., Holmes, L. H., Jr., and McGlynn, S. P., *J. Inorg. Nucl. Chem.*, 25(11), 1359, 1963.
170. Galanin, M. D. and Chizhikova, Z. A., *Opt. Spektrosk.*, 34(1), 197, 1973.
171. Savel'ev, D. A. and Sidorov, A. N., *Zh. Fiz. Khim.*, 43(5), 1080, 1969.
172. Sartori, G. and Ercolani, C., *Ric. Sci. R. Sez. A*, 3(3), 323, 1963.
173. Dale, B. W., *Trans. Faraday Soc.*, 65(2), 331, 1969.
174. Stillman, M. J. and Thomson, A. J., *J. Chem. Soc. Faraday Trans. 2*, 70(5), 790, 1974.
175. Bundina, N. I., Kaliya, O. L., Lebedev, O. L., Luk'yanets, E. A., Rodionova, G. N., and Ivanova, T. M., *Koord. Khim.*, 2(7), 940, 1976.
176. Butenin, A. V., Kogan, B. Ya., Luk'yanets, E. A., and Molchanova, L. I., *Opt. Spektrosk.*, 37(4), 696, 1974.
177. Keen, I. M. and Malerbi, B. W., *J. Inorg. Nucl. Chem.*, 27(6), 1311, 1965.
178. Kobayashi, T., Uyeda, N., and Suito, E., *Bull. Inst. Chem. Res. Kyoto Univ.*, 52(4), 605, 1974.
179. Kobayashi, T., Uyeda, N., and Suito, E., *Bull. Inst. Chem. Res. Kyoto Univ.*, 47(4), 401, 1969.
180. Zalesskii, I. E., Kotlo, V. N., Solov'ev, K. N., and Shkirman, S. F., *Zh. Prikl. Spektrosk.*, 20(6), 1010, 1974.
181. Lebedev, O. L. and Nasonov, V. S., *Opt. Spektrosk.*, 23(2), 318, 1967.
182. Markova, I. Ya., Popov, Yu. A., and Shaoulov, Yu. Kh., *Zh. Fiz. Khim.*, 44(10), 2636, 1970.
183. Akopov, A. S., Berezin, B.D., and Klyuev, V. N., *Izv. Vyssh. Ucheb. Zaved. Khim. Khim. Tekhnol.*, 15(8), 1190, 1972.
184. Mathur, S. C., Singh, J., and Krupnick, A. C., *Indian J. Phys. 1971—1972*, 44(12), 657, 1970.
185. Esposito, J. N., Sutton, L. E., and Kenney, M. E., *Inorg. Chem.*, 6, 1116, 1967.
186. Berezin, B. D. and Akopov, A. S., *Zh. Obshch. Khim.*, 44(5), 1089, 1974.
187. Hush, N. S. and Woolsey, I. S., *Mol. Phys.*, 21(3), 465, 1971.
188. Ukei, K., *Bussei*, 15(13), 134, 1974.
189. Kroenke, W. J. and Kenney, M. E., *Inorg. Chem.*, 3(5), 696, 1964.
190. Sutton, L. E. and Kenney, M. E., *Inorg. Chem.*, 6, 1869, 1967.
191. Johnson, F. M., *Mém. Soc. R. Sci. Liège 6ᵉ Série*, 3, 391, 1972.
192. Blinov, L. M. and Kirichenko, N. A., *Fiz. Tverd. Tela*, 12(5), 1574, 1970.
193. Blinov, L. M. and Kirichenko, N. A., *Fiz. Tverd. Tela (Leningrad)*, 14(8), 2490, 1972.
194. Blinov, L. M., Kirichenko, N. A., and Duninin, N. V., *Zh. Prikl. Spektrosk.*, 25(3), 548, 1976.
195. Masahiko, M. and Yatabe, K., *J. Phys. Soc. Jpn.*, 33(4), 1176, 1972.
196. Gurevich, M. G. and Solov'ev, K. N., *Dokl. Akad. Nauk Belorussk. S.S.R.*, 5, 291, 1961.
197. Solov'ev, K. N., *Vestsi Akad. Navuk Belarusk. S.S.R. Ser. Fiz. Tekhn. Navuk*, 3, 27, 1962.
198. Solov'ev, K. N., Tsvirko, M. P., and Kachura, T. F., *Opt. Spektrosk.*, 40(4), 684, 1976.
199. Kobyshev, G. I., Lyalin, G. N., and Terenin, A. N., *Dokl. Akad. Nauk S.S.S.R.*, 148, 1294, 1963.
200. Lyalin, G. N. and Kobyshev, G. I., *Dokl. Akad. Nauk S.S.S.R.*, 148, 1053, 1963.
201. Lyalin, G. N. and Kobyshev, G. I., *Opt. Spektrosk.*, 15(2), 253, 1963.
202. Personov, R. I., *Opt. Spektrosk.*, 15(1), 61, 1963.

203. Gradyushko, A. T., Sevchenko, A. N., Solov'ev, K. N., and Tsvirko, M. P., *Photochem. Photobiol.*, 11(6), 387, 1970.
204. Zalesskii, I. E., Kotlo, V. N., Sevchenko, A. N., Solov'ev, K. N., and Shkirman, S. F., *Dokl. Akad. Nauk. S.S.S.R.*, 207(6), 1314, 1972.
205. Assour, J. M. and Harrison, S. E., *J. Am. Chem. Soc.*, 87(3), 651, 1965.
206. Shkirman, S. F. and Solov'ev, K. N., *Izv. Akad. Nauk S.S.S.R. Ser. Fiz.*, 29(8), 1378, 1965.
207. Gurinovich, G. P., Pateeva, M. V., and Shul'ga, A. M., *Izv. Akad. Nauk S.S.S.R. Ser. Fiz.*, 27(6), 777, 1963.
208. Kotlo, V. N., Solov'ev, K. N., and Shkirman, S. F., *Izv. Akad. Nauk. S.S.S.R. Ser. Fiz.*, 39(9), 1972, 1975.
209. Solov'ev, K. N., Gradyushko, A. T., and Tsvirko, M. P., *Izv. Akad. Nauk S.S.S.R. Ser. Fiz.*, 36(5), 1107, 1972.
210. Iogansen, L. V., *Dokl. Akad. Nauk S.S.S.R.*, 205(2), 390, 1972.
211. Strelkova, T. I., Gurinovich, G. P., and Sinyakov, G. N., *Zh. Prikl. Spektrosk. Akad. Nauk Belorussk. S.S.R.*, 4(5), 429, 1966.
212. Sevchenko, A. N., *Izv. Akad. Nauk S.S.S.R. Ser. Fiz.*, 26, 53, 1962.
213. Sevchenko, A. N., Solov'ev, K. N., Gradyushko, A. T., and Shkirman, S F., *Dokl. Akad. Nauk S.S.S.R.*, 169(1), 77, 1966.
214. Kosonocky, W. F. and Harrison, S. E., *J. Appl. Phys.*, 37(13), 4789, 1966.
215. Solov'ev, K. N., Mashenkov, V. A., and Kachura, T. F., *Zh. Prikl. Spektrosk.*, 7(5), 773, 1967.
216. Shmuttsler, K., Lyalin, G. N., and Terenin, A. N., *Dokl. Akad. Nauk S.S.S.R.*, 174(1), 147, 1967.
217. Pine, A. S., *J. Appl. Phys.*, 39(1), 106, 1968.
218. Rieckhoff, I., Klaus, E., and Voigt, E. M., *in Mol. Lumin. Int. Conf.*, Lim, E. C., Ed., W. A. Benjamin, New York, 1969, 295.
219. Ake, R. L. and Gouterman, M., *Theor. Chim. Acta*, 15(1), 20, 1969.
220. Personov, R. I. and Korotaev, O. N., *Dokl. Akad. Nauk S.S.S.R.*, 182(4), 815, 1968.
221. Personov, R. I. and Korotaev, O. N., *Vop. Radiofiz. Spektrosk.*, 3, 71, 1967.
222. Personov, R. I., Godyaev, E. D., and Korotaev, O. N., *Fiz. Tverd. Tela*, 13(1), 111, 1971.
223. Bajema, L., Gouterman, M., and Meyer, B., *J. Mol. Spectrosc.*, 27(1—4), 225, 1968.
224. Gurinovich, G. P. and Strelkova, T. I., *Fiz. Probl. Spektroskopii, Akad. Nauk S.S.S.R., Mater. 13-go [Trinadtsatego] Soveshch., Leningrad, 1960*, 1, 305, 1962.
225. Solov'ev, K. N., Mashenkov, V. A., and Kachura, T. F., *Opt. Spektrosk.*, 27(1), 50, 1969.
226. Personov, R. I. and Al'shits, E. I., *Chem. Phys. Lett.*, 33(1), 85, 1975.
227. Yoshino, K., Keneto, K., Kyokane, J., and Inuishi, Y., *J. Phys. Soc. Jpn.*, 31(5), 1594, 1971.
228. Vincett, P. S., Voigt, E. M., and Rieckhoff, K. E., *J. Chem. Phys.*, 55(8), 4131, 1971.
229. Morita, M., *J. Phys. Soc. Jpn.*, 33(3), 863, 1972.
230. Stadelmann, H. R., *J. Lumin.*, 5(3), 171, 1972.
231. Menzel, E. R., Rieckhoff, K. E., and Voigt, E. M., *Chem. Phys. Lett.*, 13(6), 604, 1972.
232. Menzel, E. R., Rieckhoff, K. E., and Voigt, E. M., *J. Chem. Phys.*, 58(12), 5726, 1973.
233. Tsvirko, M. P., Sokol'ev, K. N., and Sapunov, V. V., *Opt. Spcktrosk.*, 40(5), 843, 1976.
234. Zalesski, I. E., Kotlo, V. N., Sevchenko, A. N., Solov'ev, K. N., and Shkirman, S. F., *Dokl. Akad. Nauk S.S.S.R.*, 210(2), 312, 1973.
235. Ermolaev, V. L., Krasheninnikov, A. A., and Shablya, A. V., *Opt. Spektrosk.*, 34(5), 1011, 1973.
236. Yoshino, K. Hikida, M., Tatsuno, K., Kaneto, K., and Inuishi, Y., *J. Phys. Soc. Jpn.*, 34(2), 441, 1973.
237. Kaneto, K., Yoshino, K., Hikida, M., and Inuishi, Y., *Technol. Rep. Osaka Univ.*, 23 (1121—1154), 493, 1973.
238. Kaneto, K., Yoshino, K., and Inuishi, Y., *J. Phys. Soc. Jpn.*, 35(2), 621, 1973.
239. Yoshino, K., Hikida, M., Kaneto, K., and Inuishi, Y., *Technol. Rep. Osaka Univ.*, 23 (1090—1120), 171, 1973.
240. Tsvirko, M. P., Sapunov, V. V., and Solov'ev, K. N., *Opt. Spektrosk.*, 34(6), 1094, 1973.
241. Kachura, T. F., Sevchenko, A. N., Solov'ev, K. N., and Tsvirko, M. P., *Dokl. Akad. Nauk S.S.S.R.*, 217(5), 1121, 1974.
242. Tsvirko, M. P., Solov'ev, K. N., and Sapunov, V. V., *Opt. Spektrosk.*, 36(2), 335, 1974.
243. Korotaev, O. N. and Personov, R. I., *Opt. Spektrosk.*, 37(5), 886, 1974.
244. Stoilov, Yu. Yu. and Trusov, K. K., *Kvantovaya Elektron. (Moscow)*, 1(6), 1458, 1974.
245. Zalesskii, I. E., Kotlo, V. N., Sevchenko, A. N., Solov'ev, K. N., and Shkirman, S. F., *Dokl. Akad. Nauk S.S.S.R.*, 218(2), 324, 1974.
246. Gorokhovskii, A. A. and Kaarli, R., *Izv. Akad. Nauk S.S.S.R. Ser. Fiz.*, 39(11), 2326, 1975.
247. Gorokhovskii, A. A., *Opt. Spektrosk.*, 40(3), 477, 1976.
248. Gorokhovskii, A. A. and Kaarli, R., *Tezisy Dokl. Vses. Soveshch. Lyumin.*, 89, 1975.
249. Zalesski, I. E., Kotlo, V. N., Solov'ev, K. N., and Shkirman, S. F., *Opt. Spektrosk.*, 38(5), 917, 1975.

250. Kaneto, K., Yoshino, K., and Inuishi, Y., *J. Phys. Soc. Jpn.*, 37(5), 1297, 1974.
251. Kaneto, K., Ido, Y., Yoshino, K., and Inuishi, Y., *Technol. Rep. Osaka Univ.*, 25(1230—1253), 43, 1975.
252. Kaneto, K., Ido, Y., Yoshino, K., and Inuishi, Y., *J. Phys. Soc. Jpn.*, 38(4), 1042, 1975.
253. Huang, T.-H., Rieckhoff, K. E., and Voigt, E. M., *Can. J. Phys.*, 54(6), 633, 1976.
254. Personov, R. I. and Bykovskaya, L. A., *Dokl. Akad. Nauk S.S.S.R.*, 199(2), 299, 1971.
255. Kotlo, V. N., Solov'ev, K. N., Shkirman, S. F., and Zalesskii, I. E., *Vestsi Akad. Navuk Belarus. S.S.R. Ser. Fiz. Mat. Navuk*, 3, 99, 1974.
256. Solov'ev, K. N., Zalesskii, I. E., Kotlo, V. N., and Shkirman, S. F., *Pis'ma Zh. Eksp. Teor. Fiz.*, 17(9), 463, 1973.
257. Gorokhovskii, A. A., Kaarli, R., and Rebane, L., *Pis'ma Zh. Eksp. Teor. Fiz.*, 20(7), 474, 1974.
258. Gorokhovskii, A. A., Kaarli, R., and Rebane, L., *Opt. Commun.*, 16(2), 282, 1976.
259. Aleksandrov, I. V., Bobovich, Ya. S., and Maslov, V. G., *Pis'ma Zh. Eksp. Teor. Fiz.*, 19(5), 264, 1974.
260. Aleksandrov, I. V., Bobovich, Ya. S., Maslov, V. G., and Sidorov, A. N., *Opt. Spektrosk.*, 37(3), 467, 1974.
261. Ksenofontova, N. M., Aleksandrov, I. V., Bobovich, Ya. S., Solov'ev, K. N., Shkirman, S. F., and Kachura, T. F., *Zh. Prikl. Spektrosk.*, 20(5), 834, 1974.
262. Aleksandrov, I. V., Belyaevskaya, N. M., Bobovich, Ya. S., Bortkevich, A. V., and Maslov, V. G., *Zh. Eksp. Teor. Fiz.*, 68(4), 1274, 1975.
263. Bobovich, Ya. S., Aleksandrov, I. V., Maslov, V. G., and Sidorov, A. N., *Pis'ma Zh. Eksp. Teor. Fiz.*, 18(3), 175, 1973.
264. Robertson, J. Monteath, in Proc. Robert A. Welch Foundation Conf. Chem. Res. IV. Mol. Struct. Org. React., Houston, Tex., November 7 to 9, 1960.
265. Gieren, A. and Hoppe, W., *J. Chem. Soc. Sect. D. Chem. Commun.*, 8, 413, 1971.
266. Rogers, D. and Osborn, R. S., *J. Chem. Soc. Sect. D. Chem. Commun.*, 15, 840, 1971.
267. Kobayashi, T., Kurokawa, F., Ashida, T., Uyeda, N., and Suito, E., *J. Chem. Soc. Sect. D. Chem. Commun.*, 24, 1631, 1971.
268. Kirin, I. S., Kolyadin, A. B., and Lychev, A. A., *Zh. Strukt. Khim.*, 15(3), 486, 1974.
269. Sokol, V. I., Porai-Koshits, M. A., Butman, L. A., Rozanov, I. A., and Berdnikov, V. R., *Izv. Akad. Nauk S.S.S.R. Ser. Khim.*, 2, 485, 1974.
270. Kirner, J. F., Dow, W., and Scheidt, W. R., *Inorg. Chem.*, 15(7), 1685, 1976.
271. Mooney, J. R., Choy, C. K., Knox, K., and Kenney, M., *J. Am. Chem. Soc.*, 97(11), 3033, 1975.
272. Kroenke, W. J., Sutton, L. E., Joyner, R. D., and Kenney, M. E., *Inorg. Chem.*, 2(5), 1064, 1963.
273. Ukei, K., *Phys. Lett. A.*, 55A (2), 111, 1975.
274. Mitchell, G. R., *J. Chem. Phys.*, 37, 216, 1962.
275. Okamoto, N., Kajikawa, M., and Hasegawa, K., *Nippon Kagaku Zasshi*, 87(4), 363, 1966.
276. Levina, S. D., Astakhov, I. I., Lobanova, K. P., and Surikov, V. V., *Elektrokhimiya*, 4(11), 1380, 1968.
277. Murata, H., Yamada, T., and Matsukawa, H., *Shikizai Kyokaishi*, 43(8), 384, 1970.
278. Miyake, S., Fujiwara, K., Tokonami, M., and Fujimoto, F., *Jpn. J. Appl. Phys.*, 3(5), 276, 1964.
279. Stabenow, J., *Ber. Bunsenges. Phys. Chem.*, 72(3), 374, 1968.
280. Vertsner, V. N., Vorona, Yu. M., and Zhadanov, G. S., *Stekloobraznoe Sostoyanie, Inst. Khim. Silikatov, Akad. Nauk S.S.S.R. Gas. Optich. Inst. Zavod Khudozhestvennogo Stekla, Leningrad Elektrotrkhn. Inst., Fr. Simpoziuma, Leningrad*, 1, 81, 1963.
281. Bowden, F. P. and Chadderton, L. T., *Nature (London)*, 192, 31, 1961.
282. Bowden, F. P. and Chadderton, L. T., *Proc. R. Soc. London Ser. A*, 269, 143, 1962.
283. Labaw, L. W. J., *Ultrastruct. Res.*, 5, 409, 1961.
284. Wakabayashi, T., *Shikizai Kyokaishi*, 46(2), 129, 1973.
285. Uyeda, N., Kobayashi, T., Suito, E., Harada, Y., and Watanabe, M., *J. Appl. Phys.*, 43(12), 5181, 1972.
286. Uyeda, N. and Ishizuka, K., *J. Electron. Microsc.*, 23(2), 79, 1974.
287. Ishizuka, K. and Uyeda, N., *Bull. Inst. Chem. Res. Kyoto Univ.*, 53(2), 200, 1975.
288. Birrell, G. B., Burke, C., Dehlinger, P., and Griffith, O. H., *Biophys. J.*, 13 (5), 462, 1973.
289. Burke, C. A., Birrell, G. B., Lesch, G. H., and Griffith, O. H., *Photochem. Photobiol.*, 19 (1), 29, 1974.
290. Dam, R. J., Griffith, O. H., and Rempfer, G. F., *J. Appl. Phys.*, 47(3), 861, 1976.
291. Riggs, W. M. and Beimer, R. G., *Chemtech.*, November, p. 652, 1975.
292. Frost, D. C., McDowell, C. A., and Woolsey, I. S., *Mol. Phys.*, 27(6), 1473, 1974.
293. Betteridge, D., Carver, J. C., and Hercules, D. M., *J. Electron. Spectrosc. Relat. Phenomena*, 2(4), 327, 1973.
294. Niwa, Y., Kobayashi, H., and Tsuchiya, T., *Inorg. Chem.*, 13(12), 2891, 1974.

295. Larsson, R., Folkesson, B., and Schon, G., *Chem. Scr.*, 3(2), 88, 1973.
296. Kramer, L. N. and Klein, M. P., *J. Chem. Phys.*, 51(8), 3618, 1969.
297. Bernard, C., Legras, C., Magner, G., and Savy, M., *C. R. Acad. Sci. Ser. C*, 277(18), 829, 1973.
298. Johansson, L. Y. and Blomquist, J., *Chem. Phys. Lett.*, 34(1), 115, 1975.
299. Rupp, H. and Weser, U., *Biochim. Biophys. Acta*, 446(1), 151, 1976.
300. Borod'ko, Yu. G., Vetchinkin, S. I., Zimont, S. L., Ivleva, I. N., and Shul'ga, Yu. M., *Chem. Phys. Lett.*, 42(2), 264, 1976.
301. Dubois, R., Carver, J. C., and Tsutsui, M., U.S. NTIS AD Rep. AD-AO25863, National Technical Information Service, Springfield, Va., 1976.
302. Hoechst, H., Goldmann, A., Huefner, S., and Malter, H., *Phys. Status Solidi B*, 76(2), 559, 1976.
303. Zeller, M. V. and Hayes, R. G., *J. Am. Chem. Soc.*, 95(12), 3855, 1973.
304. Joergensen, C. K. and Berthou, H., *Kgl. Dan. Vidensk. Selsk., Mat. Fyss Medd.*, 38(15), 1972.
305. Niwa, Y., Kobayashi, H., and Tsuchiya, T., *J. Chem. Phys.*, 60(3), 799, 1974.
306. Niwa, Y., *Kagaku No Ryoiki*, 29(5), 342, 1975.
307. Epstein, L. M., *J. Chem. Phys.*, 36, 2731, 1962.
308. Hudson, A. and Whitfield, H. J., *Chem. Commun.*, 17, 606, 1966.
309. Hudson, A. and Whitfield, H. J., *Inorg. Chem.*, 6(6), 1120, 1967.
310. Dale, B. W., Williams, R. J. P., Edwards, P. R., and Johnson, C. E., *Trans. Faraday Soc.*, 64(3), 620, 1968.
311. Dale, B. W., *Mol. Phys.*, 28(2), 503, 1974.
312. Dale, B. W., Williams, R. J. P., Edwards, P. R., and Johnson, C. E., *J. Chem. Phys.*, 49(8), 3445, 1968.
313. Dale, B. W., Williams, R. J. P., Edwards, P. R., and Johnson, C. E., *Trans. Faraday Soc.*, 64(11), 3011, 1968.
314. Moss, T. H. and Robinson, A. B., *Inorg. Chem.*, 7(8), 1692, 1968.
315. Dezsi, I., Balazs, A., and Molnar, B. *J. Inorg. Nucl. Chem.*, 31(6), 1661, 1969.
316. Dudreva, B. Ts. and Pirinchieva, R. K., *Bulg. J. Phys.*, 2(2), 126, 1975.
317. Blomquist, J. and Moberg, L. C., *Phys. Scr.*, 9(6), 350, 1974.
318. Shapiro, N. I., Suzdalev, I. P., Gol'danskii, V. I., Sherle, A. I., and Berlin, A. A., *Teor. Eksp. Khim.*, 11(3), 330, 1975.
319. Srivastava, T. S., Przybylinski, J. L., and Nath, A., *Inorg. Chem.*, 13(7), 1562, 1974.
320. Grenoble, D. C. and Drickamer, H. G., *J. Chem. Phys.*, 55(4), 1624, 1971.
321. Champion, A. R. and Drickamer, H. G., *Proc. Natl. Acad. Sci. U.S.A.*, 58(3), 876, 1967.
322. Dolphin, D., Sams, J. R., Tsin, Ts. B., and Wong, K. L., *J. Am. Chem. Soc.*, 98(22), 6970, 1976.
323. Stein, G. E., Thesis, Univ. Microfilms, Order No. 73—5175, Ann Arbor, Mich., 1972; *Diss. Abstr. Int. B*, 33(8), 3587, 1973.
324. Thompson, J. L., Ching, J., and Fung, E. Y., *Radiochim. Acta*, 18(1), 57, 1972.
325. Nath, A., Harpold, M., Klein, M. P., and Kuendig, W., *Chem. Phys. Lett.*, 2(7), 471, 1968.
326. Stoeckler, H. A., Sano, H., and Herber, R. H., *J. Chem. Phys.*, 45(4), 1182, 1966.
327. Herber, R. H. and Stoeckler, H. A., *Tech. Rep. Ser. Int. At. Energ. Agency*, 50, 110, 1966.
328. Stoeckler, H. A. and Sano, H., *Nucl. Instr. Methods*, 44(1), 103, 1966.
329. O'Rourke, M. and Curran, Brother C., *J. Am. Chem. Soc.*, 92(6), 1501, 1970.
330. Friedel, M. K., Hoskins, B. F., Martin, R. L., and Mason, S. A., *J. Chem. Soc. D*, 7, 440, 1970.
331. Krivoglaz, M. A., *Zh. Eksp. Teor. Fiz.*, 46(2), 637, 1964.
332. Fluck, E., *Katal. Phthalocyaninen, Symp.* Kropf. H. and Steinbach, F., Eds., Thieme, Stuttgart, Germany, 1973, 37.
333. Johansson, L. Y., Larsson, R., Blomquist, J., Cederstrom, C., Grapengiesser, S., Helgeson, U., Moberg, L. C., and Sundbom, M., *Chem. Phys. Lett.*, 24(4), 508, 1974.
334. Larsson, R., Mrha, J., and Blomquist, J., *Acta Chem. Scand.*, 26(8), 3386, 1972.
335. Graham, W. R., Hutchinson, F., and Reed, D. A., *J. Appl. Phys.*, 44(11), 5155, 1973.
336. Swanson, L. W. and Crouser, L. C., *Surf. Sci.*, 23(1), 1, 1970.
337. Schuett, W., Koester, H., and Zuther, G., *Surf. Sci.*, 45(1), 163, 1974.
338. Kleint, Ch. and Moeckel, K., *Surf. Sci.*, 40(2), 343, 1973.
339. Leger, A., Klein, J., Belin, M., and Defourneau, D., *Solid State Commun.*, 11(10), 1331, 1972.
340. De Cheveigne, S., Leger, A., and Klein, J., *Proc. 14th Int. Conf. Low Temp. Phys.*, Vol. 3, North-Holland, Amsterdam, 1975, 491.
341. Kawai, T., Soma, M., Matsumoto, Y., Onishi, T., and Tamaru, K., *Chem. Phys. Lett.*, 37(2), 378, 1976.
342. Betteridge, D. and Jones, D., *Proc. Anal. Div. Chem. Soc.*, 12(1), 31, 1975.
343. York, , *Ap. J.*, 166, 65, 1971.
344. Blinov, L. M. and Kirichenko, N. A., *Phys. Solid State (Russia)*, 21, 1246, 1970.

Chapter 13

MASS SPECTRA

Mass spectra of pcs continue to be obtained[1-8] as a means to characterize them and their mixtures.

"Pc pigments were among the earliest compounds whose mass spectra were determined using a heated probe which could vaporize the sample in the ionisation chamber directly into the path of the electron beam. The value of the direct insertion probe in structural studies of natural products by mass spectrometry is well established and mass spectra of the structurally related porphyrin class of natural pigments have been frequently examined. Correlations of the mass spectra to structure of the pcs are relatively rare. Lester interpreted the mass spectra of four metal pcs in terms of the electronic structure of large ring systems."[1]* The electronic aspects of the structure and mass spectra of large ring systems including pcs are the subject of a discussion and review.[2]

A novel method of sample introduction at high temperatures is used to determine the mass spectra of Cu pc, Cu 3,6-octachloro pc, Cu dodecachloro pc, Cu 3,6-hexachloro pc, and Cu monochloro pc.[3] The spectra of the highly substituted pcs have proved especially difficult to obtain.[4] The mass spectrum of Cu tetrachloro pc permits detection of impurities in the sample such as the penta and trichloro analogs.[5] Another chlorinated pc, chloroaquoindium pc, is also characterized by mass spectroscopy.[6] Another mass spectrometric investigation of chlorinated pcs shows the molecular ion for chloroaluminum-, chloroindium-, and tribromoniobium pc.[7] However, the mass spectra of chloroscandium-, chloroyttrium-, and tribromotantalum pc "showed neither peaks of the molecular mass nor metal containing fragments of the pc structure."[7]

An electron attachment mass spectrographic study of chlorinated Cu and Co pcs reveals that the samples were not uniform but were mixtures of compounds containing various amounts of chlorine.[8]

Field desorption mass spectrometry can be readily employed to obtain mass spectra of pcs, and provide a rapid method for the qualitative examination of mixtures of pcs[4] which often occur in their preparation. For example, the field desorption mass spectrum of Cu tetra-(3,6)-mono-(4)-chloro pc readily discerns the presence of di-, tri-, tetra-, penta-, hexa-, hepta-, and octachloro pcs.

Mass spectrometry is used to determine the threshold energy of transition metal pcs for ionization in the gas phase.[1] "The average difference between the gaseous ionization potentials and the most probable surface ionization potentials for the pc molecules of 2.34 eV suggest that the polarization energies for these molecules are much greater than those observed for fused ring aromatic compounds."*

* From Eley, D. D., Hazeldine, D. J., and Palmer, T. F., *J. Chem. Sci., Faraday Trans. 2*, 69(12), 1808, 1973. With permission.

REFERENCES

1. Eley, D. D., Hazeldine, D. J., and Palmer, T. F., *J. Chem. Soc., Faraday Trans. 2,* 69(12), 1808, 1973.
2. Lester, G. R., *Mass Spectrometry,* NATO Advanced Study Institute, Glasgow, 1964, 153.
3. Hill, H. C. and Reed, R. I., *Tetrahedron,* 20(5), 1359, 1964.
4. Games, D. E., Jackson A. H., and Taylor, K. T., *Org. Mass. Spectrom.,* 9(12), 1245, 1974.
5. Beynon, J. H., Saunders, R. A., and Williams, A. E., *Appl. Spectrosc.,* 17(3), 63, 1963.
6. Yoshihara, K., Shiokawa, T., Kishimoto, M., and Suzuki, S., *Shitsuryo Bunseki,* 22(3), 231, 1974.
7. Varmuza, K., Maresch, G., and Meller, A., *Monatsch. Chem.,* 105(2), 327, 1974.
8. Starke, M. and Tuemmler, R., *Z. Chem.,* 7(11), 433, 1967.

Chapter 14

LASERS

I. Q-FACTOR MODULATION AND Q-SWITCHING

The parameters of duration, stability, and transformation coefficient, of a smooth pulse obtained during modulation of the ruby laser quality Q-factor by Cu pc vapors are discussed.[1]

A nonpoisonous Q-switch for ruby lasers consists of a plastic foil, comprised of a chloroaluminum pc incorporated in polymethyl methacrylate dissolved in ethanol.[2] Such foils with thickness 20 to 40 μm and transmission of 40 to 85% give reproducible laser pulses of 1 to 25 MW output with energies of 0.03 to 0.75 Wsec and allowed out-coupling of $>$ 100 pulses of 20 MW at intervals of 20 sec or out-coupling of $>$ 100 pulses of 15 MW at 10-sec intervals.

Pc-ethanol solutions at various concentrations are studied as a Q-switch for a resonance ruby laser system.[3] The system consists of an MFT 524M Xe tube, a pc solution cell with 1-cm optical path, a ruby rod (Cr^{3+} 0.05%, 6 mm diameter, 5 cm length), an LSD 39 Å photodiode, and an energy meter. Giant pulses with 1.1 MW are observed in the concentration range 3.5×10^{-5} to 2×10^{-4} *M.*

The optical transmission of some pc solutions increases with the intensity of visible light. The effect, which is very fast, can be used in the passive Q-switching of ruby lasers.[4] Experiments are presented which give insight into the behavior of this process.

The saturable action of pc dyes in the vapor phase is studied and successful Q-switching of the ruby laser by using H_2 pc vapor is accomplished.[5]

The feasibility of increasing the intensity of the single pulse radiation of a ruby laser by using passive and active modulators of the resonator losses is studied oscillographically with a GaClpc solution as the passive modulator and a KH_2PO_4 crystal or a rotating prism as the active modulator.[6] The development of the single pulse in lasers is a function of the modulator: pulse development times are several microseconds for the passive modulator and 200 to 300 nsec for the active modulator.

A laser is constructed, based on dyes with distributed feedback.[7] A low-mode single pulsed Nd laser, the Q modulation of which is realized by a solution of brominated pc, is the pumping source.

The methods and results of an experimental study of the spectral width of the radiation from five ruby rods under free generation conditions and also during the modulation of the Q-factor by using solutions of Ga and Zr pcs in ethanol, Cu pc sulfate in water, chlorophyll a + b in ethanol, along with a rotary prism, are studied.[8]

II. SHUTTERS AND FILTERS

The effect of spectral luminescent properties of transparent filters on the parameters of generated monopulses is studied with a ruby laser, 15 mm diameter and 60 mm in length, and pc filters.[9] The energy of the pulses is lowest with Mg pc and Zn pc in pyridine and highest in GaClpc in chlorobenzene.

A systematization is made of phototropic solutions of pcs based on their absorption spectra from the point of view of their possible use as passive shutters of ruby lasers.[10] Absorption maxima, λ_m, of solutions of pcs in 23 solvents are given. Besides λ_m of the dyes, the choice of the solvent is relevant because of self-focusing and scattering in the solvent or its possible association with pc molecules. The solutions of pcs of metals from Groups III, IV, and V seem to be the most suitable.

The effect of thermal distortion or thermal operating conditions of a passive shutter, based on a chlorobenzene solution of GaClpc, on the generation characteristics of a pulsed ruby laser show that, in the construction of lasers with phototropic shutters operating at a high pulse frequency, it is essential to compensate for the thermal lens in the dye solution and active rod and to eliminate causes of nonuniform heating of the cuvette of the light beam.[11]

Energies of single ruby laser pulses are measured for various shutters, GaClpc, AlClpc, and cryptocyanine, and for varying laser energy outputs to determine the optimum operating conditions for the passive shutters.[12]

The physical properties of a solvent for stable laser passive shutters are described.[13] The eutectic mixture $AlBr_3$ + CH_3COCl, 1.7 to 1, is suggested to be a suitable solvent for the pc passive shutter. This mixture has a sp gr 2.48 ± 0.01 g/cm^3 at 20°C, relative coefficient of viscosity 45.3 ± 0.2 Ostwald, heat capacity 0.154 ± 0.001 cal/g at 25 to 50°C, volume expansion coefficient $6.5 \pm 0.1 \times 10^{-4}$ degree^{-1} at 18 to 77°C, n = 1.5640 ± 0.0002 at 20°C, temperature derivative for the D-line $-(4.1 \pm 0.2) \times 10^{-4}$ degree^{-1} at 22 to 32°C, and freezing point − 84 ± 2°C. This mixture is single phase at − 30 to + 65°C.

A peakless sytem of ruby laser generation using dark screening solutions is described.[14] The solutions of pc complexes of aluminum and gallium in a eutectic are used as dark screening solutions in the ruby laser generator in order to obtain a peakless generation. The dependences of generation kinetics on the pumping value (2500 to 6300 J) and concentration of the filter are studied. The generation pulse duration increases with increased pumping. The peakless laser generation can be realized with a passive dark screening filter.

The dependency of the lasing spectral width on the transmittency T_o of various filters is studied and as an opaque deflector, a convex spherical mirror is used.[15] This results in formation of a very narrow spectrum of $< 10^{-3}$ cm^{-1}. The lasing spectral width also depends on the filter material, the optimal effect being obtained with GaClpc.

Dynamic filtering is a means of laser eye protection.[16] "With the growing uses and new developments in the field of laser radiation it has become increasingly important to develop an eye protection device for the safety of industrial workers and combat troops. An ideal dynamic filter has high visible transmittance at ordinary light levels and high optical density at increased laser powers. This is accomplished with a three-level system. Absorption of laser radiation would populate an excited singlet state. If the decay to the ground state is slow and a higher excited singlet, of the proper energy separation, exists with a sufficiently large extinction coefficient, then dynamic filtering will take place. Heptaphene, AlClpc, Sudan Black B, and indanthrone have shown an increase in optical density over that of the ground state when subjected to a Q-switched laser."

Complex compounds of pcs and transition metal halogenides are applied as nonlinear filters for ruby lasers.[17] "An employment of the optical and mechanical shutters is usually connected with all kinds of technical difficulties. The mechanical shutters are especially troublesome in operation, due to the necessity of applying the suitable engines and synchronization systems, whereas the electro-optical shutters must be driven by the corresponding generators. In view of these difficulties, the use of shutters based on solutions of organic dyes appear to be particularly convenient.

"In ruby lasers the following substances are most commonly used as passive filters: solutions of cryptocyanines and pcs of various metals and glass filters doped with selenides and cadmium sulfides. A cuvette filled with the solution of the respective dye is placed between the active element of the laser and one of its mirrors. The solution strongly absorbs the light of the wavelength emitted by the laser. The laser action may be developed when (as a result of intense pumping) the population of the upper laser level becomes so numerous that the grain in the active medium exceeds the absorption loss in the dye solution. The laser action, though initially weak, is however, sufficient for a rapid bleaching of the dye, due to quick saturation of the absorption transitions, which in turn results in shifting of the absorption band beyond the spectrum of the emitted laser beam. The recovered high value of the Q-factor in the resonator enables a rapid development of the laster action and the whole energy accumulated in the active medium may be emitted in form of one or several light pulses. Thereafter, the dye molecules are quickly brought back to their initial state and the dye may be used anew.

"The so far used cryptocyanine dyes, though simple in exploitation, exhibit one serious shortcoming — being unstable chemically . . .

"The dyes of the pc group characterized by a good chemical stability do not decompose due to laser irradiation, even of great power. Their only disadvantage is a poor solubility in commonly used organic solvents. Consequently, the use of such solvents as alcohols, ether, acetone or benzene is impossible, since achievable concentrations range is between 10^{-7} mole/1 — 10^{-9} mol/1 . . . Nitrobenzene proves to be one of the best organic solvents for the pcs. The obtained concentrations range within 10^{-3} — 10^{-6} mole/1, depending on the type of pc.

"When looking for new chemical compounds to be applied as nonlinear filters for ruby lasers a series of complex compounds of pc of *p* and *d* transition metal halogenides has been obtained. The molecule of pc serves as an electron donor, while the transition metal halogenide plays a part of an electron acceptor. In polar or anhydrous media these compounds form complexes of ionic structure . . . [Thus] they exhibit a good solubility in polar organic solvents, such as nitrobenzene, chlorobenzene, chloronitrobenzene, and so on. The solubility increases by 2 to 4 orders of magnitude . . . This fact, enabling obtaining nonlinear filters of small thickness of the absorbing layers to 0.1 mm, seems to be important for the reduction in the losses of laser beam energy caused by inserting the filter into the resonator . . ."* The complex compounds obtained are spectroscopically examined in order to determine their chemical structure and to select the compounds of absorption bands overlapping the ruby laser $\lambda = 694.3$ nm, an example of which is shown in Figure 1.

The measurement system to study nonlinear filters is shown in Figure 2.

"From the performed measurements it follows that the energy loss in laser beams caused by insertion of the nonlinear filter made of an organic dye, depends both on layer thickness and concentration of the solution. Detailed results will be given in a paper to follow. Nonlinear filters for ruby laser prepared of pc complex solution being water sensitive should work in tightly closed cuvettes . . . They are resistant to the very high power laser induced damage and their chemical stability under waterless conditions — may last for years."*

The efficiency of a device for increasing the signal-to-noise ratio is evaluated and the spatial filtration of a ruby laser signal is studied by using a nonlinearly absorbing solution of GaBrpc in nitrobenzene, in a cell with a 1 mm path length.[18] A maximum value of the signal-to-noise ratio of 12 is obtained.

III. OPTICAL BLEACHING

The properties of the spectral bleaching in saturable dyes are studied.[19] It is shown by a rate equation analysis why hole burning effects in the absorption lines of certain

* From Graczyk, A., Sobczynska, J., Kozikowska, A., and Puchalski, S., *Opt. Appl.*, 4(4), 31, 1974. With permission.

organic dyes are observed only when the saturable absorber is within a laser cavity. A review of spectroscopic properties of complex molecules is used to show that a simple two-level scheme is inadequate to describe the optical bleaching of dye molecules. Experimental data are reported for the transmission of intense ruby laser radiation by several types of dyes.[20]

IV. OPTICAL PUMPING AND OPTICAL GENERATION

The optical characteristics of Mg pc in quinoline, pc in sulfuric acid, and cryptocyanine in methanol are observed.[21] The ruby laser is perpendicular to the observed generation in the solution. Lines generated in the 7000 to 10,000 Å interval can be obtained by varying the concentration of the solution and the reflection coefficient of the mirrors used.

A laser employing AlClpc dissolved in ethanol lases at 7560 Å when excited with a 6942 Å laser pulse as shown in the following diagram:[22]

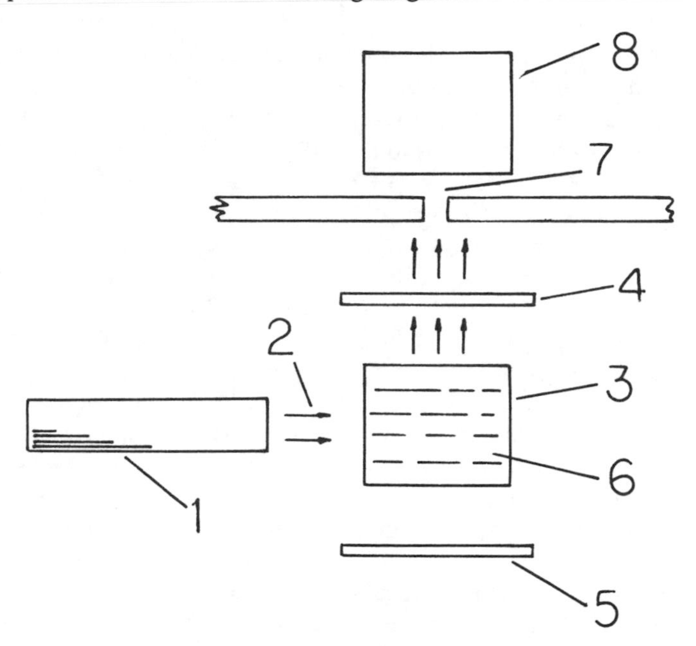

Ruby laser *1* emits a giant pulse *2* of light having a wavelength of 6943 Å and an intensity in the range 1 to 100 mw/cm². This pulse is directed at a glass cell *3* which is a cylinder 1 in. in length and ¾ in. in diameter. Parallel mirrors *4* and *5* are treated by conventional coating techniques so that they are highly reflecting and a few percent transmitting for the wavelength corresponding to 7560 Å. Adjacent to mirror *4* is a spectrograph slit *7* through which the lasing beam output frequency of 7560 Å passes on its way to being recorded on the film of a spectrograph designated as *8*. The glass cell is filled with a room temperature solution of AlClpc dissolved in ethanol at a pc concentration range of about 5×10^{6} to 2×10^{17} molecules/cm³.

V. WAVEFRONT MULTIPLICATION

A 3×10^{-3} *M* AlClpc in ethanol dye solution in a cell of thickness 100 μm is used for wavefront multiplication of a ruby laser.[23] A Q-switched ruby laser with a peak power of about 2 MW in the TEM_{001} mode is used with a KDP doubler to provide simultaneous signals at 0.694 and 0.347 μm. The optical components beyond the cou-

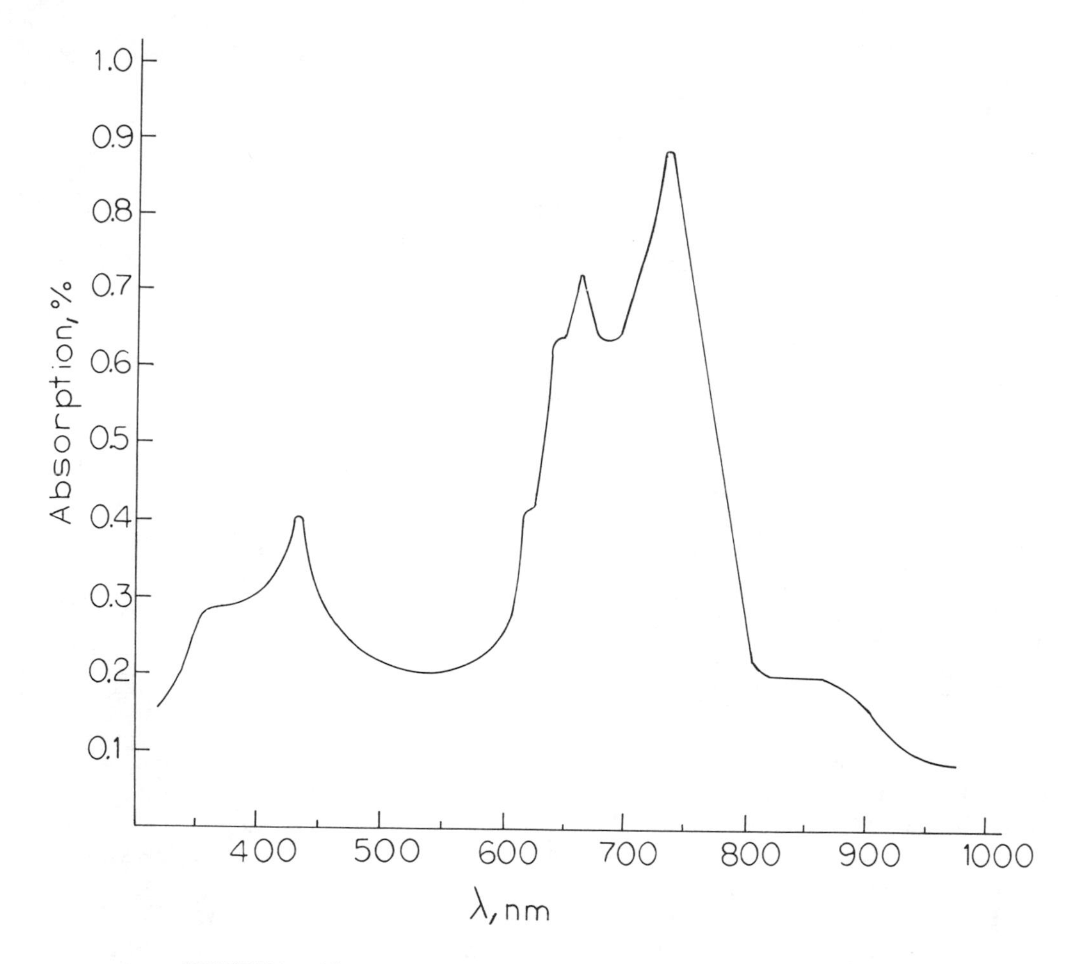

FIGURE 1. Absorption spectrum of the $Na_2PcAlCl_3$ complex in nitrobenzene.

pler are aligned so that the fundamental red beam is collimated and the second harmonic UV beam converge to a focus near mirrors M_1 and M_2. Both beams strike a reflection grating of 1000 lines per millimeter, and two first-order diffraction beams of the second harmonic are produced. These diffracted beams are reflected by mirrors M_1 and M_2 and combined at the dye cell plane for form interference fringes. The intensity distribution at the interference plane is used to pump the dye cell. The pumping establishes a gain and refractive index distribution in the dye solution which is then used to spatially modulate the amplitude of a reconstruction beam.

VI. TRANSMISSION

The intensity of radiation in the resonator of a ruby laser is estimated, necessary for illumination across the following organic solutions: vanadyl pc in $HCON(CH_3)_2$, vanadyl pc in nitrobenzene, cryptocyanine in methanol, and Zr pc in α-bromonaphthalene.[24] The transmissivity of these solutions is due to transitions from ground energy levels to singlet levels, the lifetime of which is 2 to 8×10^{-9} sec.

The formation of a transparent channel is investigated experimentally in colloidal liquid media during the passage through them of ruby laser radiation, the laser operating under Q modulation.[25] Colloidal solutions of dyes, presumably including an Alpc complex, in organic solvents are used as the media of interest.

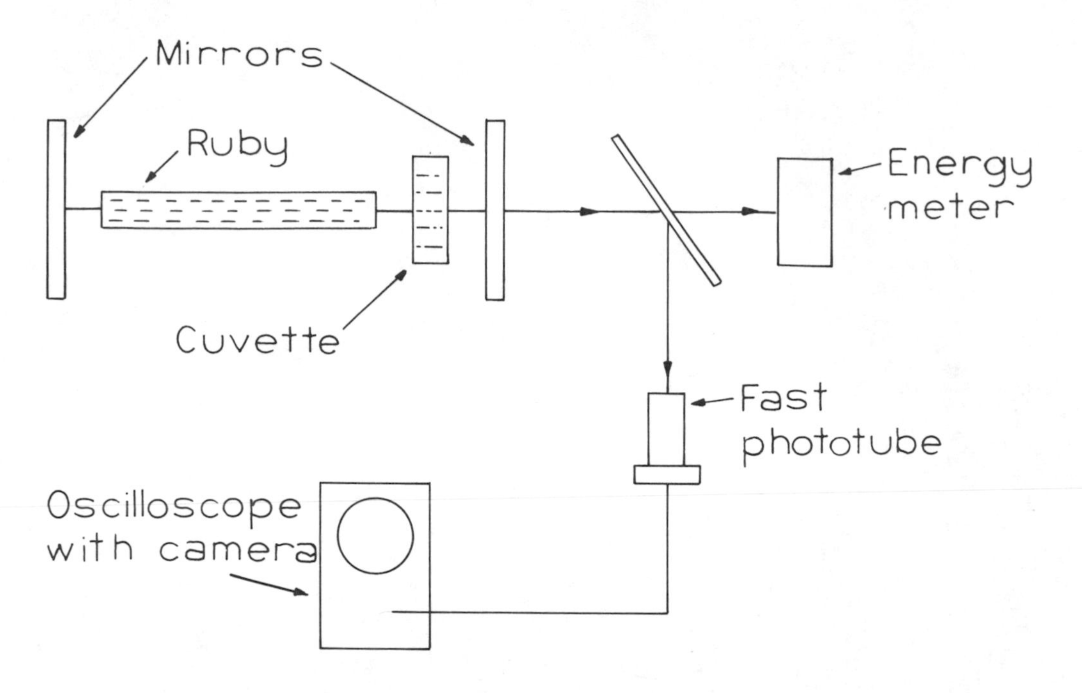

FIGURE 2. Scheme of the measuring system for examination of nonlinear filters.

Passage of radiation from a free-running Q-modulated, with a passive shutter, ruby laser through the semicolloidal solutions of AlClpc in *o*-dichlorobenzene, VO pc in $CHCl_3$ and Al-1,2-naphthalocyanine in *o*-dichlorobenzene and quinoline are studied, where the dimensions of the micelles and inhomogeneities around them are smaller than the wavelength of light.[26]

The passage of weak nonmonochromatic radiation through a resonance medium in the presence of a strong monochromatic, presumably laser, field is discussed, taking into account transverse and longitudinal relaxations.[27] In the case of long relaxation times, the frequency-dependent amplification curves are obtained for vanadium pc.

The kinetics of transillumination processes of AlClpc in ethanol and Mg pc in quinoline are studied.[28] The kinetics are evaluated from the intensity change of probing light during an irradiation of the cuvette by a single pulse of a ruby laser.

VII. ABSORPTION

Absorption of laser radiation in a low temperature plasma is studied.[29] Optical density values are obtained for chlorocryptocyanine and Ga pc complexes in chlorobenzene.[29]

Absorption losses in organic dye lasers are due to the absorption of the pumping light in the excited singlet and triplet states of the dye.[30] The absorption coefficient can be expressed as a sum of three terms that correspond to the transitions $S_0 \rightarrow S_1$, $S_1 \rightarrow S_2$, and $T_1 \rightarrow T_2$, "It is very difficult to obtain the absorption coefficient for the $S_1 \rightarrow S_2$ transition because of the very short lifetime of the S_1 level ($10^{-8} - 10^{-9}$ second). The absorption that can be ascribed to this process is observed as a temporary darkening of the AlClpc solution in ethanol or pyridine, and of the Mg pc solution in quinoline."

Induced losses and their effect on the output efficiency of several pc dyes excited by a single ruby laser pulse are studied by comparing the kinetics of development of nonsteady-state losses.[31]

"Solutions of many organic substances deviate from the Beer-Lambert's absorption law when illuminated by an intense laser pulse. These deviations, either positive or negative (namely, either bleaching or coloring) can give a wealth of information regarding the dynamics of the excited levels, that is, the various relaxation rates and cross sections coupling these states."* This nonlinear behavior of solutions illuminated by a ruby laser is studied. Simultaneous temporal and spatial equations governing the light intensity and electronic molecular level population in a laser illuminated medium are solved numerically for the cases of H_2pc and AlClpc.

VIII. SATURABLE OPTICAL ABSORPTION

The saturation of optical absorption at 6943 Å of AlClpc dissolved in chloronaphthalene and of vanadium pc dissolved in nitrobenzene is studied.[33] A giant ruby laser is used as the source of intense resonance radiation. The absorption transition occurs between the singlet ground state and the lowest lying excited singlet and has a molar absorptivity of the order of 300,000.

IX. SUBPICOSECOND PULSES

"The passively modelocked CW dye laser is a unique source of optical pulses for investigating ultrafast processes on picosecond or even subpicosecond time scale . . . The generation of optical pulses as short as a few tenths of a picosecond at high repetition rates [is studied]. Techniques are described for the application of these pulses to perform time resolved spectroscopy on a subpicosecond time scale."[34] The absorption recovery of Malachite Green in methanol and glycerin and AlClpc in methanol and ethylene glycol are investigated.

X. POLARIZATION OF INDUCED LASER RADIATION

The polarization of induced radiation of organic dyes is studied.[35] A solution of AlClpc in ethanol, pumped with a ruby laser single pulse (about 4 J, duration of about 20 to 30 sec), is included in the study. "The effects of an angle between the electric vector of exciting radiation and the normal to the resonator axis, exciting radiation intensity, and the value of losses in resonator on the degree of polarization, DP, of induced radiation is described."

XI. FLUORESCENCE

The triplet state lifetime of H_2pc varies from 10^{-6} sec at room temperature to 150×10^{-6} sec at 77 K.[36] Decay from the first excited state proceeds via the triplet state. Excitation by a giant pulse ruby laser completely empties the ground state. Luminescence is quenched in metallic pcs by the extremely rapid intersystem crossing.

Absorption spectra and fluorescence spectra excited by the He-Ne laser line at 633 mμ are reported for H_2pc, Mg pc, and Zn pc in the vapor.[37] The yields of fluorescence are temperature dependent but relatively independent of the atmospheres Ar, N_2, and CF_2Cl_2.

Spectra of pcs of Mg, Zn, AlBr, GaBr, GaCl, VO, and Cu are recorded in the region 1200 to 1700 cm^{-1}. Gain as well as generation are possible on singlet-singlet transitions of luminescent metallo pcs under excitation by a ruby laser.[38]

Experiments on the fluorescence excitation by monochromatic laser radiation are

* From Zunger, A. and Bar-Eli, K., *IEEE J. Quantum Electron.*, 10(1), 29, 1974. With permission.

carried out to determine the causes of the large width (about 200 cm^{-1}) of the spectral bands in the luminescence and absorption spectra of organic molecules, presumably including Zn pc, in solid solutions.[39]

The fluorescence spectra of Zn pc and 3,4,8,9-dibenzopyrene are determined at 4.2 K under laser excitation in different media.[40] The nature of the multiplets rising in the fluorescence spectra in the range of the vibronic transitions is studied.

The dependence of fluorescence spectra of pure pc and Zn pc on laser excitation frequency is studied.[41] The spectra are measured at 4.2 K at various excitation wavelengths.

XII. PHOSPHORESCENCE

Phosphorescence emission and its decay time in Pt pc single crystals are investigated in the near IR region at about 971 nm by a Q-switched ruby laser.[42] The intensity of the phosphorescence emission at about 971 nm at 4.2 K tends to saturate and its decay time begins to decrease with increasing excitation intensity greater than the incident photon flux of 1×10^{23} photons/cm^2 sec. This dependence is explained in terms of triplet exciton-triplet exciton (T-T) collisions.

Mutual annihilation of triplet excitons in a Pt pc single crystal at 4.2 K is studied under magnetic fields of $\leqslant$ 100 kG through the measurement of the phosphorescence decay by Q-switched ruby laser excitation.[43] "In Pt pc, singlet excitons convert rapidly to triplet excitons through a highly efficient intersystem crossing due to the effect of a heavy metal, resulting in the appearance of enhanced phosphorescence without fluorescence. The zero field splitting, ZFS, of the triplet spin state of Pt pc has a fairly large value of about 8.5 cm^{-1}. Therefore, the measurement of the phosphorescence emission instead of EPR and delayed fluorescence should be a unique method of studying the behavior of triplet excitons in Pt pc crystals."*

XIII. EMISSION SPECTRA

Spontaneous and stimulated emission spectra of organic dye molecules of cyanine and pc groups and kinetics of the emission are studied.[44] A ruby laser in the regime of free oscillation is used as the excitation source.

XIV. LONG-LIVED PHOTOINDUCED EXCITED FORMS OF ADDUCTS OF BROMINATED PHTHALOCYANINE

The long-lived photoinduced excited forms of $Br_{12}pcAlBr \cdot (n-m)AlBr_3$ of the adducts $Br_{12}pcAlBr \cdot nAlBr_3$, where $m < n < 16$, formed when brominated pc dissolved in a eutectic containing $AlBr_3$ and CH_3OCl in a mole ratio 1.5 to 2.2:1 is irradiated by a neodymium laser are investigated.[45] The light transmitted through the cell containing the sample and through a monochromator is registered on an oscillograph. The laser radiation induces the formation of the long-lived adducts which changes the optical density and thus the transparency of the solutions. The relaxation times of these changes varies from 0.1 to 10 min.

XV. PHOTOCONDUCTIVITY

In an investigation of the photoconductivity of Zn pc induced by ruby lasers and

* From Kaneto, K., Yoshino, K., and Inuishi, Y., *Chem. Phys. Lett.*, 40(3), 505, 1976. With permission.

glass lasers, dependence of induced current on ruby and glass laser intensity show that the triplet state plays an important role in the carrier generation in the near IR.[46,47] The phosphorescence emission in Zn pc and Cu pc suggests this result.

XVI. DIFFRACTION GRATING

An experimental investigation is described for diffraction gratings obtained in a layer of an absorbing substance.[48] A detailed optical scheme indicates the pulsed laser and auxiliary apparatus used. Experiments are performed for an aqueous solution of $CuSO_4$ and for solutions of pcs of Ga and Mg chloride in pyridine. A solution of the Ga complex gives a diffraction two orders of magnitude greater than the $CuSO_4$.

XVII. THERMAL ACTION

Decomposition of naphthoquinone diazide by ruby laser radiation is accomplished on a transparent substrate by addition of 0.1 wt% pc to the photoresist area.[49]

REFERENCES

1. Klochkov, V. P. and Bogdanov, V. L., *Zh. Prikl. Spektrosk.*, 19(6), 1014, 1973.
2. Leupold, D., Konig, R., and Rosler, K., (to VEB Carl Zeiss Jena), British Patent 1,298,539, (December 6, 1972.
3. Yamauchi, T., Sasaki, H., and Sugahara, M., *Yamaguchi Daigaku Kogakubu Kenkyu Hokoku*, 22, 6, 1971.
4. Gires, F. and Combaud, F., *J. Phys. (Paris)*, 26(6), 325, 1965.
5. Bradley, D. J., U.S. NTIS AD Rep. No. 750789, National Technical Information Service, Springfield, Va., 1972; *Govt. Rep. Announce. (U.S.)*, 72(24), 183, 1972.
6. Kovalev, A. A., Pilipovich, V. A., and Razvin, Yu. V., *Kvantovaya Elektron, (Moscow)*, 1(5), 1191, 1974.
7. Efendiev, T. Sh. and Rubinov, A. N., *Kvantovaya Elektron. (Moscow)*, 2(4), 858, 1975.
8. Sharlai, S. F. and Priyateleva, L. P., *Tr. Leningr. Mat. Tochnoi Mekh. Opt.*, 65, 21, 1968.
9. Kovalev, A. A., Pilipovich, V. A., Bogdanovskaya, L. A., and Murashko, L. S., *Zh. Prikl. Spektrosk.*, 9(1), 71, 1968.
10. Sharlai, S. F., *Tr. Leningr. Inst. Tochnoi Mekh. Opt.*, 67, 110, 1970.
11. Ivanov, V. A. and Lebedev, V. I., *Zh. Prikl. Spektrosk.*, 18(3), 400, 1973.
12. Ivanov, V. A., Kovalev, A. A., and Lebedev, V. I., *Zh. Prikl. Spektrosk.*, 20(4), 597, 1974.
13. Gryaznov, Yu. M., Kirsanova, T. I., and Savelova, V. K., *Zh. Prikl. Spektrosk.*, 20(4), 716, 1974.
14. Gryaznov, Yu. M. and Chastov, A. A., *Zh. Prikl. Spektrosk.*, 16(4), 658, 1972.
15. Morgun, Yu. F., Muravitskii, M. A., and Lavrovskii, L. A., *Zh. Prikl. Spektrosk.*, 19(1), 33, 1973.
16. Blumenthal, A. H., Anderson, R. W., Jr., and Mikula, J. J., U.S. NTIS AD Rep. No. 779555/2GA, National Technical Information Service, Springfield, Va., 1973; *Govt. Rep. Announce. (U.S.)*, 74(15), 59, 1974.
17. Graczyk. A., Sobczynska, J., Kozikowska, A., and Puchalski, S., *Opt. Appl.*, 4(4), 31, 1974.
18. Gryaznov, Yu. M. and Chastov, A. A., *Kvantovaya Elektron. (Moscow)*, 2(12), 2610, 1975.
19. Degiorgio, V., *Appl. Phys. Lett.*, 10(6), 175, 1967.
20. Giuliano, C. R. and Hess, L. D., *IEEE J. Quantum Electron.*, 3(8), 358, 1967.
21. Stepanov, B. I., Rubinov, A. N., and Mostovnikov, V. A., *Zh. Eksp. Teor. Fiz. Pis'ma Redaktisyu.*, 5(5), 144, 1967.
22. Sorokin, P. P., (to International Business Machines Corporation), U.S. Patent 3,679,995, July 25, 1972.
23. Sawatari, T. and Shupe, D. M., *Appl. Phys. Lett.*, 24(2), 95, 1974.
24. Dovger, L. S., Ermakov, B. A., Lukin, A. V., and Shklover, L. P., *Opt. Spektrosk.*, 20(5), 903, 1966.
25. Chastov, A. A., *Zh. Prikl. Spektrosk.*, 15(6), 997, 1971.
26. Chastov, A. A. *Zh. Prikl. Spektrosk.*, 16(4), 649, 1972.

27. Shakhnazarova, N. V., *Izv. Akad. Nauk Arm. S.S.R. Fiz.*, 7(3), 173, 1972.
28. Kovalev, A. A., Pilipovich, V. A., and Razvin, Yu. V., *Zh. Prikl. Spektrosk.*, 17(3), 536, 1972.
29. Burakov, V. S., *Kvantovaya Elektron. Lazernaya Spektrosk.*, 316—345, 502—504, 1974.
30. Razvin, Yu. V., Tyushkevich, B. N., and Kovalev, A. A., *2nd Mater. Resp. Konf. Molodykh Uch. Fiz.* Vol. 2, Akad. Nauk Beloruss. S.S.R. Fiz., Minsk, S.S.R., 1972, 14.
31. Kovalev, A. A., Pilipovich, V. A., and Razvin, Yu. V., *Vestsi Akad. Navuk Belarus. S.S.R. Ser. Fiz. Mat. Navuk*, 6, 83, 1973.
32. Zunger, A. and Bar-Eli, K., *IEEE J. Quantum Electron.*, 10(1), 29, 1974.
33. Armstrong, J. A., *J. Appl. Phys.*, 36(2), 471, 1965.
34. Shank, C. V. and Ippen, E. P., *Lect. Notes Phys.*, 43, 408, 1975.
35. Kovalev, A. A., Pilipovich, V. A., and Razvin, Yu. V., *Zh. Prikl. Spektrosk.*, 16(4), 654, 1972.
36. Kosonocky, W. F., Harrison, S. E., and Stander, R., *J. Chem. Phys.*, 43(3), 831, 1965.
37. Eastwood, D., Edwards, L., Gouterman, M., and Steinfeld, J., *J. Mol. Spectrosc.*, 20(4), 381, 1966.
38. Stepanov, B. I., Rubinov, A. N., and Mostovnikov, V. A., *Zh. Prikl. Spektrosk.*, 6(3), 300, 1967.
39. Personov, R. I., Al'shits, E. I., Bykovskaya, L. A. and Kharlamov, B. M., *Zh. Eksp. Teor. Fiz.*, 65(5), 1825, 1973.
40. Al'shits, E. I., Personov, R. I., Pyndyk, A. M., and Smogov, V. I., *Opt. Spektrosk.*, 39(2), 274, 1975.
41. Al'shits, E. I., Personov, R. I., and Stogov, V. I., *Izv. Akad. Nauk S.S.R. Ser. Fiz.*, 39(9), 1918, 1975.
42. Kaneto, K., Yoshino, K., and Inuishi, Y., *J. Phys. Soc. Jpn.*, 37(5), 1297, 1974.
43. Kaneto, K., Yoshino, K., and Inuishi, Y., *Chem. Phys. Lett.*, 40(3), 505, 1976.
44. Melishchuk, M. V., Tikhonov, E. A., and Shpak, M. T., *Ukr. Fiz. Zh. (Russian Ed.)*, 16(3), 451, 1971.
45. Gryaznov, Yu. M., *Zh. Prikl. Spektrosk.*, 23(6), 1014, 1975.
46. Yoshino, K., Keneto, K., Tatsuno, K., and Inuishi, Y., *J. Phys. Soc. Jpn.*, 35(1), 120, 1973.
47. Yoshino, K., Kaneto, K., and Inuishi, Y., in *Energy Charge Transfer Org. Semicond., Proc. U.S.-Japan Semin.*, Masuda, K. and Silver, M., Eds., Plenum Press, New York, 1974, 37.
48. Rubanov, A. S. and Ivakin, E. V., *Kvantovaya Elektron. Lazernaya Spectrosk.*, 407, 507, 1974.
49. Veiko, V. P., Kotov, G. A., and Libenson, M. N., *Fiz. Khim. Obrab. Mater.*, 2, 16, 1973.

Chapter 15

MAGNETIC PROPERTIES

I. MAGNETIC CIRCULAR DICHROISM (MCD)

"The electronic absorption spectra of metal pcs have provoked the interest of both experimentalist and theoretician. The main features of the absorption spectra of closed-shell metal pcs have been satisfactorily accounted for by the work of Longuet-Higgins, Rector and Platt [H. C. Longuet-Higgins, C. Rector and J. R. Platt, *J. Chem. Phys., 18,* 1174, 1950]. Gouterman and his collaborators have considerably extended both the collection of experimental data and their interpretation [L. Edwards and M. Gouterman, *J. Mol. Spectr.*, 33, 293, 1970]. However, in the spectra of pcs containing certain transition-metal ions, especially iron (II), cobalt (I) and (III), and manganese, there are extra features that have been attributed to charge transfer transitions between metal and pc ring orbitals [P. S. Braterman, R. C. Davies and R. J. P. Williams, *Adv. Chem. Phys.*, *7,* 359, 1963]. Since the unambiguous assignment of these bands would yield information about the spin state and the oxidation state of the complex in solution it would be of value to place such assignments on a firm basis. The technique of magnetic circular dichroism (mcd) spectroscopy provides an assignment criterion, the polarisation of an electronic transition in a magnetic field, and can conveniently be applied to solutions [A. D. Buckhingham and P. J. Stephens, *Ann. Rev. Phys. Chem.*, 17, 399, 1966]. We report here the results of mcd studies on solutions of iron (II) pc and cobalt (I), (II), and (III) pcs."[1] The technique provides "a valuable aid to the assignment of charge-transfer bands simply by a comparison of the observed polarizations with those predicted by group theory."[1]*

"It is also possible to use mcd spectra to furnish quantitative information about the magnitudes of the orbital angular momentum of excited states. In this account[2] we describe the use of mcd spectroscopy to measure the variation in the angular momentum of one excited state of the pc ring as a function of the central metal ion and also as a function of axial substituents on the metal."**

"The mcd spectra of Mn and Fe pcs were found to be different from that of Co pc, the latter being characteristic of that of the remaining pcs [Ni→Zn] of the first transition series. A possible explanation is suggested in terms of a greater participation of metal 3d orbitals in the bonding of the Mn and Fe complexes. The Soret (B) band was also investigated for Zn pc in indan. No distinct mcd spectrum was found despite the fact that it is expected to show an A-component. "The techniques of mcd measurement have been described previously [P. N. Schatz, A. J. McCaffery, W. Suetaka, G. N. Henning, A. B. Ritchie and P. J. Stephens, *J. Chem. Phys.*, 45, 722, 1966]. In the present work, a superconducting magnet and a Durrum Jasco ORD/CD UV-5 spectropolarimeter were employed."***

Electrochromism is observed in Lu di pc[4] deposited as a thin film on an electrode surface. Electrochromism exists in all rare earth di pcs and also in those of Y and Sc.[4]

The electronic structure of the α, β, and X polymorphs of H_2 pc "are poorly understood in spite of the considerable interest aroused by their photoconductive properties

* From Stillman, M. J. and Thomson, A. J., *J. Chem. Soc. Faraday Trans. 2,* 70(5), 790, 1974. With permission.

** From Stillman, M. J. and Thomson, A. J., *J. Chem. Soc., Faraday Trans. 2,* 70(5), 805, 1974. With permission.

*** From Gall, C. and Simkin, D., *Can. J. Spectrosc.*, 18(5), 124, 1973. With permission.

[S. E. Harrison and K. H. Ludwig, *J. Chem. Phys.*, 45, 343, 1966.] In recent studies of the intense transitions in the visible spectrum of the dimeric form [J. H. Sharp. and M. Lardon, *J. Chem. Phys.*, 72, 3230, 1968,] the splitting into two main bands is attributed to a large Davydov interaction. The same interaction is postulated for the smaller splittings observed in the α and β forms, reflecting the greater separation of the molecules in these lattices. The use of polarized crystal spectra to confirm these assignments is difficult because of the high intensity of the bands. It is virtually impossible for the α form for which sufficiently large crystals have not been grown. The necessary polarization data can be obtained, however, from magnetic circular dichroism spectroscopy [A. D. Buckingham and P. J. Stephens, *Ann. Rev. Phys. Chem.*, 17, 399, 1966] on the sublimed films of this polymorph. The absorption spectra of pcs in solution show very little dependence on the central metal ion or axially coordinated ligands but the MCD spectra are very sensitive to such changes in the environment of the basic ring system. Two such studies have been reported[2,3] but in one case[3] instrumental observations prevented observation of the complete band system. In extending this work to studies of the solid phase of pcs we have observed the temperature dependent mcd and absorption spectra of α metal free pc, α H_2 pc."[5]*

An apparatus is described for measuring the MCD spectra of matrix-isolated species over the temperature range 14 to 50 K. The MCD of H_2 pc and Cu pc under these conditions are measured with a Cary 61 dichrograph.[6]

From an electro dichroic study of dilute suspensions of macroparticles such as Cu pc crystals in an aqueous dispersion, "the directions of their electronic transition moments can be determined. In certain cases the presence of additional peaks may be inferred from the data which were not obvious from conventional absorption spectra on the solution of the randomly oriented particles."[7]

II. NUCLEAR MAGNETIC RESONANCE

"During the past few years a variety of nonlanthanide reagents have been developed. One group that is noteworthy in terms of its potential is made up of species characterized by the presence of metal or metalloid porphyrin or porphyrin-like ring systems (Figure 1). During the same time, a number of lanthanide shift reagents have also been developed. All the members of this promising group function in the same general fashion. They all bind the substrates or the significant portions of the substrates to their ring centers by forming, as appropriate, addition or condensation bonds with them. Similarly they all shift the protons brought under their influence by enveloping them in nonhomogeneous local ring current generated fields (in the functioning reagents there are no paramagnetic atoms). These reagents are used in one or both of two ways. Some are used to spread and simplify the spectra of substrates, that is, as ordinary shift reagents. The others, as well as two of those used as ordinary shift reagents, are used in work that has as its objective the determination of substrate bond parameters.

A. New Techniques and Spectra

The reagents used to simplify spectra are best classified according to the atoms in their ring centers. Those of current interest have iron, cobalt, silicon, or germanium as ring center atoms . . . One of the most important reagents that has iron as a central atom is iron (II) pc, Fe pc. This reagent is applicable to unhindered amines . . . Another reagent that has iron as a central atom is bis-*n*-deuteriobutylamine iron (II) pc, Fe pc $(ND_2C_4D_9\text{-}n)_2$. It is also applicable to unhindered amines . . . A third reagent that has iron as a central atom is the aniline analog of the butylamine reagent, Fe pc

* From Hollebone, B. R. and Stillman, M. J., *Chem. Phys. Lett.*, 29(2), 284, 1974. With permission.

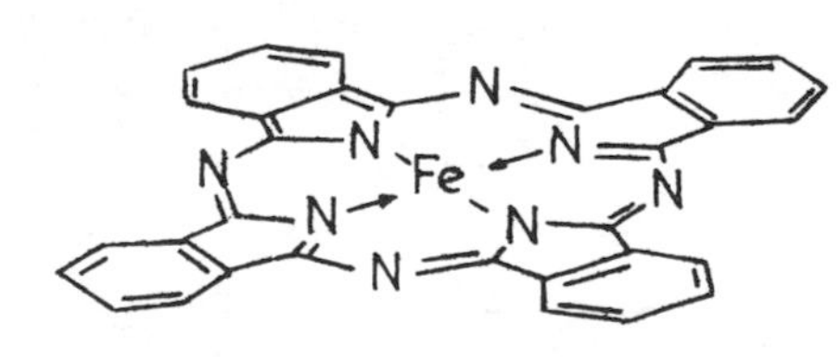

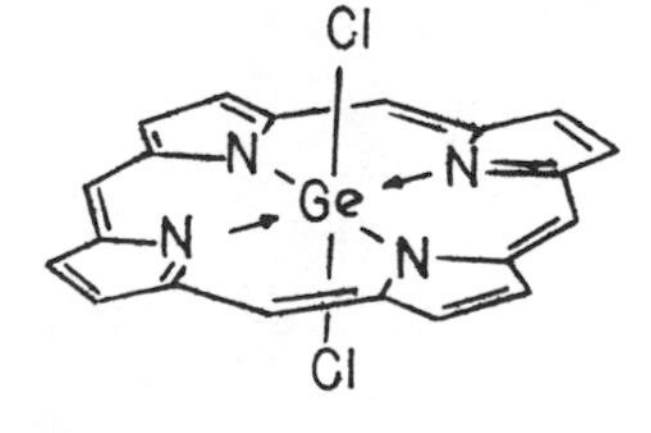

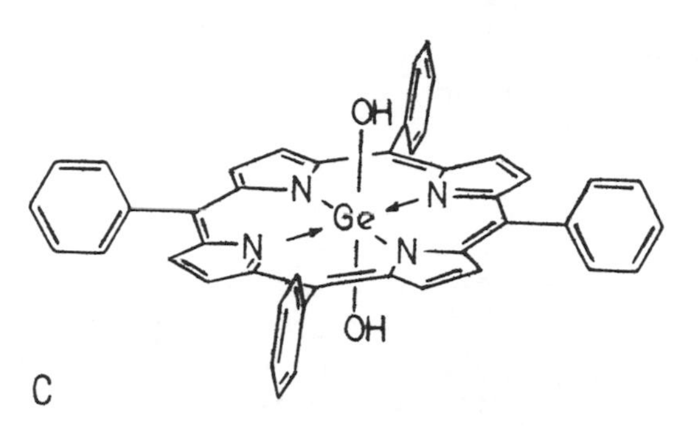

FIGURE 1. Three typical macrocyclic shift reagents. (A) Iron(II) phthalocyanine; (B) dichlorogermanium (IV) porphine; and (C) dihydroxygermanium (IV) tetraphenylporphine.

$(ND_2C_6D_5)_6$. This reagent is applied in the same way as its butylamine analog and is particularly easy to use . . .

"The reagents to which most of the attention has been given in the bond parameter work are dihydroxysilicon pc, pc, $Si(OH)_2$, and the four closely related oligomeric oxygen-bridged hydroxy silicon pcs, HO $(pcSiO)_{2-5}H$ . . . The ring current equation upon which the work has been based is of the superconducting loop type. This equation was developed with the aid of data pertaining to some oligomeric siloxysilicon pcs and some siloxymethylsilicon pcs."[8]*

The use of Fe^{2+} as a shift reagent is illustrated by work carried out on methylamine and *n*-butylamine.[9]

NMR shift reagents that are an alternative to donor-acceptor pseudocontact-shift NMR shift reagents are described. They rely on the ring current effect of certain germanium-porphyrin systems for field asymmetry and on covalent bonds between the germanium atoms of the porphyrin and the compound of interest for holding the reagent in place.[10]

In another work designed to develop NMR shift reagents that fills at least a portion of the hiatus of the β-diketone lanthanide shift reagents, "the reagents dealt with are dichlorogermanium tetraphenylporphine, dihydroxygermanium tetraphenylporphine, dichlorogermanium porphine, dihydroxygermanium porphine, dichlorogermanium pc, and Fe^{2+} pc. The substrates dealt with include hexanols, phenols, acetic acid, bromides, and amines."[11]

Qualitative and quantitative aspects of Fe^{2+} and Si^{4+} pcs as NMR shift reagents are the subject of a thesis.[12]

NMR studies on germanium hemiporphyrazines and pcs indicate that the hemiporphyrazine ring, in contrast to the pc ring, is not aromatic.[13]

* Kenney, M. E., Nuclear magnetic resonance, in *McGraw-Hill Yearbook of Science and Technology*, McGraw-Hill, New York, 1975. With permission.

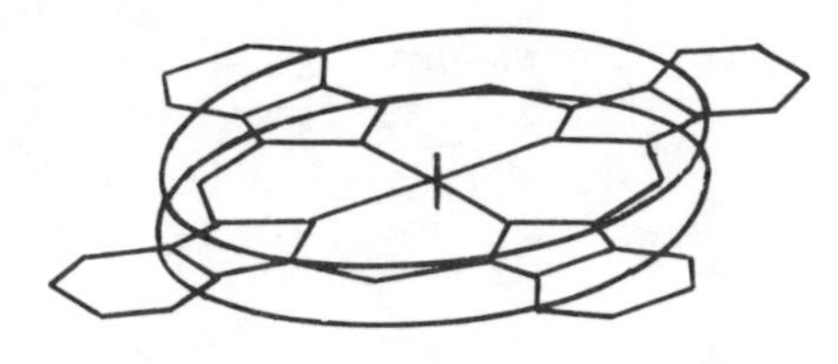

FIGURE 2. Single-loop-pair model.

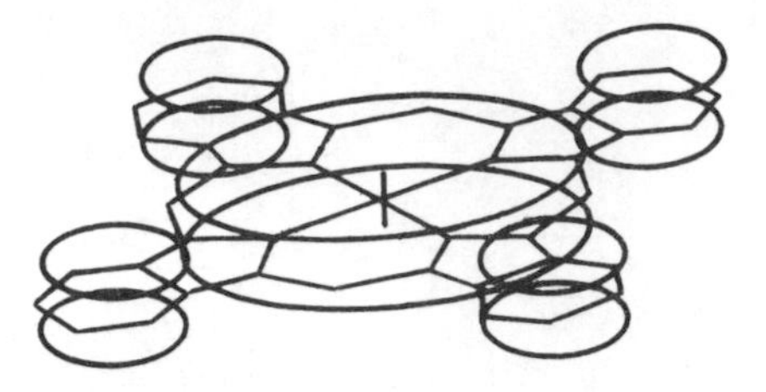

FIGURE 3. Five-loop-pair model.

The NMR spectra of pc Si $(OSi(C_2H_5)_3)_2$, pc $Sn(OSi(C_2H_5)_3)_2$, and hp $Sn(OSi(C_2H_5)_3)_2$ provide further evidence for the nonaromaticity of the hemiporphyrazine ring.[14]

The NMR spectra of pc $Si(OSi(C_6H_5)_3)(OSi(CH_3)(OSi(CH_3)_3)_2)$, pc $Si(CH_3)(OSi(C_3H_7)_3)$, pc$Si(OSi(C_3H_7)_3)_2$, pc $Si(C_3H_7)(OSi(CH_3)(OSi(CH_3)_3)_2)$, pc $Si(CH_3)$ $(OSi(CH_3)(OSi(CH_3)_3)_2)$ and pc $Si(CH_3)(OSi(CH_3)_2OSi(CH_3)(OSi(CH_3)_3)_2)$ are also examined.[15] "For the first four compounds particular attention has been given to the simplification of the spectra of the side groups caused by the ring current of the pc ring. For the last two compounds attention has been given to a determination of the ^{29}Si-C-H coupling constants of the side groups."

The NMR spectra of porphine, Zn porphine, Pd porphine, Zn tetrabenzoporphine, and Zn pc are measured in neutral deuterated chloroform, piperidine, and CH_3SOCH_3 with a detailed discussion.[16]

The proton magnetic resonance spectra of 2,9,16,23-tetra-*tert*-butyl pc and its Zn complex are studied in carbon tetrachloride.[17]

NMR techniques are also used to investigate the mechanism of the phase transition in the α-H_2pc probably to the ferroelectric state at about 17°C.[18]

"The anion of Co tetrasulfo-pc is reduced by a variety of materials to a diamagnetic species that is most reasonably considered to be a derivative of the spin paired Co (I) ion." Proton magnetic resonance measurements indicate that the alternate formulation as a hydride of Co (III) is incorrect.[19]

B. Ring-Current Effect of the Phthalocyanine Ring

Simple classical current loop calculations are carried out on the pc ring.[20,21] NMR data for the calculations are supplied by single ring and multi ring silicon and germanium pcs. The single loop and five loop pair models are shown in Figures 2 and 3.

III. ESR SPECTRA

Electron spin resonance, ESR, is also known as electron magnetic resonance, EMR, and electron paramagnetic resonance, EPR.

A. Cobalt Phthalocyanine

ESR line shapes of polycrystalline samples of $S' = 1/2$ transition element ions are calculated for several compounds including Co pc complexes.[22]

ESR is observed for α and β Co pc.[23] The spectra reveal the difference of the structures of the two polymorphs.

A study of solvent effects on the electronic structure of Co^{2+} in the Co pc molecule is made with several heterocyclic amines.[24] One notable feature of the ESR spectra is the superhyperfine (SHF) structure arising from the magnetic interaction between the Co^{2+} unpaired electron and the out-of-plane solvating molecules.

The effects of the extraplanar ligands $(CH_3)_2SO$, C_5H_5N, imidazole, and CN^- on the ESR spectra of Co complexes of tetrasulfonated pc are studied.[25] It is possible to deduce the relative d-orbital energies of Co porphyrin systems using ESR data.[26]

The ESR of Co pc is the subject of a thesis.[27]

B. Vanadyl Phthalocyanine

An ESR study of vanadyl pc magnetically diluted in metal-free and Zn pc salts, and in sulfuric acid solution, confirms the placement of the 3d unpaired electron in the d_{xy} singlet orbital of the vanadyl ion.[28]

The ESR spectrum of vanadyl pc is recorded at 77 K in dilute glassy matrices of concentrated sulfuric acid, α-chloronaphthalene, and quinoline.[29] Nitrogen SHF structure is observed in all solvents. The unpaired electron in the V $3d_{xy}$ orbital interacts with the ligand-N nuclei in vanadyl pc, similar to vanadyl porphyrin.

Exchange effects on the ESR of Cu or vanadyl pc-H_2 pc solid solutions indicate that the intermolecular exchange interaction is stronger in β metal pcs than in the α polymorphs.[30] In dilute solutions, no change in the ESR spectrum accompanies the $\alpha \rightarrow \beta$ phase transition.

An ESR study of vanadium complexes in petroleum indicate "the ESR spectra of vanadyl ion have been investigated in natural oils. By computer simulation the anisotropic *g* values and hyperfine constants have been determined: $g_{\parallel} = 1.965$, $g_{\perp} = 1.987$, $A_{\parallel} = 171.0$ Oe and $A_{\perp} = 59.2$ Oe. The ESR parameters reveal that vanadyl porphyrin or vanadyl pc may be present in the investigated oil samples."[31]

Another ESR study is referenced.[32]

C. Copper Phthalocyanine

An expression is derived for the total ESR absorption intensities of polycrystalline substances with large *g* anisotropy.[33] "The formula is checked by the investigation of the resonance intensities of frozen sulfuric acid solutions of Cu and Co pcs."

ESR measurements on crystals of Cu pc which are considerably diluted by Zn pc are summarized.[34] "In this way it has been possible to reduce the line width of the Cu resonance so that not only is the hyperfine structure from the Cu isotopes observed, but also that from the four surrounding nitrogen atoms. An analysis of this superhyperfine structure then enables some idea to be obtained on how the electron orbit is spread over the molecule."

The ESR in Cu pc diluted in H_2 pc allows the hyperfine structure of ^{65}Cu to be distinguished from that of ^{63}Cu; the values of hyperfine tensors and g-factor tensors are given for each isotope.[35]

α Cu pc is studied in αH_2 pc matrices by ESR at room temperature.[36] The SHF spin Hamiltonian used in analyzing earlier results of ESR investigations on square bonded Cu complexes is reconsidered and a new SHF term is derived. This ESR spectrum is also analyzed by powder line shape and MO calculations.[37]

In an investigation of dimer formation, the ESR data do not provide evidence for dimerization in Cu pc 3,3,4″,4‴ tetrasulfonic acid in frozen aqueous solution.[38]

The degree of molecular association of a Cu pc dye in benzene and THF is studied by electronic spectroscopy and ESR.[39] "To deduce the manner in which aggregation in these solvents influence the detailed features of the ESR envelope spectrum, the evolution of envelope lineshape is followed in Cu pc-H_2 pc solid solutions as the concentration of Cu pc is progressively increased."

The ESR spectrum of isotopically pure powered ^{63}Cu pc magnetically diluted in H_2 pc is measured.[40] "Much improved resolution of the hyperfine spectrum has allowed more accurate measurement of the nitrogen hyperfine tensor."

An ESR study is made of molecular interactions in aqueous solutions containing transition metal ion chelates of 4,4′,4″,4‴ tetrasulfo pc.[41] A typical solution is frozen aqueous DMF containing Cu tetrasulfo pc.

D. A Variety of Phthalocyanines

The four negative ions, corresponding to successive 1-electron addition of a series of metal pcs, where the metal is sodium, magnesium, chloro-Al, Zn^{2+}, Cu^{2+}, Ni^{2+}, and Co^{2+} are prepared in THF solution.[42] The results of ESR measurements on these solutions are discussed and the parameters are tabulated.

The hyperfine structure in the ESR spectra of reduced chromium and iron pcs are observed.[43] For example, when Cr pc is reduced by Na THF, a well-resolved 9-line ESR spectrum is obtained centered at g = 1.975, with a spacing of 3.07 G. This spectrum is attributed to the interaction of the unpaired electron with the four inner nitrogen atoms of the pc.

Cr, Fe, Ni, Cu, and H_2 pcs are reduced by sodium in THF and in hexamethylphosphoramide. Nitrogen hyperfine structure is observed in reduced Fe pc, whereas both Cr and nitrogen hyperfine structure is observed in reduced Cr pc.[44] A qualitative energy-level scheme is used to systematize the information.

The ESR of Mn(II) pc is studied in glasses at 90K.[45] "The Mn appears to be in its rare low-spin state ($S = ½$) and shows anisotropic hyperfine and spectroscopic splitting tensors: $A = 151 \times 10^{-4}$, $B = 25 \times 10^{-4}$ cm^{-1}; $g\| = 1.90$, $g\perp = 2.16$. The unpaired electron is probably in the d_{xy} orbital of Mn. The energy ordering of levels is believed to be d_{xz}, $d_{yz} < d_{yz} < d^2_z < d_{x2-y2}$.

The EPR spectrum of Ag(II) pc is determined in 1-chloronaphthalene at 77 K and in the solid state at room temperature. The N hyperfine structure is observed and its spacing shows that the odd electron can be found in Ag d_{x2-y2} orbital about 54% of the time. The g values ($g\| = 2.093$; $g\perp = 2.017$) and Hamiltonian parameters are compared to those found for Cu pc and Ag porphyrins.[46]

The EPR spectra of pc, Fe^{3+}, Co^{2+}, Cu^{2+}, Ni^{2+}, $2Na^{1+}$, $2K^{1+}$, Ag^{2+}, Be^{2+}, Mg^{2+}, Zn^{2+}, Cd^{2+}, Al^{3+}, Sn^{2+}, Pt^{2+} and Ce^{3+} pcs are also scrutinized.[47] The magnetic behavior of transition metal pcs-square planar Cr(II), Cr(III), Mn(II), Fe(II), and octahedral Cr(II), Cr(III), and Mn(IV) pcs are studied at 90 to 300 K. "The results, together with those of other authors, are discussed, with the help of ESR data, in terms of the relative energies of the metal ion 3d orbitals. It is suggested that Cr(II) pc and chloro Fe(III) pc have structures involving metal-metal bonds."[48]

The EPR signals in crystalline pcs are discussed[49] and observed.[50] The product of the dark reduction of pcs by metallic sodium using the EPR method are also investigated.[51] An EPR study is made of the molecular complexes of Mg pc and other pcs with iodine.[52]

The interaction of pcs with such transition metal halides as $AlCl_3$, $AlBr_3$, and $GaCl_3$ yield charge transfer complexes characterized by an ESR signal.[53] The nature of paramagnetic centers in H_2 pc and Ni pc are studied from EPR signals.[54] The EPR spectra of Co, Ni, Cu, and H_2 pcs obtained at 77 and 298 K[55] are the subject of a thesis.[56]

The absence of hyperfine structure in the EPR spectra of Cu pc and Sn pc is due to the interaction of the unpaired electrons with the metal, whereas the presence of hyperfine structure in the EPR spectra of H_2 pc is explained by hyperconjugation of the molecular orbital with which the unpaired electron is described with sp^2-hybrid orbitals of the four inner nitrogen atoms.[57] The EPR spectra of low spin iron porphins and their derivatives are recorded and discussed in terms of crystal field.[58] The EPR parameters of pc are given.

A second order theory, including quadrupole effects, for the calculation of EPR spectra of polycrystalline samples of $s = ½$ transition metal ion complexes possessing axial symmetry is developed.[59] With this theory, the EPR spectrum of cobalt pc sodium tetrasulfonate in dimethylsulfoxide can be explained.

The reaction of tetrapyrrole pigments with pyridine and anthracene negative molecular ions is studied by EPR spectroscopy. Zn tetraphenylporphyrin, Mg, Zn, and Co pc, chlorophyll a, among other pigments interact with these molecular ions in tetrahydrofuran (THF) solution producing EPR singlets, characteristic of the pigment monoions. The systems studied are sensitive to visible light and the reduction involved is reversible in the presence of an oxidizing agent.[60]

The temperature dependence of the EPR spectrum of dichlorosilicon pc is studied. After prolonged storage at about 0°C and cooling to −70°C, dichlorosilicon pc shows an anomalous temperature dependence of the EPR-line parameters.[61] A study of the EPR spectra of Zn pc-I_2 vs. temperature shows that at low temperature the EPR signal obeys the Curie law but at > 180 K thermal excitation to the paramagnetic state takes place.[62] The temperature dependence of the EPR line parameters of polycrystalline Ge pc Cl_2 shows anomalies that are interpreted as a disorder-order phase transition.[63] At room temperature the angular dependence of the EPR spectra of single crystalline pc (H_2 pc) and Cu pc and polycrystalline Sn pc are obtained.[57]

Pc doped with iodine shows strong ESR absorption at $g = 2.0036$ with a line width of 5.03 G.[64]

In α pc an amplification of up to two times the EPR signal is attained at saturation of the forbidden transitions at 77 K.[65] A concentrated solution of Ni pc in sulfuric acid gives a quadrupole amplification at the same temperature. "The appearance of the effect in both cases indicates a statistical dipole interaction between the paramagnetic centers and the spins of the hydrogen nuclei."

Metal pcs form in oleum cations or dications with localization of positive charges either on the metal atom, or on the ligand, or on both of them simultaneously.[66] During localization of the charge on the ligand, the cations have spectral similarities independent of the type of metal and in general independent of its presence in the molecule and are characterized by IR absorption bands in the 1100 nm region. The EPR spectra of the ions with unpaired electrons on the ligand exhibit singlet absorption lines with resolved hyperfine structure. For ions with unpaired electrons on the metal and on the ligand, EPR signals are not observed. With dilution of cation solutions in oleum protonated metal pcs are formed.

IV. APPLICATIONS

"The process of dyeing wool involves the transfer of dye from aqueous solution by means of absorption by and diffusion into the wool structure. An understanding of the process requires a knowledge of, amongst other factors, the molecular status of the dye both in solution and in the fibre structure. While there is a wealth of information which may be gathered concerning the solute-solute and solute-solvent interactions of the dye molecules in solution, it has proved to be more difficult to determine the interactions of the dye molecules when incorporated into the fibre structure. If the dye contains a suitable paramagnetic transition metal ion, ESR measurements can be used to assist in determining the environment of the metal ion, and hence may be able to provide information about the molecular state of the dye molecule. A dye suitable for such studies is the Cu(II) chelate of 4,4′,4″,4‴ tetrasulfo pc (tspc), a wholly acid anionic dye which, under appropriate conditions, diffuses deep into the wool structure."[67]* "The ESR results suggest that the regions of the wool structure which contain the dye molecules have the following characteristics: there are cavities in the wool structure large enough to accommodate dimeric forms of the planar dye molecules. These cavities must be quite large, as the tspc molecules are approximately 15Å × 15Å in size, and the molecules of a pair are separated by about 4.5Å." Also, the

* From De Bolfo, J. A., Smith, T. D., Boas, J. F., and Pilbrow, J. R., *J. Chem. Soc., Faraday Trans.*, 72(2), 495, 1976. With permission.

ESR spectra "show that the dye molecules are found in the amorphous or plastic phase of the wool structure and are not attached to specific sites."[68]

The ESR spectra of vanadyl ion are studied in mineral oil.[69] By computer simulation the anisotropic *g* values and hyperfine constants are determined: $g\| = 1.965$, $g\perp = 1.987$, $A\| = 171.0$ Oe, and $A\perp = 59.2$ Oe.

"The ESR parameters reveal that vanadyl porphyrin or vanadyl pc may be present in the investigated oil samples."

V. X-RAY PHOTOELECTRON SPECTRA (XPS) OF PARAMAGNETIC COBALT COMPLEXES

X-ray photoelectron spectra of a number of Co(II) compounds, presumably including Co pcs, are studied and a correlation is shown between the intensity of the Co $2p_{3/2}$ satellite and the magnitude of its magnetic moment.[70] The appearance of satellites is explained by a spin exchange in the process of photoionization.[71]

VI. PARAMAGNETIC ANISOTROPY

"The study of planar transition metal compounds has generated much interest in their electronic structures, mainly because an *a priori* determination of the ground state is often not possible. This is in contrast to complexes with octahedral or tetrahedral geometries, where the total spin multiplicity uniquely determines the ground state. Apart from their intrinsic interest, investigation of transition metal ion materials of lower than cubic symmetry, particularly those with a square-planar configuration, are of increasing importance because of the relation between these types of molecules and those of biological interest.

"Apart from the measurement of the average magnetic moment, the only widely used techniques for investigating the electronic properties of planar complexes have been optical absorption and electron spin resonance (ESR) spectroscopy. Although much useful information is often obtained from these techniques, they suffer from several limitations, for example assignment problems involving the d-d and charge transfer transitions in the former and relaxation phenomena in the latter.

"Recently we have been interested in applying the paramagnetic anisotropy technique to problems of electronic structures and in particular to planar complexes containing first row transition metal ions. The method is very useful, a determination of the sign of the molecular magnetic anisotropy often being enough to characterize, or at least indicate, a particular ground state. Also, paramagnetic anisotropy data are essential for describing quantitatively the isotropic shifts observed in the NMR spectra of paramagnetic compounds."[72]*

The paramagnetic anisotropy and ground state of β-cobalt (II) pc are determined, and the results agree with those determined by ESR.[73] The paramagnetic anisotropy and ferromagnetism in spin $s = 3/2$ Mn(II) pc are also found.[74] The paramagnetic anisotropy, magnetization, and magnetic susceptibility are measured for Fe(II) pc.[75] The paramagnetic anisotropy is used to deduce the electronic structure of the square-planar β-Cu (II) pc.[76]

VII. DIAMAGNETIC ANISOTROPY

A system for calculating the diamagnetic susceptibility of organic molecules is detailed, based on the addition of susceptibility for individual atoms, bond types, and

* From Gregson, A. K., Martin, R. L, and Mitra, S., *J. Chem. Soc., Dalton Trans.*, 15, 1458, 1976. With permission.

increments for resonant structure.[77] Especially noteworthy is the very large ring increment in porphyrin and pc molecules, where it is concluded that the electron delocalization involves a path through all of the ring systems.

Diamagnetic susceptibilities and anisotropies of H_2 pc, Ni pc, and Zn pc are measured. The results are interpreted such that only part of the total anisotropy can be attributed to the π electron macrocyclic ring.[78,79] "The contributions from σ-electrons and other local paramagnetic and diamagnetic terms are also very significant."

VIII. MAGNETIC MOMENTS AND SUSCEPTIBILITIES

Natural and synthetic organic substances, which contain the structure of a broad-sensed system of conjugated double bonds, may exhibit appreciable paramagnetism.[80] The paramagnetic properties and magnetic susceptibilities of several conjugated polymers are tabulated.

The susceptibility and magnetic moments of a Cr(III) pc arc measured.[81] The effective magnetic moment corresponds to a nuclear spin of two unpaired electrons for a Cr(O) complex.

The magnetic moments and susceptibilities of Mn, Fe, Co, Ni, and Cu 4,4′,4″,4‴-tetrasulfo pcs are determined in the solid state and in solution.[82] The results are

Molar Susceptibilities and Magnetic Moments of Metal Tetrasulfopcs (MSpc) in the Solid State[a]

Compound	Field strength (kG)	(Complex) $\chi_m \times 10^6$	μ_{eff} (Bohr Magneton)
Co(II)S pc	7.8	+1225	2.07
	6.4	+1449	2.20
	3.2	+1763	2.36
FeSpc[b]	7.8	+3369	3.08
FeSpc[c]	7.8	+8986	4.80
Mn(II)S pc	7.8	+2972	2.94
	6.4	+3516	3.17
	3.2	+5653	3.90
Cu(II)S pc	7.8	+1449	2.21
	6.4	+1903	2.44
	3.2	+2737	2.83
Ni(II)S pc	7.8	− 358	. . .

[a] At room temperature.
[b] In the presence of dry air after drying 24 hr at 100°C *in vacuo.*
[c] In the presence of dry, oxygen-free nitrogen after drying 24 hr at 100°C *in vacuo.*

Molar Susceptibilities and Magnetic Moments of Metal Tetrasulfo pcs (MSpc) in Solution[a]

Compound	$\chi_m \times 10^6$	μ_{eff} (Bohr Magneton)
Co(II) Spc	+ 940	1.88
Fe(III)Spc	+ 824	1.80
Mn(II)Spc[b]	+1050	1.94
Cu(II)Spc	+ 768	1.77
Ni(II)Spc	− 441	. . .

[a] 0.025 *M* aqueous solutions at 20°C and 19 kG.
[b] Same μ_{eff} at 19 and 12 kG.

The influence of such extra planar ligands as pyridine, $NO_2^{\ominus}$, $SCN^{\ominus}$, $CN^{\ominus}$, imidazole, benzimidazole, 2-methylimidazole, and histidine on the magnetic properties of 4,4′,4″,4‴-tetrasulfo pcs are also observed.[82]

The magnetic and spectral properties of metal tetrasulfo pcs are the subject of a thesis.[83]

The magnetic moments of Mn and Fe pc formates are also studied.[84]

The magnetic properties of Fe(II) pc are studied at 1.25 to 20 K and 100 to 300 K. Between 100 and 300 K the magnetic susceptibility obeys the Curie-Weiss law, but the measured value of μ_{eff} at 25°C of 3.71 Bohr magnetons is intermediate between the theoretical spin-only values for $s = 1$ and $s = 2$ states. However, at 1.25 to 20 K the susceptibility is virtually independent of temperature and with this result it is shown that the ground state of the central iron atom is an orbital singlet, with a real spin triplet state.[85]

The susceptibilities of Cu, Co, and Fe(II) pcs are measured and explained "as those of substantially polar or covalent complexes and conform with the ligand field theory."[86]

"Low spin cobalt (II) complexes generally show remarkably high magnetic moments (1.9—2.8 B.M.) compared with the spin only value (1.73 B.M.)." The magnetic and optical properties of low spin cobalt (II) complexes including the pcs are elucidated.[87] "It has been found that the ^2E_g state is the lowest excited state and is mainly responsible for the large orbital contribution to the magnetic moment through spin-orbit interaction with the $^2A_{1g}$ ground state."*

"X-ray structural data on the transition metal pcs show that the magnetic interaction occurs through its developed π-system via a 90° metal-ligand-metal pathway, and the molecules are stacked in the lattice in such a way that it forms approximately discrete linear chains. Since the electronic structures of these compounds are now known, the low temperature magnetic studies on the metal pcs promise to offer good scope to test the theoretical results on the linear chains and the 90° super exchange mechanism."[88]

Magnetic susceptibility measurements are made at 1.5 to 293 K on Co, Cu, and Mn pc, using a null-coil pendulum magnetometer. Magnetization measurements are made at 1 to 15 kOe. Cu pc has a magnetic moment of 1.82 B.M. at room temperature, which remains constant down to 1.5 K. Co pc shows evidence of antiferromagnetic interaction below about 60 K. The magnetization data on Mn pc at 4.2 and 1.5 K confirm the occurrence of ferromagnetic interaction.[88]

The magnetic susceptibilities of Mn(II), Co(II), and Cu(II) pcs are studied.[89] The magnetic susceptibility of Mn pc is measured in the paramagnetic region between 13.8 and 280 K. The deviation from the Curie-Weiss law is found at about 70 K and the Weiss constant is found at about 70 K, and the Weiss constant θ is obtained at 23 K. The magnetic susceptibilities of Co(II) pc and Cu(II) pc are measured at 1.8 to 77 K. In Co pc a broad maximum is observed and the data are analyzed in terms of the theory of the antiferromagnetic Ising-chain model. Cu pc obeys the Curie law down to 1.8 K.

IX. MAGNETIZATION

Magnetization curves are given for systems consisting of powdered nickel and cobalt covered by a thin film of pc. The magnetization, J_{satn}, for the Co pc Co system is higher than for the Nipc Ni system.[90]

* From Nishida, Y. and Kida, S., *Inorg. Nucl. Chem. Lett.*, 7(4), 325, 1971. With permission.

X. MAGNETO OPTICAL ROTATION (MOR) SPECTROSCOPY

"Magneto optical rotation (MOR) spectroscopy is based on Faraday's observation [M. Faraday, *Phil. Trans. Roy. Soc. London,* 3, 1, 1841] that any substance will rotate the plane of polarized light when a magnetic field is applied parallel to the light beam."*

Magneto optical rotatory dispersion (MORD) data of porphyrins and pcs of the visible and near UV absorption bands are analyzed.[92] It is shown that the magnetic moments of the excited states can be obtained from the data.

A discussion of MOR spectra includes porphyrins and pcs which "as a group have large magnetic rotations with several anomalous dispersion features in their MOR spectra."*

XI. ENERGY LEVELS

Energy levels or eigenvalues for the Cu^{2+} ion in Cupc in a transverse static magnetic field varying from 1 to 1000 G are calculated, making use of $g\perp = 2.045$, $A = 208 \times 10^{-4}$ cm^{-1}, and $B = 31 \times 10^{-4}$ cm^{-1} for Cu^{2+} ion in Cupc.[93]

REFERENCES

1. Stillman, M. J. and Thomson, A. J., *J. Chem. Soc. Faraday Trans. 2,* 70(5), 790, 1974.
2. Stillman, M. J. and Thomson, A. J., *J. Chem. Soc. Faraday Trans. 2,* 70(5), 805, 1974.
3. Gall, C. and Simkin, D., *Can. J. Spectrosc.,* 18(5), 124, 1973.
4. Moskalev, P. N. and Kirin, I. S., *Zh. Fiz. Khim.,* 46(7), 1778, 1972.
5. Hollebone, B. R. and Stillman, M. J., *Chem. Phys. Lett.,* 29 (2), 284, 1974.
6. Douglas, I. N., Grinter, R., and Thomson, A. J., *Mol. Phys.,* 28 (6), 1377, 1974.
7. Jennings, B. R. and Foweraker, A. R., *Spectrosc., Lett.,* 7 (8), 371, 1974.
8. Kenney, M. E., Nuclear magnetic resonance, *McGraw-Hill Yearbook of Science and Technology,* McGraw-Hill, New York, 1975.
9. Maskasky, J. E., Mooney, J. R., and Kenney, M. E., *J. Am. Chem. Soc.,* 94(6), 2132, 1972.
10. Maskasky, J. E. and Kenney, M. E., *J. Am. Chem. Soc.,* 93, 2060, 1971.
11. Maskasky, J. E. and Kenney, M. E., *J. Am. Chem. Soc.,* 95, 1443, 1973.
12. Mooney, J. R., Qualitative and Quantitative Aspects of Iron (II) and Silicon (IV) Pcs as NMR Shift Reagents, thesis, avail. Univ. Microfilms, Order No. 74-2550, Ann Arbor, Mich., 1973; *Diss. Abstr. Int. B,* 34(8), 3640, 1974.
13. Esposito, J. N., Sutton, L. E., and Kenney, M. E., *Inorg. Chem.,* 6 (6), 1116, 1967.
14. Sutton, L. E. and Kenney, M. E., *Inorg. Chem.,* 6 (10), 1869, 1967.
15. Kane, A. R., Sullivan, J. F., Kenney, D. H., and Kenney, M. E., *Inorg. Chem.,* 9 (6), 1445, 1970.
16. Solov'ev, K. N., Mashenkov, V. A., Gradyushko, A. T., Turkova, A. E., and Lezina, V. P., *Zh. Prikl. Spektrosk.,* 13 (2), 339, 1970.
17. Andronova, N. A. and Luk'yanets, E. A., *Zh. Prikl. Spektrosk.,* 20 (2), 312, 1974.
18. Dudreva, B. and Grande, S., *J. Phys. (Paris) Colloq.,* 2, 183, 1972.
19. Busch, D. H., Weber, J. H., Williams, D. H., and Rose, N. J., *J. Am. Chem. Soc.,* 86 (23), 5161, 1964.
20. Janson, T. R., Kane, A. R., Sullivan, J. F., Knox, K., and Kenney, M. E., *J. Am. Chem. Soc.,* 91(19), 5210, 1969.
21. Janson, T. R., Ring Current Effect of the Pc Ring, thesis avail. Univ. Microfilms, Ann. Arbor, Mich., Order No. 72—51; *Diss. Abstr. Int. B,* 32(6), 3237, 1971.
22. Vanngard, T. and Aasa, R., *Paramagnetic Resonance, Proc. 1st Int. Conf., Jerusalem, 1962,* Vol. 2, 1963, 509.
23. Assour J. M. and Kahn, W. K., *J. Am. Chem. Soc.,* 87(2), 207, 1965.

* From Shashoua, V. E., *Methods Enzymol.,* 27, 796, 1973. With permission.

24. Assour, J. M., *J. Am. Chem. Soc.*, 87(21), 4701, 1965.
25. Chan, S. I. and Rollmann, L. D., *Inorg. Chem.*, 10(9), 1978, 1971.
26. Lin, W. C., *Inorg. Chem.*, 15(5), 1114, 1976.
27. Assour, J. M., Electron Spin Resonance of Co Pc, Univ. Microfilms, Ann Arbor, Mich., Order No. 65-10,560; *Diss. Abstr. B*, 27(10), 3627, 1967.
28. Assour, J. M., Goldmacher, J., and Harrison, S. E., *J. Chem. Phys.*, 43(1), 159, 1965.
29. Sato, M. and Kwan, T., *J. Chem. Phys.*, 50(1), 558, 1969.
30. Abkowitz, M. and Chen, I., *J. Chem. Phys.*, 54(2), 811, 1971.
31. Galambos, G., Korosi, G., Siklos, P., and Tudos, F., *Magy. Kem. Foly.*, 79(8), 364, 1973.
32. Sato, M. and Kwan, T., *Chem. Pharm. Bull. (Tokyo)*, 16(12), 2517, 1968.
33. Aasa, R. and Vanngard, T., *Z. Naturforsch.*, 19a, 1425, 1964.
34. Deal, R. M., Ingram, D. J. E., and Srinivasan, R., *Proc. Colloq. Amphere (Atomes Mol. Etudes Radio Elec.,)* 12, 239, 1963.
35. Aoyagi, Y., Masuda, K., and Yamaguchi, J., *J. Phys. Soc. Jpn.*, 23(5), 1188, 1967.
36. Abkowitz, M., Chen, I., and Sharp, J. H., *J. Chem. Phys.*, 48(10), 4561, 1968.
37. Chen, I., Abkowitz, M., and Sharp, J. H., *J. Chem. Phys.*, 50(5), 2237, 1969.
38. Boas, J. F., Pilbrow, J. R., and Smith, T. D., *J. Chem. Soc. A*, 5, 721, 1969.
39. Abkowitz, M. and Monahan, A. R., *J. Chem. Phys.*, 58(6), 2281, 1973.
40. Guzy, C. M., Raynor, J. B., and Symons, M. C. R., *J. Chem. Soc. A*, 15, 2299, 1969.
41. De Bolfo, J. A., Smith, T. D., Boas, J. F., and Pilbrow, J. R., *J. Chem. Soc. Faraday Trans. 2*, 72(2), 481, 1976.
42. Clack, D. W., Hush, N. S., and Yandle, J. R., *Chem. Phys. Lett.*, 1(4), 157, 1967.
43. Guzy, C. M., Raynor, J. B., and Stodulski, L. P., *Nature (London)*, 221 (5180), 551, 1969.
44. Guzy, C. M., Raynor, J. B., Stodulski, L. P., and Symons, M. C. R., *J. Chem. Soc. A*, 6, 997, 1969.
45. Phillips, L. K., U. S. AEC UCRL-17853, U.S. Atomic Energy Commission, Oak Ridge, Tenn., 1967.
46. MacCragh, A. and Koski, W. S., *J. Am. Chem. Soc.*, 85 (16), 2375, 1963.
47. Kholmogorov, V. E. and Glebovskii, D. N., *Opt. Spektrosk.*, 12, 728, 1962.
48. Lever, A. B. P., *J. Chem. Soc.*, p. 1821, 1965.
49. Kholmogorov, V. E., *Opt. Spektrosc.*, 14, 303, 1963.
50. Tikhomirova, N. N. and Chernikova, D. M., *Zh. Strukt. Khim.*, 3(3), 335, 1962.
51. Kholmogorov, V. E. and Shablya, A. V., *Opt. Spektrosk.*, 17(2), 298, 1964.
52. Sharoyan, E. G., Dubrov, Yu. N., Tikhomirova, N. N., and Blyumenfel'd, L. A., *Teor. Eksp. Khim. Akad. Nauk Ukr. S.S.R.*, 1(4), 519, 1965.
53. Bialkowska, E., Graczyk, A., and Leibler, K., *Uniw. Adama Mickiewicza Poznaniu Wydz. Mat. Fiz. Chem. [Pr]. Ser. Fiz.*, 19, 205, 1975.
54. Sharoyan, E. G. and Markosyan, E. A., *Izv. Akad. Nauk Arm. S.S.R. Fiz.*, 4(3), 169, 1969.
55. Rollman, L. D. and Iwamoto, R. T., *J. Am. Chem. Soc.*, 90(6), 1455, 1968.
56. Rollmann, L. D., Electrochemistry Electron Paramagnetic Resonance and Visible Spectra of Cobalt, Nickel, Copper, and Metal-Free Pcs in Dimethyl Sulfoxide, thesis, avail. Univ. Microfilms, Order No. 68-6939, Ann Arbor, Mich., 1967; *Diss. Abstr. B*, 28(11), 4481, 1968.
57. Martynenko, A. P., Shorin, V. A., Borodkin, V. F., and Al'yanov, M. I., *Izv. Vyssh. Ucheb. Zaved. Khim. Khim. Tekhnol.*, 17(4), 528, 1974.
58. Peisach, J., Blumberg, W. E., and Adler, A., *Ann. N. Y. Acad. Sci.*, 206, 310, 1973.
59. Rollmann, L. D. and Chan, S. I., *J. Chem. Phys.*, 50(8), 3416, 1969.
60. Kholmogorov, V. E., *Biofizika*, 15(6), 983, 1970.
61. Martynenko, A. L., Markova, I. Ya., Shaulov, Yu. Kh., and Popov, Yu. A., *Zh. Fiz. Khim.*, 45(9), 2331, 1971.
62. Sharoyan, E. G. and Samuelyan, A. A., *Dokl. Akad. Nauk Arm. S.S.R.*, 54 (3), 154, 1972.
63. Martynenko, A. P., Markova, I. Ya., Il'gasov, V. P., and Kuznetsova, S. I., *Elektron. Tekh. Nauch. Tekh. Sb., Upr. Kach. Stand.*, No. 5, 104, 1972.
64. Aoyagi, Y., Masuda, K., and Namba, S., *J. Phys. Soc. Jpn.*, 31(2), 524, 1971.
65. Schulze, H. and Burkersrode, W., *Z. Phys. Chem.*, 233(5—6), 419, 1966.
66. Bobrovskii, A. P. and Sidorov, A. N., *Zh. Strukt. Khim.*, 17(1), 63, 1976.
67. De Bolfo, J. A., Smith, T. D., Boas, J. F., and Pilbrow, J. R., *J. Chem. Soc. Faraday Trans. 2*, 72(2), 495, 1976.
68. De Bolfo, J. A., Smith, T. D., Boas, J. F., and Pilbrow, J. R., *Proc. 18th Ampere Congr. Nottingham, North-Holland, Amsterdam*, 1974, 277.
69. György, G., Gâbor, K., Pâl, S., and Ferenc, T., *Magy. Kem. Foly.*, 79(8), 364, 1973.
70. Borod'ko, Yu. G., Vetchinkin, S. I., Zimont, S. L., Ivleva, I. N., and Shul'ga, Yu. M., *Chem. Phys. Lett.*, 42(2), 264, 1976.

71. Borod'ko, Yu. G., Vetchinkin, S. I., Zimont, S. L., Ivleva, I. N., and Shul'ga, Yu. M., *Dokl. Akad. Nauk S.S.S.R.*, 228(4), 873, 1976.
72. Gregson, A. K., Martin, R. L., and Mitra, S., *J. Chem. Soc. Dalton Trans.*, 15, 1458, 1976.
73. Martin, R. L. and Mitra, S., *Chem. Phys. Lett.*, 3(4), 183, 1969.
74. Martin, R. L., Mitra, S., and Sherwood,, R. C., *J. Chem. Phys.*, 53(5), 1638, 1970.
75. Barraclough, C. G., Martin, R. L., Mitra, S., and Sherwood, R. C., *J. Chem. Phys.*, 53(5), 1643, 1970.
76. Martin, R. L. and Mitra, S., *Inorg. Chem.*, 9(1), 182, 1970.
77. Haberditzl, W., *Sitzungsber. Dtsch. Akad. Wiss. Berlin, Kl. Chem. Geol. Biol.*, 2, 1964.
78. Barraclough, C. G., Martin, R. L., and Mitra, S., *J. Chem. Phys.*, 55(3), 1426, 1971.
79. Mitra, S. *Indian J. Chem. Sect. A*, 14A(4), 267, 1976.
80. Liu, T.-M., *Hua Hsueh Tung Pao*, No. 7, 34, 1962.
81. Taube, R. and Lunkenheimer, Kl., *Z. Naturforsch.*, 19b(7), 653, 1964.
82. Weber, J. H. and Busch, D. H., *Inorg. Chem.*, 4(4), 469, 1965.
83. Weber, J. H., Magnetic and Spectral Properties of Metal-Tetrasulfo Pc Complexes, thesis, avail. Univ. Microfilms, Order No. 64-6974. Ann Arbor, Mich; *Diss. Abstr.* 24(11), 4394, 1964.
84. Walter, H. P., *Monatsber. Deut. Akad. Wiss. Berlin*, 11, (11-12), 873, 1969.
85. Dale, B. W., Williams, R. J. P., Johnson, C. E., and Thorp, T. L., *J. Chem. Phys.*, 49 (8), 3441, 1968.
86. Havemann, R., Haberditzl, W., and Mader, K. H, *Z. Physik. Chem. (Leipzig)*, 218, 71, 1961.
87. Nishida, Y. and Kida, S., *Inorg. Nucl. Chem. Lett.*, 7 (4), 325, 1971.
88. Mitra, S., Barraclough, C. G., Martin, R. L., and Sherwood, R. C., *Proc. 15th Nucl. Phys. Solid State Phys. Symp.*, Vol. 3, Phys. Comm., Dept. of Atomic Energy, Bombay, India, 1971, 599.
89. Miyoshi, H., *Bull. Chem. Soc. Jpn.*, 47 (3), 561, 1974.
90. Levina, S. D., Bryukhatov, N. L., Lobanova, K. P., and Deeva, N. T., *Zh. Fiz. Khim.*, 44 (11), 2903, 1970.
91. Shashoua, V. E., *Methods Enzymol.*, 27, 796, 1973.
92. Stephens, P. J., Suetaak, W., and Schatz, P. N., *J. Chem. Phys.*, 44(12), 4592, 1966.
93. Matta, M. L., Sukheeja, B. D., and Narchal, M. L. *J. Phys. Chem. Solids*, 35(9), 1339, 1974.

Chapter 16

CONDUCTION

I. INTRODUCTION

A variety of topics are discussed in this chapter. It is important to note, however, that the dichotomy into topics does not necessarily mean that only a given topic is discussed in a given reference. At times, a number of subjects are the concern of the authors. It is not possible to weight each topic in each reference, such that the reader may find pertinent information in which he is interested by reading the literature on conduction over a wide spectrum of topics. As examples, two references cited here relate to a number of matters concerning conduction.

The dark conductivity, photoconductivity, as well as the Hall effect, thermoclectric power, space charge limited currents, and rectification phenomena of pcs and poly pcs are reviewed. The influence of doping on the conductivity and photoconductivity of pcs and the theory of conduction are discussed.[1]

"To investigate the influence of the central metal on electrical and optical properties of pc . . . we report the dark conductivity, activation energy, photoconduction spectra, fluorescence spectra and laser induced photoconductivity . . . The dark current of pc was measured by vibrating reed electrometer under electric field strength between 10^2 V/cm and about 10^4V/cm."* Dark conductivity, activation energy, photoconduction spectra, fluorescence spectra and laser induced photoconductivity of pc single crystals (H_2pc, Cu pc, Co pc . . .) are also studied.[2]

II. PURIFICATION OF PHTHALOCYANINES PRIOR TO MAKING CONDUCTIVITY MEASUREMENTS

In the introduction to their discussion of their work on the conduction of pcs, given at a symposium on catalysis, the authors stress their concern relating to the purity of pcs the conductivities of which are being measured. "There is an obvious point in examining the metal pcs, containing metals of the first transition period. Some results on the room temperature dark conduction in this series became available at the start of this research. In taking the study further, we have laid some stress on purification procedures for trace metals, for organic impurities which may charge complex with the pcs and for polymers, all factors which may enhance the conductivity above that for the pure compounds. We have also studied the activation energies, or 'energy gaps' for semiconduction and the electron spin resonance spectra of the solids. Although problems of purity are far from solved, an interesting picture has emerged as to the extent of involvement of the central metal atom in pc semiconduction."[3]

III. CONDUCTION IN PHTHALOCYANINES

The bulk current in single crystals of H_2pc exhibits ohmic behavior up to fields of 10^4 V/cm and exhibits square law dependence on voltage for higher fields.[4]

Semiconduction in pcs is the subject of a thesis.[5]

"Several recent papers present data purporting to show that a definite correlation exists between the optical absorption spectra and dark conduction in the various pcs

* From Yoshino, K., Kaneto, K., Kyokane, J., and Inuishi, Y., *Technol. Rep. Osaka Univ.*, 21 (995—1026), 549, 1971. With permission.

. . . the above correlation probably does not exist as the agreements between quantities derived from electrical and optical measurements are fortuitous."[6,7]

The temperature dependence of electrical conductivity at 25 to 230°C and 10^{-3} torr are determined for Ti, Zr, Hf, Pd, and Pt chloro pcs and Ni pc to determine the effect of the metal atoms on the activation energy of electrical conductivity of pcs. The activation energy ranges from 1.38 eV for Ti to 1.60 eV for Pt.[8]

Uranyl pc in its electrical properties might be expected to be quite different from pc complexes with transition metals. In the dark, it is one of the most conductive of the pcs but its activation energy is 0.87 eV, about the same as metal-free pc.[9]

Transient current measurements in pc are studied in a thesis.[10]

In a study of the crystal structure of pc and of the conductivity of systems consisting of a metal covered with a pc film it is shown that the existence of other factors is required to obtain conductivity of the p or n type besides the crystal modification.[11]

The mechanism of conductivity in thin pc films is studied by vacuum coating glass plates with Pt filament electrodes with Ni, followed by a Cu pc or Ni pc film, which is contacted with Mg or saturated Hg(Tl). Current-voltage curves up to ± 10 V across this film show essentially a quadratic dependence of current on applied voltage.[12]

Bulk and surface conduction in metal-free pc is the subject of a thesis.[13]

The electrical conductivity of pc semiconductors in the ferro electric state is measured.[14]

The temperature dependence of conductivity and the differential thermal EMF of chlorinated Cu pcs with chlorine contents from 2 to 50% is determined.[15]

Voltage-current characteristics are studied in metal-free pc.[16]

"Electrical conductivity and electron spin resonance of metal free pc doped with iodine were investigated. The electrical resistivity and its activation energy of the pc doped with iodine were largely decreased as compared with the undoped metal free pc. The change of the conductivity is attributed to the change of the activation energy."[17]*

"The dark conductivities in metal and metal free pcs have been measured by several authors and the effects of the central metal ion to the conductivity have been discussed. However, the difference of the conductivity and the activation energy of each pc has not yet been explained. In this report, the electrical conductivities and the magneto resistance effects of Ni, Cu, Zn and metal free pc are measured and the conductivity and the activation energy are measured" and discussed.[18]**

The temperature dependence of the conductivity of highly chlorinated Cu pc above room temperature in dry NH_3 can be fitted to the expression $\sigma = \sigma'_o \exp(E/2kT_o)\exp(-E/2kT)$ characterized by the three parameters σ'_o, E, and T_o rather than to the more usual expression $\sigma = \sigma_o \exp(-E/2kT)$.[19]

The electrical conductivity of pc derivatives containing OH and CH_3O groups and their complexes with Cu, Ni, and Co is studied at 293 to 493 K. The conductivity is 10^5 times that of unsubstituted pcs.[20]

The electrical conductivity of Os pc Cl_2, Os pc Br_2, Os pc I_2, and Os pc SO_4 at 300 K is 1.1×10^{-6}, 2.1×10^{-6}, 4.5×10^{-5}, and 3.4×10^{-7}, respectively. Activation energies over the range $10^3/T$ K = 2.7 to 3.3 are 0.44, 0.61, 0.85, and 1.09 eV, respectively.[21]

Electrical conductivity measurements of Co, Cr, and Mn pcs are made at 160 to 330 K, from which data the activation energies of conductivity are calculated. Decrease of activation energy at certain temperatures indicate structural changes of the compounds.[22]

The effect of preparation parameters on current-voltage characteristics measured in

* From Aoyagi, Y., Masuda, K., and Namba, S., *J. Phys. Soc. Jpn.*, 31 (2), 524, 1971. With permission.
** From Aoyagi, Y., Masuda, K., and Namba, S., *J. Phys. Soc. Jpn.*, 31 (1), 164, 1971. With permission.

vacuum at 37°C is studied for Cu pc films 0.5 μm thick on quartz substrates with Pt and Al electrodes. The samples were annealed for 1 hr at 10^{-4} torr.[23]

"The study of the temperature dependence of the electrical conductivity of a semiconductor is of considerable importance since it is capable of providing information about the energy differences between free and bound electron states.

"The dark electrical conductivity of β H_2pc single crystals is investigated over the temperature range 273—600°K., at a reduced pressure of 10^{-7} torr.[24]

Metal pcs of the first transition period elements are prepared and their dc conductivity and ESR spectra are measured and discussed in terms of the purity and extent of involvement of the central metal atom in pc semiconduction.[25] The activation energy or energy gap for semiconduction is also discussed.

The mechanism of formation of p-n junctions and the nature of barriers in the variation of the diode properties of Al-pc-Ag during doping and subsequent heat treatment are studied.[26] The activation energy of direct current flow is 0.25 to 0.30 eV.

IV. RESISTIVITY

Resistance measurements of pressed pellets of metal-free pc range from 2×10^{11} at room temperature to 8×10^6 ohm-cm at 225°C. The calculated activation energy of conduction is 1.385 eV and the Seebeck coefficient is -1680 μV/°C at 220°C.[27]

The resistivity of Nd H pc at different temperatures is determined in an atmosphere of argon in an article entitled "Pcs With High Electrical Conductivity".[28]

The specific resistance of pc siloxane, $C_{32}H_{14}N_8SiH_2O$ is 8×10^5 ohm-cm.[29]

The temperature dependence of the electrical resistivity and of the magnetic susceptibility of Pb pc is studied down to liquid helium temperatures, above and below 1 K, "as a step toward finding a one-dimensional superconductor," because "the existence of a one-dimensional superconductor has not yet been reported."[30]

V. ENERGY BAND MODEL TO EXPLAIN CONDUCTION

The band model is used to explain conduction in H_2 pc. Evidence for the band model is suggested by the equality of the thermal activation energy for electrical conduction, the energy for first optical absorption, the energy for the initiation of photoconductivity, and from Hall measurements.[31]

The energy band structure of H_2 pc is calculated.[32] The authors' purpose in making this calculation is related to the observation that "Pcs are the most important class of organic photo- and semiconductors. Also, the similarities of their molecular structures with that of chlorophyll and hemoglobin, the two most important biological molecules, add to the importance of pcs. Hence, the determination of the energy band structure of pcs is expected to give valuable information which may be extrapolated to get some insight into the complex biological systems."[32]

The mechanism of charge carrier transport in β H_2 pc is studied with the aid of the Munn and Siebrand theory. It is shown that the band model is not applicable to H_2 pc except for electron transport along the crystallographic b axis.[33]

VI. ONE-DIMENSIONAL CONDUCTOR

Lead pc is considered to be a one-dimensional conductor.[34] In crystalline Pb pc, the molecules are stacked linearly in the direction of the c axis; the interatomic distance between lead atoms within a molecular column is slightly larger than that in lead metal; there is X-ray scattering evidence for a Kohn anomaly in Pb pc, all of which support the concept of one-dimensional conductivity in Pb pc. Experiments on electrical resistivity of Pb pc films and a model for the conduction mechanism are presented.

VII. EFFECT OF PRESSURE ON CONDUCTIVITY

The electrical conductivity of pc and of Cu pc are determined at 25 to 200°C and over a pressure range of 1250 to 50,000 bars. The conductivity increases markedly with increasing pressure at constant temperature and with increasing temperature at constant pressure. Linear plots of conductivity versus 1/T are obtained.[35]

Vanadyl, thorium, lead, copper, zinc, and metal-free pcs become appreciably conductive under the static pressure of up to several hundred kilobars. The least resistivity observed is 0.1 ohm-cm for vanadyl pc. The high pressures are generated in a pyrophyllite octahedron placed in the central chamber of a split-sphere-type appratus.[36]

VIII. FREQUENCY DEPENDENCE OF CONDUCTIVITY

The electrical conductivity of β pc is measured at 2.5 to 50 kbar and at 25 to 200°C over a frequency range of 1 to 20 kc/sec.[37] The resistance decreases with increasing frequency and reaches a constant value. The decrease in resistance from low to high frequencies is lower, the higher the pressure.

Low frequency conductivity is studied of vacuum evaporated polycrystalline metal-free pc films. Current-voltage measurements, carried out at 10^{-4} torr on films 1 μm thick at 135°C, are nonlinear.[38]

The dielectric permittivity and conductivity of Cu pc are measured in the frequency range 10^2 to 10^{10} Hz. At low frequencies the conductivity is approximately constant while at high frequencies the conductivity is proportional to f^n, where $n < 1$. "A very likely explanation of the behavior in the higher frequency range is in terms of electronic carriers hopping either within or between molecules."[39]

The frequency dependence of conductivity and capacitance of evaporated thin films of Cu pc, studied by the alternating current method, are determined.[40] The activation energy for dc conductivity in α and β Cu pc are also determined, and with the films *in vacuo,* there is no difference in electrical properties between phases. The frequency dependence of conductivity and capacitance is explained in terms of hopping and band conductors. Adsorbed oxygen has a pronounced effect on increasing DC conductivity but not on AC conductivity of α Cu pc film.

The frequency dependence of the electrical capacitance and conductivity of compressed powder discs of Cu pc is determined at 30 Hz to 3 MHz and 20 to 120°C.[41] The electrical properties of the highly purified discs prepared by repeated sublimation of the impure sample *in vacuo* at 380°C are consistent with hopping charge transport between localized states.

IX. EFFECT OF ADSORBED ATMOSPHERES ON CONDUCTIVITY

The electrical conductivity of metal-free and Cu pcs as single crystals is measured *in vacuo,* nitrogen, and oxygen. The decrease in thermal activation energies in oxygen indicate increased surface conduction due to the reversibly adsorbed oxygen, although the interactions of metal-free and Cu pcs are not identical.[42]

The conductive behavior of metal-free pcs of Fe, Co, Ni, Cu, and Zn is examined on thin layers under the influence of oxygen, NO, NO_2, and NH_3. The dark conductivity is strongly increased by oxygen, NO, and NO_2; the activation energies are reduced to 10% of the original value *in vacuo.* The conductivity increases caused by the oxidizing gases are reversed by NH_3.[43]

Dyes can be divided into n- and p-type conductors according to the effect of oxygen. In some dyes (triphenylmethane derivatives, rhodamines) the conductivity decreases when oxygen is adsorbed (n-type conduction); others (merocyanines, pcs) exhibit increased conductivity (p-type conduction) under these conditions.[44]

The surface conductivity of organic semiconductors such as pcs can vary appreciably in the presence of adsorbed gases, suggesting the possibility of using them as a method for the detection of air contaminants. For example, in a vacuum, the surface conductivity of metal-free pc thin film deposited on quartz substrates depends strongly on the type of gas brought into contact with the sample.[45]

"A number of studies have indicated that conductivity in the dark and under illumination of certain organic semiconductors are significantly affected with oxygen . . . Oxygen molecules were found to enter lead pc single crystals by thermal diffusion when they were annealed in an oxygen gas of atmospheric pressure above 230°C. for more than ten hours . . . Dark conductivity σ increased by three orders of magnitude at room temperature with the heat treatment. Correspondingly, activation energy of conductivity . . . decreased from 0.85 to 0.54 eV."[46]*

The effect of thermal treatment of U and Th di pcs in oxygen and in vacuum on the conductivity and energy of activation is studied.[47]

X. EFFECT OF α AND β CRYSTAL PHASES ON CONDUCTIVITY

Pcs, namely, H_2 pc, and Cu, Ni, and Zn pcs in the α form are prepared by precipitation from concentrated sulfuric acid and sublimation to the β form, pressed at 4000 atm, and contacted with Cu or Ag powder. Resistivity (ohm-cm), activation energy (eV), and thermoelectric power (μV/degree) are measured in nitrogen.[48]

The α form of H_2pc, and Cu, Ni, and Zn pcs give resistivity of 790, 1, 14, and 4.7 × 10^7 ohm-cm, respectively; activation energy 0.71, 0.30, 0.60, and 0.43, respectively; thermoelectric power (μV/degree) 1250, 910, 970, and 960, respectively. The β form gives 42, 2, 40, and 0.043 × 10^{13}; 0.87, 0.78, 1.14, and 0.72; 670, 1170, 1280, and 1750, respectively.

The temperature dependence of the electrical conductivity of pc is measured in a nitrogen atmosphere, with $\sigma = \sigma_o \exp(-E/kT)$, with E = 0.24 eV for α and 0.91 eV for β. The conductivity of the α form is 2.4 × 10^{-6} ohm^{-1} cm^{-1} at 25°C; above 160°C the sample transforms to the β form with a drop in conductivity of $> 10^{-5}$.[49]

Some authors believe that the crystal phase transformation from α to β and vice versa has no influence on conductivity.[50] "The stronger propensity of the α phase to absorb oxygen is the source of the apparently higher conductivity of the α phase."

XI. UNPAIRED ELECTRONS AND CONDUCTION

H_2 pc single crystals contain 10^{17} unpaired electrons per cubic centimeter. Experiment indicates that the unpaired electrons may be either in the bulk or on the surface of the pc.[51]

The electrical conduction mechanism in single crystals of pc and its complexes with Ni, Cu, and Zn is studied.[52] The activation energy for hopping conduction is calculated. Hopping conduction is dependent on the exchange effect of the unpaired electron in free radicals.

XII. POLYMERS AND CONDUCTION

An invention relates to the preparation of metal-free and metal poly pcs which are "useful as pigments and as semiconductor materials."[53]

"Several poly metal pcs and also a metal-free poly pc have been prepared as part of

* From Yasunaga, H., Kojima, K., Yohda, H., and Takeya, K., *J. Phys. Soc. Jpn.*, 37(4) 1024, 1974. With permission.

a project for the synthesis of semi-conducting organic and organometallic polymers and study of their electrical properties, including their tendency towards superconduction. The resistivities and dielectric constants at different pressures and temperatures were measured for the polymers . . . The polymeric hydrogen form of pc is found to be more conductive than the metallic derivatives, in contrast to the behavior of the monomers. In addition, the polymers were found to be much more conductive than the corresponding monomers with the resistivity of the polymers ranging from 7 ohm-cm to about 3×10^6 ohm-cm."*

Semiconducting polymers for gas detection include conjugated polyenes and polyesters containing pc in their backbone.[55]

XIII. TRAPPING LEVELS, TRAPPING MEASUREMENTS, DRIFT MOBILITY, AND CHARGE CARRIERS

Drift mobilities are studied in the pcs. The Hall effect is observed in metal-free pc and in Cu pc. The sign of the effect is that for the majority electrons in the former and for majority holes in the latter, with Hall mobilities of about 0.4 and 200 $cm^2/$ Vsec, respectively.[56]

The drift mobilities of electrons and holes in metal-free and Pb pc are measured by using a transient space charge limited current method and a time of flight method. Fast trapping of carriers in the bulk and emptying of traps at the surface under illumination are observed.[57] Traps with a density of $10^{14}/cm^3$ and an activation energy of 0.36 eV are found in both metal-free and Pb pc single crystals.[58] A high density of trapping levels is found at a depth of 0.37 eV in Cu pc, H_2 pc, and Ni pc.[59] A shallower (about 0.3 eV) trap is found in Zn pc. The concentration of traps in polycrystalline films of Cu, Ni, and Co pcs is determined from stationary injected currents. For Cu, Ni, and Co pcs with injected holes the electron traps are 2×10^{20}, 5×10^{19}, and $5 \times 10^{16}/cm^3$, for hole traps 2×10^{19}, 2×10^{20}, and $3 \times 10^{18}/cm^3$, respectively.[60] Injection currents are studied in pc films with various types of traps.[61]

Evidence is given showing that for a vacuum of 5×10^{-9} torr the current in the square law region is space charge limited, 10^{19} to $10^{20}/cm^3$ trapping centers exist 0.38 eV below the conduction band edge, and below 90°C the ohmic conduction is extrinsic owing to the presence of 10^{12} to $10^{13}/cm^3$ donors at the trapping level.[62]

From temperature-dependence measurements of the space charge limited current and ohmic current densities, the electron trapping levels in metal-free pc are determined in vacuum, and in oxygen and hydrogen.[63] The electrical conductivity of β metal-free pc single crystals is investigated at 273 to 600 K at a reduced pressure of 10^{-7} torr. At temperatures below about 410 K, the conduction process is consistent with the presence of an electron trapping level located 0.32 eV below the conduction band edge, with a density of $7 \times 10^{16}/cm^3$, and a donor level of $2 \times 10^7/cm^3$ at the same energy. Above about 410 K, evidence suggests that the conduction is intrinsic.[64] By use of pulsed photoconductivity with a Q-switched ruby laser, drift mobilities of electrons and holes at 290 to 600 K are measured in β metal-free pc single crystals. The temperature dependence of the drift mobilities and the electron trapping time are calculated.[65]

Thin films of Cu pc are investigated relative to the trap distribution in the forbidden band. Space charge limited current measurements indicate that localized states exist, uniformly distributed in energy within a region of the forbidden band, below the equilibrium Fermi level.[66] The temperature dependence of the charge carrier drift mobility

* From Norrell, C. J., Pohl, H. A., Thomas, M., and Berlin, K. D., *J. Polymer Sci.*, 12(5), 913, 1974. With permission.

in single crystals of metal-free pc in the β phase is determined.[67] Thermally stimulated current in various pc single crystals, H_2 pc, Zn pc, Co pc, Ni pc, and Cu pc and the drift mobilities in Zn pc and Cu pc are studied. Trap depth is estimated to be 0.37 ± 0.03 eV. The drift mobility of electrons in Zn pc is 0.1 $cm^2/Vsec$ at room temperature by the time of flight method with ruby laser excitation.[68] The drift mobility of holes in Cu pc is observed to be 0.2 to 0.5 $cm^2/Vsec$. The mobility increases exponentially with temperature.[69]

The drift mobility in holes of Cu pc are observed to be 0.2 to 0.5 $cm^2/Vsec$. The mobility increases exponentially with temperature, with an activation energy of 0.1 eV.[69] The influence of vanadyl pc impurity on the energetic trap distribution of Cu pc is studied.[70] Vanadyl pc (VO pc) is chosen because it is very similar to Cu pc. VO pc molecules on crystal surfaces give a strong increase in the trap concentration.[70]

"For crystals with low conductivity the existence of localized electron traps has an extremely strong influence on the electrical conduction properties. The levels of traps, their concentration, their discrete or continuous distribution, and the ratio of concentration of traps distributed in both types determine the carrier concentration in the bands very sensitively. If the band model is applicable this influence may be described by the position of the Fermi level, which gives the energetical distribution of the charge carriers in the equilibrium state."*

The position of the Fermi level is investigated in crystals and in thin films of Cu pc. The position of the Fermi level has a great influence on the conductivity, on the existence of p or n-type conduction, and on the population of traps.[73] A model for the electrical transport phenomena in imperfect crystals of Cu pc is propsed in terms of the surface state[72] and of bulk defects.[71] Measurements are made of space charge limited currents (SCLC) and thermally stimulated currents, (TSC) in sublimed layers of pc. The characteristic temperature of the distribution and the concentration of the hole traps, about $10^{17}/cm^3$, are determined.[74] By measurements of the temperature dependences of SCLC and TSC the presence of discrete trap levels 0.4 and 0.1 eV above the top of the valence band is detected.

The charge carrier mobility in microcrystalline or amorphous H_2 pc layers, determined by the pulsed light technique, is of the order of 10^{-3} to 10^{-2} $cm^2/Vsec$, and the positive holes are the majority carriers.[75]

Charge carriers transfer integrals in H_2 pc single crystals are the subject of a discussion.[76]

Transient field effect measurements indicate the presence of three sets of fast surface states having densities of 10^{10} to $10^{11}/cm^2$ within 0.17 eV of midgap. The parameters of these surface states are not appreciably affected by ambient gases. The thickness of the space charge region of metal-free pc crystals is 10^{-5} to 10^{-4} cm.[77]

XIV. SANDWICH STRUCTURES

Rectification is observed on polycrystalline samples of metal-free, Cu, Ni, and Mo pcs sandwiched between different metal electrodes. The highest rectification ratio observed is 500 and is obtained with Cu pc between a Pt and a Ag electrode and also between a Ag and an Al electrode.[78]

The current-voltage characteristics of metal/metal-free pc/metal sandwich structures with different electrode metals, with varying thicknesses and crystal modifications of the organic layer are studied.[79] Pc layers are 1 to 5 μm thick and the current-voltage relationships are explained in terms of the SCLC in the pc layer. The effect of the $\alpha \rightarrow \beta$ transition on the current-voltage parameters is also measured.

* From Lehmann, G. and Hamann, C., *Phys. Status Solidi B*, 63(1), 341, 1974. With permission.

Au-pc-Al sandwich structures show two kinds of switching properties. Reproducible bistable switching is observed in pc layers 1500 Å thick at room temperature at 1 atm as well as at 1×10^{-7} torr pressure.[80] "The precise nature of the memory switching in pure organic films remains a subject of future investigations."[80]

XV. DIELECTRIC EFFECTS

A. A Variety of Effects

Dielectric relaxation is studied in sublimed thin layers of metal-free pc in the α form in the ordered state at low temperatures. The dielectric characteristics of pc in the ordered state are similar to the case of absorption of the Debye type.[81]

Capacitance measurements are made of β H_2 pc at a frequency range of 0.05 c/sec to 300 mc/sec.[82]

The conductivity, dielectric losses, and capacity of pressed powder tablets and films of H_2 pc and Cu pc are studied at 20 to 90°C as a function of frequency in the range 10^2 to 10^7 Hz.[83]

Because the pc molecule is planar, symmetrical, and does not contain polar groups, its dielectric constant is small and independent of temperature. However, at temperatures close to room temperature and somewhat lower, spontaneous slow increase in ε and tan δ are observed. In this state, pc is characterized by sharp temperature dependence of ε and tan δ, and pc has increased electrical conductivity by two to three orders of magnitude.[84]

The magnitude of the pyroelectric effect in pc increases with increased polarization voltage and at the maximum it also increases with increasing heat rate. The maximum pyrocurrent is at 17 to 20°C, which corresponds to the position of the maximum dielectric constant.[85]

Dielectric constant and loss measurements are made on α, β, and X metal-free and Cu pcs at microwave frequencies.[86] Dielectric conductivities of these pcs appear to be consistent with dc conductivities.

"Alternating current measurements carried out in thin films (about 0.7 μm) of a Cu pc semiconductor with the use of aluminum electrodes have shown that in the $T > 30$°C. region the capacitance and the dielectric loss factor exhibit complex frequency and temperature dependences typical of relaxation phenomena. Such a behavior proves to be due to the nonuniform conductivity over the film thickness because of the presence of blocking contacts on the Al-Cu pc interfaces. These studies have shown that the presence of traps with a depth of 0.55 ev and a concentration of 3×10^{22} m^{-3} in Cu pc leads to formation of Schottky barriers having widths of about 0.1 μm."*

The lifetime of Diflon polycarbonate films with 0.4% titania and 0.01% pc dye is an inverse function of the mean applied partial discharge fields, or the dielectric fields, raised to some power, depending on the geometrical dimensions of the film.[88]

B. Photodielectric Effect

A thin pc film condensed *in vacuo* from vapor onto quartz substrate, between a semitransparent Pt electrode and an Al electrode is exposed to white light from an incandescent lamp. Capacity and dielectric loss angle (tan δ) are measured *in vacuo* at 10^{-4} to 10^{-5} torr at room temperature. The capacity of the pc film increases 25 to 30% from the dark value when exposed to light.[89] After transition of pc into the ferroelectric state, its dielectric constant exhibits sensitivity to light. On illumination in the neighborhood of the phase transition, the dielectric constant decreases, but independently

* From Vidadi, Yu. A., Kocharli, K. Sh., Barkhalov, B. Sh., and Sadredinov, S. A., *Phys. Status Solidi, A*, 33(1), K67, 1976. With permission.

of the intensity of illumination.[90] The photodielectric effect is observed in the beryllium and copper complexes of pc at a frequency of 10^{10} Hz.[91]

At sufficiently high temperature the photodielectric effect is observed in pc films.[92]

XVI. SECONDARY ELECTRON EMISSION

"Studies on the secondary electron emission (SEE) from solids yields valuable information on the nature of motion of electrons and the dissipation of their energy. Such information is of great interest, especially for organic molecular crystals as the application of other direct methods for investigating the motion of slow electrons in these systems is very restricted."[93]

"The coefficients of secondary electron emission (SEE) in Cu pc and free pc have been determined. They are found to be close to the values determined earlier for the SEE coefficients in aromatic crystals . . . The experimental curve of the elastic reflection coefficient (k) vs. the energy of primary electrons (Ep) for free pc exhibits a steep drop when $E_p > 200$ eV. Such a behavior $k(E_p)$ is attributed to the scattering of electrons at the carbon atom, and is regarded as one of the causes for the stable position of the maximum of the SEE coefficient in organic crystals for $E_p \approx 200$ eV."*

XVII. HALL EFFECT

In 1879 E. H. Hall devised an experiment that gives the sign of the charge carriers in a conductor; it is known as the Hall effect. The experiment originally showed that only negative charge is free to move in metals.[94]

"The first Hall measurement ever made with organic semiconductor crystals is reported for metal free pc single crystals $5 \times 0.5 \times 0.3$ mm with Cu 200—2000, Na 1—10, Ca 1—10, Fe 6—60, Si 0.1—1, Mg 3—30, Al < ppm."[95] The carriers are negative electrons.

A highly sensitive device is described for measuring the photoconductivity and the Hall effect of pc and other substances.[96]

A measurement of the Hall effect in pc is the subject of a thesis.[97]

Formulas for the Hall coefficient and Hall mobility are derived.[98] Comparison of the experimental and theoretical values for pc shows good agreement.

XVIII. THERMOELECTRIC POWER AND SEEBECK COEFFICIENT

Some modifications of metal-free pc show characteristic differences in dark conductivity and thermoelectric power, depending on the preparation procedure.[99]

The thermoelectric or Seebeck coefficient and electrical conductivity are tabulated for $AlCl_2 \cdot 2H_2O$ pc, $AlCl(HSO_4)$ pc, AlCl(OH) pc, AlCl pc, $Al(OH)H_2 \cdot$pc, Al_2O di pc, $SnCl_2$pc, and VO pc.[100]

Seebeck measurements made on Mn(II) pc indicate that Mn pc has a negative Seebeck coefficient, that is, it decreases with temperature. "The dark resistivity, about 2×10^{-6} ohm-cm, is the lowest of the metal pcs investigated so far."[101]

Measurements are made of the thermoelectric power of β Cu pc single crystals, and the Seebeck coefficient is discussed using a model which considers trap-free solids and some trap distributions. "For many years we have been engaged in the measurement of the thermoelectric power of organic semiconductors, especially of pc and its metal complexes. Studying metal complexes (H_2 pc, Cu pc, Ni pc, Zn pc) we have found that in any case the β modification has a higher positive thermo emf than the α modification."[102]

* From Bubnov, L. Ya. and Frankevich, E. L., *Phys. Status Solidi B*, 69(1), 195, 1975. With permission.

XIX. CONTACT POTENTIAL DIFFERENCE AND WORK FUNCTION

"The evaluation of work function of organic crystals has been required for many years. Its importance in recent studies of electrical conduction in organic crystals is increasing, particularly in view of its relation to the carrier injection at a metallic contact. The measurement of contact potential difference (c.p.d.) is applicable to organic crystals and gives their work functions. The contact potential difference is a surface effect and is affected by the condition of the surface or by the ambient gases. It can be used for the investigation of solid surfaces. However, the same reason makes it often difficult to estimate the work function or the Fermi level experimentally free from surface effects. The evaluation of work functions may lose its meaning in some cases, unless one can distinguish the surface effect from the bulk one. There are already some reports on the measurement of the contact potentials of organic crystals, but published data on their work functions are very scarce."[103]* Using the vibrating condenser method, the contact potential difference between pc and Cu pc and reference metals or graphite is measured.[103]

The contact potential difference between gallium and mercury on the basis of SCLCs in pcs is determined.[104] Also, the contact potential difference between mercury and alkali metal amalgams is measured by the pc film method coupled with recording of voltammetric curves.[105]

XX. PCS AS ELECTRODES

Pcs, organic semiconductors, are studied as electrodes.[106-120] The electrode potential of pcs is studied under a variety of conditions — electrolytes, environmental gas, temperature, and light. The visible absorption spectra of the di pcs of Y, Nd, Eu, and Lu rare earths as a function of electrode potential are measured.[106] The film of Lu pc is applied on a glass sheet coated with SnO_2 acting as an electrode. An electrode is coated with Lu di pc to determine visually the electrode potential from the solution color.[107] Lu di pc is deposited on an electrode of a plate of glass coated with an electroconductive PbO_2 film. The same electrode is tested for extraction of iodine from solution.[108]

Gas diffusion type electrodes are obtained with noble metal catalysts such as palladium by depositing them on sintered nickel base substrates. The substrates "are impregnated with a mixture of a metal pc such as Co pc and fluorocarbon polymer emulsion, and heat treated at a temperature near the melting point of the fluorocarbon resin to give gas diffusion type electrodes. The electrodes have excellent catalytic activity even when small amounts of noble metals are used, and they are useful as air battery electrodes."[109]

Metal pcs are used in doping single crystal n-type semiconductor electrodes.[110] of ZnO and CdSe. "Artificial surface states were generated by the electrodeposition of noble metals e.g. Pt or Au, in submonolayer amounts or by precipitation of metal pcs or porphyrins on the surface. Various common oxidation reduction reactions were studied."

A number of possible active organometallic substances are tested for ion selective electrodes,[111] including metal pcs. The metal pc electrodes are responsive to anions rather than to cations and some anion selectivity is observed. "The metal pcs offer a series of water insoluble complexes where the electrical conductivity, although very low, is adequate for constructing electrodes by means of the 'Selectrode' technique."** "Solid state electrodes prepared from magnesium, lead, copper and iron (II) pcs were studied. For the first three of these compounds the response curves

* From Kotani, M. and Akamatsu, H., *Discuss. Faraday Soc.*, 51, 94, 1971. With permission.

** From Sharp, M., *Anal. Chim. Acta*, 76(1), 165, 1975. With permission.

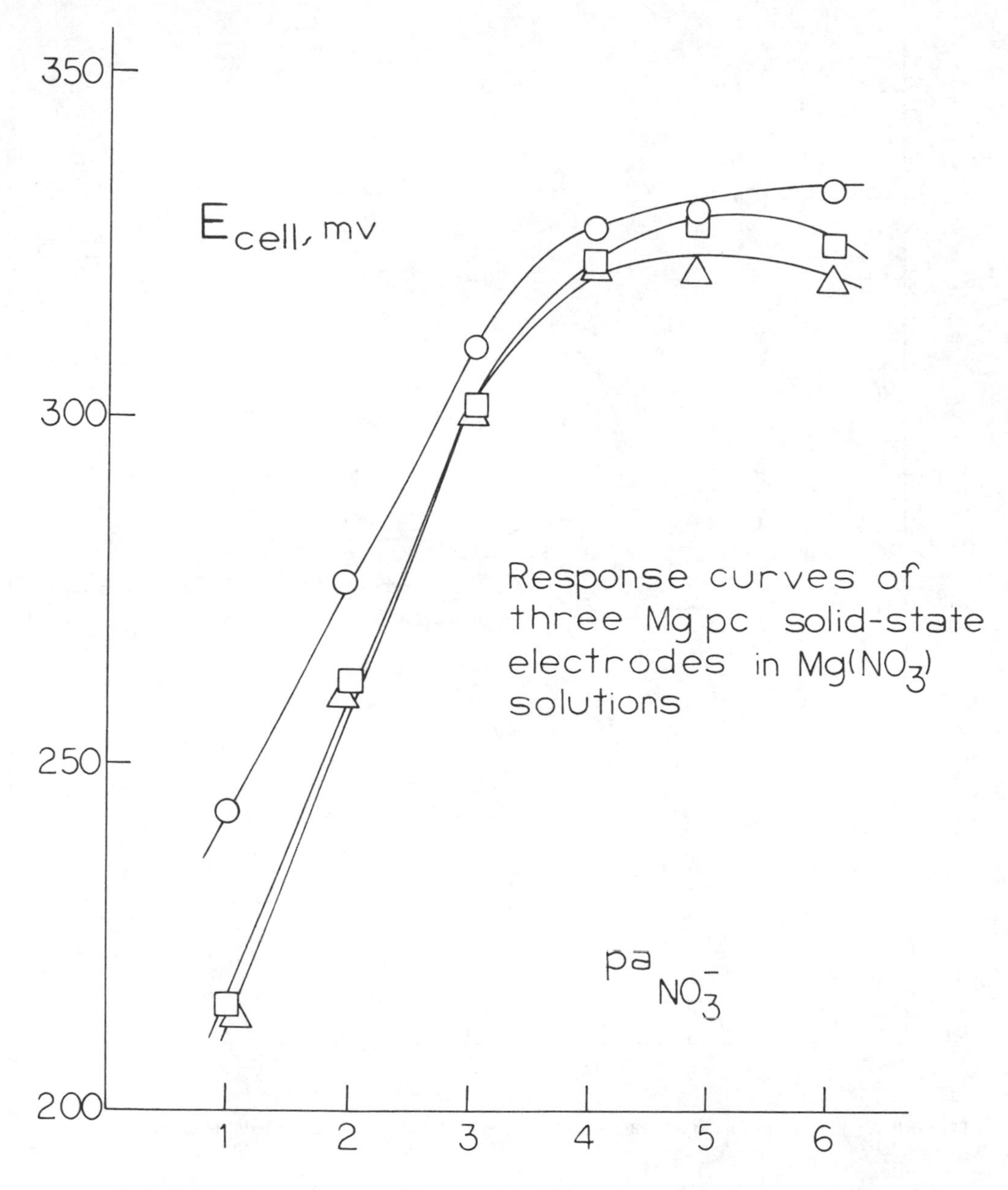

FIGURE 1. Response curves of three MgPc solid-state electrodes in $Mg(NO_3)$ solutions.

measured in the respective metal nitrate solutions were characteristic of anion rather than cation sensitivity. The results for Mg pc are typical. Figure 1 shows the response curves obtained for three Mg pc electrodes. It is apparent that sensitivity to nitrate activity variations prevails. Figure 2 illustrates the response of a Mg pc electrode measured in sodium nitrate solutions. A close correspondence between the curves of Figures 1 and 2 is clear. Additional measurements in sodium chloride and sulfate solutions, included in Figure 2, were also characteristic of anion response, and comparison between the nitrate, chloride, and sulfate curves reflected some degree of anion selectivity. Somewhat poorer reproducibility and selectivity were observed for the Pb pc and Cu pc systems. The results obtained for Fe pc electrodes were so irreproducible that

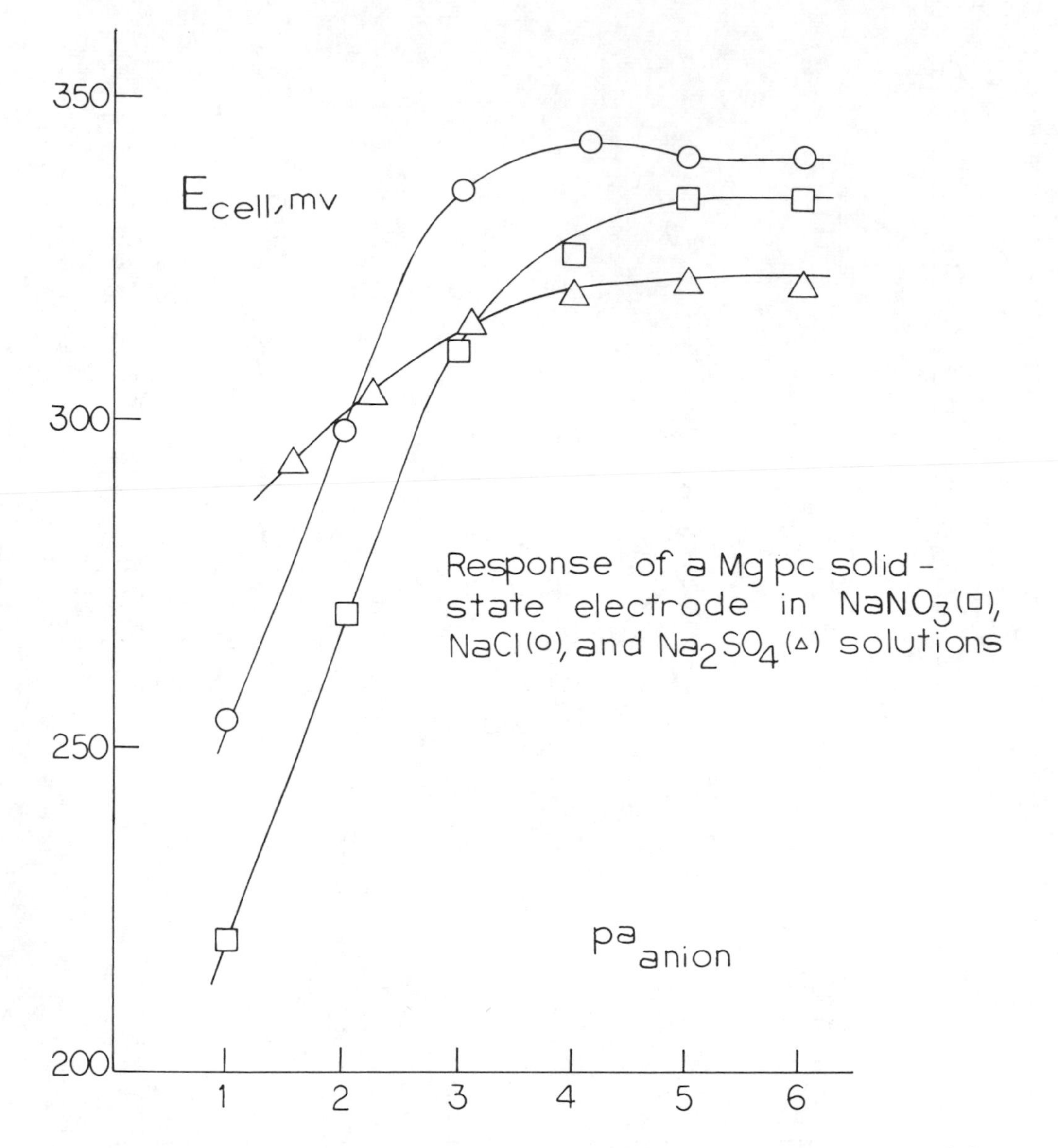

FIGURE 2. Response of an MgPc solid-state electrode in $NaNO_3$ (□) NaCl (O) and Na_2SO_4 (Δ) solutions.

no definite trends in response behavior or selectivity characteristics could reliably be discerned.''*

The photovoltaic effect of Cu pc films on a Pt electrode is dependent on electrolyte composition, both potential and current increasing with increase in pH. More effective cells of this type should be capable of generating oxygen from water.[112] Photocurrent kinetics for Pt electrodes covered with films of chlorophyll and pc are also described.[113] A simplified equivalent circuit is given to illustrate the dependence on electrode potential. Comparative photovoltaic characteristics of Co, Fe, Mg, Cu, and Be pcs deposited on Pt, Pd, Au, Ta, Nb, and Ni substrates as a function of electrolyte pH show that the photoconductivity of pcs increases in the sequence Co pc $<$ Cu pc $<$ Fe pc $<$ Be pc $<$ Mg pc.[114]

* From Sharp, M., *Anal. Chim. Acta,* 76(1), 165, 1975. With permission.

A Pt electrode carrying a film of pc or chlorophyll is placed in an electrolyte.[115] An applied potential on the electrode affects the value and the sign of the photopotential, arising on exposure of the film to light.

Continuous illumination and an oscillographic technique are used to study the potential change during illumination of a Pt electrode with a film of pc or chlorophyll in an electrolyte solution (usually 1 to 2 *M* KCl).[116]

The electrochemical activity of Cu pc electrodes deposited on Au was studied in contact with an 0.1 *M* K_2SO_4 electrolyte in the absence and presence of oxygen. The photocurrents induced by illumination of the electrodes at different wavelengths allow detection of the appearance of two different phenomena — photodesorption followed by oxygen reduction.[117]

A short circuit photocurrent produced by a semitransparent Pt electrode covered by a thin film of pc and its variation with light intensity was studied.[118] A model is defined for the acts of photoreaction.

At low voltages, the current relating to the photovoltaic and rectification properties of Al-Mg pc/Ag Schottky-barrier cells varies exponentially with voltage.[119] At 690 nm, with light incident on the Al side, the photovoltaic efficiency is about 0.1%, "one of the highest ever reported for organic photovoltaic cells."

Polarization measurements are made of cathodes containing various carbon fluorine compounds, CF_n (n = 0.24 to 1.5) as the active depolarizer material, in 1 *M* $LiClO_4$-propylene carbonate, at 298 K.[120] Included as depolarizer materials are graphite fluorides and the oxygen cross-linked polymers perfluoro pc $C_{32}N_8F_{56}$.

XXI. PHOTOCONDUCTION

The relaxation processes of photoconductivity of pcs are studied by the "τ-meter" procedure.[121] The results show that the kinetics of photoconductivity of pcs and their Cu and Mg complexes take place in an interval of 10^{-5} to 10^{-2} sec, and the oscillograms show that the photocurrent has a complicated character.

The activation energy for dark conduction of metal-free pc is 1.9 eV at 0 to 100°C. The activation energy for photoconduction is 0.34 eV.[122]

The spectral response of photoconduction in metal-free pc crystals is obtained in the region 2500 to 25,000 Å.[123] Photoconductivities of α and β forms of metal-free pc are observed. Activation energies for dark conduction are 1.9 eV for the β form and 1.4 eV for the α form, while the corresponding activation energies for photoconduction are 0.34 and 0.42 eV, respectively.[124] These differences are interpreted as due to the disordered structure of the α form.

Electric and photoelectric properties of several high ohmic semiconductor layers including H_2 pc are measured in terms of a variety of parameters including the kinetics of the photoelectric effect.[125]

The addition of *p*-chloranil, iodine, and tetracyanoethylene to the surface of films of metal-free α pc increases the dark conductivity of such films by 5×10^2, 2×10^6, and 2×10^3 times, respectively, and increases the steady-state photoconductivity by as much as 4×10^2, 8×10^4, and 9×10^2 times, respectively.[126]

The influence of vibronic coupling on one- and two-particle dissociation processes of Frenkel excitons is investigated.[127]

With respect to the pc crystal, "the probability of exciton dissociation with simultaneous annihilation of an intramolecular phonon coupled with the exciton increases with temperature with an activation energy of 0.187 eV, and this is in agreement with experimental observation" [D. R. Kearns, G. Tollin, and M. Calvin, *J. Chem. Phys.*, 32, 1020, 1960].[139] The first suggestion that in the pc crystal the exciton dissociates with inclusion of phonons is stated.[139] We conclude that this dissociation is connected with the annihilation of one quantum of the totally symmetrical intramolecular vibration, which gives vibronic structure in the spectrum of a free molecule and a dimer."*

Photoelectromotive force of several dyes, presumably including pcs, is studied in various atmospheres — vacuum, air, oxygen, and ethanol vapor.[128]

Samples of 1 to 10-μm-thick film of α Cu pc are irradiated with the light of a Hg or W lamp, filtered through a monochromator to study the internal photoeffect. The results can be explained quantitatively by assuming an exponential distribution of electron traps.[129] The photoconductivity kinetics of the barrier photoeffect of pc layers containing no metal atoms are studied.[130,131]

Photoconductivity of 6480 and 7850 Å of β form Zn pc single crystals is studied at 0 to 120°C. The photocurrent increases exponentially with light intensity.[132]

The effect of oxygen on the dark conductivity and on the stationary and pulse photoconductivity of α and β H_2 pc and α and β Mg pc is studied.[133,134]

The differences in the response of oxygen to α and β pc, α and β Mg pc are probably due to differences in molecular packing and perfection of the crystal structure.

DC conductivity measurements are made for crystals of metal-free and Cu pcs at 5000 to 20,000 cm^{-1}.[135] Photoconductivity in the visible region results from traps formed by the interaction of oxygen with free radical impurities near the surface of the crystals.

The paper "Spectra and Photoconduction of Pc Complexes" is the subject of a discussion.[136]

Photoconduction response curves are measured at 300 to 1500 mμ for single crystals of Co, Ni, Cu, Zn, and H_2 pcs.[137] H_2, Ni, and Cu pcs also have considerable photosensitivity in the near IR, peaking at 900 to 1100 mμ.

The photocurrent of metal-free, Mn, Fe(II), Co, Ni, Cu(II), and Zn is depressed markedly by oxygen, nitrogen, and water vapor.[138]

Drift mobilities of electrons (0.43 to 0.70 cm^2/Vsec) and holes (0.24 to 0.56 cm^2/Vsec) are measured by illumination with short light pulses.[139]

The photoconductivity of pcs in ultrahigh frequency fields is measured.[140] Capacitance photoemf of Cu pc films of about 2 μm thickness is measured in the near IR at 0.8 to 1.2 μm at room temperature.[141]

Photoelectric properties of CdS-Cu pc heterojunctions are studied.[142]

The photovoltaic effect of Cu pc pigment films in contact with electrolytes is measured. A variety of parameters are tested. Of the gases tested it is the presence of only oxygen and air that induces the occurrence of photoresponse.[143,144]

A review, with several references, primarily of the authors' work, of electron phototransport, presumably including Cu pc complex solutions, is presented.[145]

Photoelectric properties of an aluminum-chloro-pc complex[146] and of bis(aluminum pc) oxide[147] are studied.

The activation energy for photoconductivity for aluminum pcs increases in the order pcAlOK < pcAlOH < pcAlCl < pcAlF < $pcAlOC_6H_5$.[148] Both dark and photoconductivity are strongly activated by oxygen, for pcAlF and pcAlOH, weakly activated for pcAlOK, and unaffected for pcAlCl and pcAlOH.

The nature of the photovoltaic effect in films of pcs is the subject of a discussion.[149]

Thermally stimulated currents (TSC) in metal and metal-free pc single crystals (H_2pc, Zn pc, Co pc, Cu pc, Ni pc) and the carrier drift mobilities are studied.[150] The trap is estimated to be 0.37 ± 0.03 eV for these pcs. The drift mobility of electrons in Zn pc is found to be nearly 0.1 cm^2/Vsec at room temperature by the time of flight method. Also, photoconductivity of Zn pc induced by ruby and glass lasers shows that

* From Zgierski, M. Z., *Phys. Status Solidi B*, 59(2), 589, 1973. With permission.

the triplet state plays an important role in the carrier generation in the near IR. The phosphorescence emission in Zn pc and Cu pc supports this result.[150,151]

"Al/Mg pc/Ag sandwich cells show rectification and photovoltaic response due to a Schottky barrier at the Al/Mg pc interface. The barrier width is narrow, but it is not due to high carrier density. It is possibly related to a high trap density, about $10^{18}/cm^3$. The electron diffusion length is estimated to be about 1.5×10^{-6} cm, the quantum efficiency 1.5×10^{-3}, the lifetime about 10^{-9} sec, and the mobility about 0.1 $cm^2/Vsec$. The photovoltaic action spectra can be explained by considering that only those carriers that reach the barrier contribute to the short circuit photo current . . . At 690 nm, with light incident on the Al side, the photovoltaic efficiency is about 0.01 percent, one of the highest ever reported for organic photovoltaic cells."*

Electric and photoelectric measurements on surfaces of single crystals of Cu pc are quantitatively explained by the existence of an exponential trap distribution. The negative photoeffect and the temperature dependence of the photocurrent, influenced by the light intensity, are explained.[153]

A research disclosure relates to a photoimmobilized electrophoretic recording (PIER) process.[154] "When electrically photosensitive particles, at least some of which bear a positive polarity electrostatic charge, are disposed between two spaced electrodes . . . subjected to an electric field, and imagewise exposed to activating electromagnetic radiation, image formation can be achieved by immobilizing at least a portion of the exposed photosensitive particles and causing at least a portion of the unexpected particles to undergo a net change in charge polarity. A suitable photoconductive pigment dispersion contains β Cupc C.I. No. 74160."

The conductivity, over a range of temperature, in DC and AC fields, photoconductivity, and dielectric effects are obtained for Si and Ge pcs.[155] The photoelectric sensitivity of both Si pc and Ge pc is low.

XXII. PHOTO EFFECTS

Nelson's electron beam retardation method and photoelectric emission are used to measure the electron affinities for a number of pcs. Their affinities are found to be about 4.3 eV. Photoemission thresholds for α Cu pc and α H_2 pc are 5.2 eV and 6.1 eV, respectively. The band gaps calculated from the photoemission thresholds and electron affinities are about equal to the optical threshold for photoconduction.[156]

Photopotentials and photocurrents in various combinations of n or p semiconductors, and p or n conducting dyes, presumably including pcs, are discussed in an article entitled "P-n Junctions Between Organic and Inorganic Photoconductors."[157]

The photoionization energies of Mg pc and Cu pc single crystals are determined in different ambient atmospheres by an electrostatic method. The classic electron donor, H, raises the threshold of ionization whereas oxygen does not effect any change in the threshold from its value in vacuum.[158]

"The nature of the conductivity in β Cu pc and the electron transfer through the interfaces ZnO/pc and pc/chemisorbed species charged either positively or negatively are studied by means of a micro-probe discharge technique. The dark decay and the speed of discharge under illumination of the following photoconductor systems Al/In/ZnO single crystal Cu pc resin, Al/Cu pc resin, Al/ZnO powder resin/Cu pc resin, and Al/ZnO powder and Cu pc resin, charged under a dc corona discharge, are investigated."[159] A positive surface charge on a Cu pc layer disappears rapidly when illuminated with light of wavelength $\geqslant$ 550 nm indicating that the singlet excitons generated in the surface region of the pc layer are responsible for the fast discharge. "Based

* From Ghosh, A. K., Morel, D. L., Feng, T., Shaw, R. F., and Rowe, Jr., C. A., *J. Appl. Phys.*, 45(1), 230, 1974. With permission.

on this study, layers containing only β Cu pc resin without and with ZnO exhibit excellent electrophotographic behavior and can be used as bichargeable material."

"Photoinjection of charge from some photoconductive materials into low dielectric constant fluids can be observed using light in the visible region of the spectrum. The quantum efficiency for this process is low unless a strong electric field of order 10^5 V/cm or greater, is applied."[160] Measurements are described "of negative charge photoinjection from sublimed films of metal-free pc into hydrocarbon and dimethyl siloxane fluids. Metal-free pc is chosen because it is a well characterized organic material whose photoconductive properties have been extensively studied. Films can be prepared by vacuum evaporation and are photoconductive in red light. A transient photoinjection technique is used to measure the mobility of ions released into the fluids, the rate of injection, and the injection efficiency."[160]

Photoemission and optical studies of pcs and porphyrins are the subject of a thesis.[161]

"Photoemission from a photoconductor into an insulator is important in many electrographic processes. Its investigation provides a means of understanding the mechanisms that limit current flow across the photoconductor-insulator interface . . . investigations of charge transfer from photoconductors into insulating fluids are presented and an appropriate theoretical formation is developed."*

Photoelectrophoretic pigments giving good quality reproductions at comparatively low light intensity consist of pc metal (Co and Cu) complexes sensitized by tri or pentamethine-cyanines, e.g., 3,3′-diethylthiadicarbocyanine iodide.[163]

Photoelectrophoretic image production is achieved by electrophoretic pigments, such as ZnO or pcs, suspended in an insulating fluid and subjected to an electric field. They undergo a change of charge by illumination and thereafter change the direction of their migration.[164] "The improvement over prior art lies in the addition to the suspension of small amounts of stabilizer (e.g. calcium diisopropylsalicylate, copper oleate) which interacts with the free electrons, or reacts with the positive holes formed during exposure. In this manner, images characterized by a broad grey scale and the absence of objectionable background are produced."[164]

Cu pc is a photoelectrophoretic pigment in an electrophoretic imaging process based on the photoinduced immobilization of photosensitive pigment particles.[165] The photoelectrophoretic process is especially suitable in providing a multicolor image using a mixture of $\geqslant 2$ differently colored photosensitive pigment particles, each of which is responsive to activating radiation of a different wavelength.

In an electrical double layer process in electrophotography metal-free pc is cited as an electrophotographic photoconductive toner.[166]

Results of the energy spectra of photoelectrons during photoelectronic emission from the surface of anthracene, pentacene, naphthacene, pcs of Mg, Cu, Al, Fe, and some other organic compounds, and the values of their photoelectric work functions are presented.[167] Photoemission measurements on Cu pc films are made for photon energies of 7 to 23 eV.[168]

A magnetophotoconductive effect is found in Cu pc single crystals heated at 120°C in an oxygen atmosphere and excited with a W-iodide lamp in electric fields of about 10^3V/cm. The change in photoconductivity relative to that measured at room temperature in equilibrium with air is +0.5 to 1.0% for magnetic fields H of about 8000 G.[169]

H_2pc, V_2O_5pc, and AlCl pc used in a photovoltaic cell give two to ten times higher currents as compared with other pcs.[170] The effect of electrolyte pH is measured. Photopotential of H_2 pc films and polarization curves in the dark and illuminated by a laser is shown for a H_2 pc-Nb electrode in 0.1 *N*KCl at pH 6.8.[171] The importance of

* From Hartmann, G. C. and Shmidlin, F. W., *J. Appl. Phys.*, 46(1), 266, 1975. With permission.

electrolyte concentration in optimizing the conditions for producing high photopotential and photocurrent values is emphasized for a H_2 pc film formed on a surface of a Pt electrode in KCl solution.[172] The photochemical behavior of porous films of H_2 pc, Cu pc, and Fe pc under galvanostatic conditions is studied.[173] The photoelectrochemical effects of a H_2 pc film on a semitransparent metallic electrode is investigated.[174]

Techniques involving the use of an electrolytic condenser and monochromator are applied to the study of photopotential distribution at 400 to 1500 mμ in films of pc and chlorophyll contacting the electrolyte.[175] Maximum photopotential in air is observed for pc at 610 mμ and for chlorophyll at 840 mμ.

Photoelectric phenomena are observed in chlorophyll, its derivatives, and pcs.[176]

REFERENCES

1. Witkiewicz, Z. and Dabrowski, R., *Wiad. Chem.*, 27(6), 385, 1973.
2. Yoshino, K., Kaneto, K., Kyokane, J., and Inuishi, Y., *Technol. Rep. Osaka Univ.*, 21(995—1026), 549, 1971.
3. Beales, K. J., Eley, D. D., Hazeldine, D. J., and Palmer, T. F., Katalyse an Phthalocyaninen, Symp., Hamburg, 1, 1973.
4. Heilmeier, G. H. and Warfield, G., *J. Chem. Phys.*, 38, 163, 1963.
5. Heilmeier, G. H., Semiconduction in PC: A Study in Organic Semiconduction, thesis, Univ. Microfilms, Ann Arbor, Mich., Order No. 63-526; *Diss. Abstr.*, 23, 2959, 1963.
6. Fielding, P. E. and MacKay, A. G., *J. Chem. Phys.*, 38(11), 2777, 1963.
7. Fielding, P. E. and MacKay, A. G., *Aust. J. Chem.*, 17(7), 750, 1964.
8. Nasirdinov, S. D., Shugam, E. A., Berger, L. I., Plyushchev, V. E., and Shklover, L. P., *Zh. Fiz. Khim.*, 40(3), 741, 1966.
9. Kronick, P. L., Bloor, J. E., and Labes, M. M., *Mol. Cryst.*, 1(1), 113, 1966.
10. Westgate, C. R., Transient Current Measurements in Pc, An Organic Semiconductor, thesis, Univ. Microfilms, Ann Arbor, Mich., Order No. 66-9652; *Diss. Abstr. B*, B 27(4), 1265, 1966.
11. Levina, S. D., Astakhov, I. I., Lobanova, K. P., and Rotenberg, Z. A., *Electrokhimiya*, 2(11), 1343, 1966.
12. Rotenberg, Z. A., Levina, S. D., and Korob, L. A., *Electrokhimiya*, 2(10), 1224, 1966.
13. Barbe, D. F., Bulk and Surface Conduction in Metal-Free Pc, thesis, Univ. Microfilms, Ann Arbor, Mich., Order No. 69-21,110; *Diss. Abstr. Int. B*, 30(6), 2684, 1969.
14. Vidadi, Yu. A., Rozenshtein, L. D., and Chistyakov, E. A., *Fiz. Tverd. Tela*, 12(2), 634, 1970.
15. Starke, M. and Hamann, C., *Z. Phys. Chem. (Leipzig)*, 243(3—4), 166, 1970.
16. Petrova, M. L. and Rozenshtein, L. D., *Fiz. Tverd. Tela*, 12(3), 961, 1970.
17. Aoyagi, Y., Masuda, K., and Namba, S., *J. Phys. Soc. Jpn.*, 31(2), 524, 1971.
18. Aoyagi, Y., Masuda, K., and Namba, S., *J. Phys. Soc. Jpn.*, 31(1), 164, 1971.
19. Meshkova, G. N. and Vartanyan, A. T., *Fiz. Tekh. Poluprov.*, 6(11), 2227, 1972.
20. Dabrowski, R., Witkiewicz, Z., and Waclawek,, W., *Pr. Nauk. Inst. Chem. Org. Fiz. Politech. Wroclaw*, 7, 329, 1974.
21. Koifman, O. I., Zemlyanaya, N. G., Al'yanov, M. I., and Larionov, V. R., *Izv. Vyssh. Ucheb. Zaved. Khim. Khim. Tekhnol.*, 18(10), 1644, 1975.
22. Makles, M., Przywarska-Boniecka, H., and Wojciechowski, W., *Rocz. Chem.*, 49(10), 1647, 1975.
23. Blagodarov, A. N., Lutsenko, E. L., and Khairusova, L. I., *Izv. Vyssh. Ucheb, Zaved. Priborostr.*, 16(10), 114, 1973.
24. Cox, G. A. and Knight, P. C., *J. Phys. Chem. Solids*, 34, 1655, 1973.
25. Beales, K. J., Eley, D. D., Hazeldine, D. J., and Palmer, T. F., *Ketal. Phthalocyaninen, Symp.*, Georg Thieme, Stuttgart, 1, 1972.
26. Fedorov, M. I. and Benderskii, V. A., *Fiz. Tekh. Poluprov.*, 4(10), 2007, 1970.
27. Johnson, J. W., Office of Technical Services, AD 277,109, U.S. Department of Commerce, Washington, D.C., 1961.
28. Kirin, I. S. and Moskalev, P. N., *Zh. Fiz. Khim.*, 41(2), 497, 1967.
29. Shiihara, I., Takahashi, H., Uchifuji, K., Adachi, K., and Yamaguchi, M., Japanese Kokai, 69, 05, 711, March 10, 1969.
30. Ukei, K., Takamoto, K., and Kanda, E., *Low Temperature Phys.-LT 13, Proc. 13th Int. Conf. Low Temp. Phys.*, Vol. 4, Plenum Press, New York, 1974, 301.

31. Heilmeier, G. H. and Warfield, G., *J. Appl. Phys.*, 34(8), 2278, 1963.
32. Mathur, S. C. and Ramesh, N., *Proc. 17th Nucl. Phys. Solid State Phys. Symp.*, 16C, 64, 1973.
33. Mathur, S. C. and Ramesh, N., *Chem. Phys. Lett.*, 37(2), 276, 1976.
34. Ukei, K., *J. Phys. Soc. Jpn.*, 40(1), 140, 1976.
35. Bradley, R. S., Grace, J. D., and Munro, D. C., *Trans. Faraday Soc.*, 58, 776, 1962.
36. Onodera, A., Kawai, N., and Kobayashi, T., *Solid State Commun.*, 17(7), 775, 1975.
37. Bradley, R. S., Grace, J. D., and Munro, D. C., *J. Phys. Chem. Solids*, 725, 1964.
38. Blagodarov, A. N., Lutsenko, E. L., and Rozenshtein, L. D., *Fiz. Tverd. Tela*, 11(11), 3379, 1969.
39. Fendley, J. J. and Jonscher, A. K., *J. Chem. Soc. Faraday Trans. 1*, 69(Part 7), 1213, 1973.
40. Sakai, Y., Sadaoka, Y., and Yokuchi, H., *Bull. Chem. Soc. Jpn.*, 47(8), 1886, 1974.
41. Sadaoka, Y. and Sakai, Y., *J. Chem. Soc. Faraday Trans. 2*, 72(2), 379, 1976.
42. Assour, J. M. and Harrison, S. E., *J. Phys. Chem. Solids*, 26(3), 670, 1965.
43. Kaufhold, J. and Hauffe, K., *Ber. Bunsenges. Phys. Chem.*, 69(2), 168, 1965.
44. Meyer, H., *Angew. Chem. Int. Ed.*, 4(8), 619, 1965.
45. Van Oirschot, Th. G. J., Van Leeuwen, D., and Medema, J., *J. Electroanal. Chem. Interfacial Electrochem.*, 37, 373, 1972.
46. Yasunaga, H., Kojima, K., Yoshida, H., and Takeya, K., *J. Phys. Soc. Jpn.*, 37(4), 1024, 1974.
47. Preobrazhenskii, N. I. and Kolyadin, A. B., *Zh. Fiz. Khim.*, 49(3), 692, 1975.
48. Wihksne, K. and Newkirk, A. E., *J. Chem. Phys.*, 34, 2184, 1961.
49. Hamann, C. and Storbeck, I., *Naturwissenschaften*, 50, 327, 1963.
50. Harrison, S. E. and Ludewig, K. H., *J. Chem. Phys.*, 45(1), 343, 1966.
51. Assour, J. M. and Harrison, S. E., *J. Phys. Chem.*, 68(4), 872, 1964.
52. Masuda, K. and Namba, S., *Energy Charge Transfer Org. Semicond., Proc U.S.-Japan Semin.*, Plenum Press, New York, 1973, 53.
53. Tuemmler, W. B., (to Monsanto Company), U.S. Patent 3,245,965, April 12, 1966.
54. Norrell, C. J., Pohl, H. A., Thomas, M., and Berlin, K. D., *J. Polym. Sci.*, 12(5), 913, 1974.
55. Byrd, N. R. and Sheratte, M. B., NASA Contract. Rep. NASA-CR-134885, National Technical Information Service, Springfield, Va., 1975.
56. Le Blanc, O. H., Jr., *Am. Chem. Soc. Div. Polym. Chem. Prepr.*, 4(1), 220, 1963.
57. Westgate, C. R. and Warfield, G., *J. Chem. Phys.*, 46(1), 94, 1967.
58. Westgate, C. R. and Warfield, G., *J. Chem. Phys.*, 46(2), 537, 1967.
59. Devaux, P. and Mas, J., *Solid State Commun.*, 7(16), 1095, 1969.
60. Rotenberg, Z. A., *Elektrokhimiya*, 3(10), 1269, 1967.
61. Lutsenko, E. L., Rozenshtein, L. D., and Blagodarov, A. N., *Vses. Konf. Fiz. Dielektr. Perspekt. Razvit. Sb. Ref.*, 3, 134, 1973.
62. Barbe, D. F. and Westgate, C. R., *Solid State Commun.*, 7(7), 563, 1969.
63. Barbe, D. F. and Westgate, C. R., *J. Chem. Phys.*, 52(8), 4046, 1970.
64. Cox, G. A. and Knight, P. C., *J. Phys. Chem. Solids*, 34(10), 1655, 1973.
65. Cox, G. A. and Knight, P. C., *J. Phys. Chem.*, 7(1), 146, 1974.
66. Mueller, M. A., Mihai, I. C., and Mueller, L. P., *Phys. Status Solidi A*, 4(2), 479, 1971.
67. Cox, G. A. and Knight, P. C., *Phys. Status Solidi B*, 50(2), K135, 1972.
68. Yoshino, K., Kaneto, K., Tatsuno, K., and Inuishi, Y., *Technol. Rep. Osaka Univ.*, 585, 1972.
69. Aoyagi, Y., Masuda, K., and Namba, S. *Mol. Cryst. Liq. Cryst.*, 22(3—4), 301, 1973.
70. Hamann, C., Starke, M., and Wagner, H., *Phys. Status Solidi A*, 16(2), 463, 1973.
71. Lehmann, G. and Hamann, C., *Phys. Status Solidi B*, 55(2), 585, 1973.
72. Hamann, C. and Lehmann, G., *Phys. Status Solidi B*, 60(1), 407, 1973.
73. Lehmann, G. and Hamann, C., *Phys. Status Solidi B*, 63(1), 341, 1974.
74. Blagodarov, A. N., Lutsenko, E. L., and Rozenshtein, L. D., *Phys. Status Solidi A*, 5(2), 333, 1971.
75. Kearns, D. R. and Calvin, M., *J. Chem. Phys.*, 34, 2022, 1961.
76. Devaux, P. and Delacote, G. M., *J. Chem. Phys.*, 52(9), 4922, 1970.
77. Barbe, D. F. and Westgate, C. R., *J. Phys. Chem. Solids*, 31(12), 2679, 1970.
78. Haak, F. A. and Nolta, J. P., *J. Chem. Phys.*, 38, 2648, 1963.
79. Fustoss-Wegner, M., *Thin-Solid Films*, 36(1), 89, 1976.
80. Fustoss-Wegner, M. and Ritvay, E. K., *KFKI*, 1975.
81. Vidadi, Yu. A., Rozenshtein, L. D., and Chistyakov, E. A., *Fiz. Tverd. Tela*, 11(9), 2695, 1969.
82. Huggins, C. M. and Sharbaugh, A. H., *J. Chem. Phys.*, 38, 393, 1963.
83. Vidadi, Yu. A. and Rozenshtein, L. D., *Electrokhimiya*, 3(10), 1241, 1967.
84. Vidadi, Yu. A., Chistyakov, E. A., and Rozenshtein, L. D., *Fiz. Tverd. Tela*, 11(8), 2403, 1969.
85. Chistyakov, E. A., Vidadi, Yu. A., and Rozenshtein, L. D., *Fiz. Tverd. Tela*, 12(9), 2772, 1970.
86. Abkowitz, M. A. and Lakatos, A. I., *J. Chem. Phys.*, 57(12), 5033, 1972.
87. Vidadi, Yu. A., Kocharli, K. Sh., Barkhalov, B. Sh., and Sadredinov, S. A., *Phys. Status Solidi A*, 33(1), K67, 1976.

88. Il'chenko, N. S., Kirilenko, V. M., Okis, Ya. V., Mikhal'chenko, V. I., and Sverbii, N. A., *Dielektriki,* 1, 44, 1971.
89. Vidadi, Yu. A. and Rozenshtein, L. D., *Dokl. Akad. Nauk S.S.S.R.,* 168(5), 1041, 1966.
90. Chistyakov, E. A., Vidadi, Yu. A., and Rozenshtein, L. D., *Fiz. Tverd. Tela,* 11(11), 3383, 1969.
91. Ionov, L. N. and Akimov, I. A., *Fiz. Tekh. Poluprov.,* 5(10), 2017, 1971.
92. Blagodarov, A. N., Lutsenko, E. L., and Rozenshtein, L. D., *Fiz. Tverd. Tela,* 12(5), 1549, 1970.
93. Bubnov, L. Ya. and Frankevich, E. L., *Phys. Status Solidi B,* 69(1), 195, 1975.
94. Halliday, D. and Resnick, R., *Physics, Part II,* John Wiley & Sons, New York, 1962.
95. Heilmeier, G. H. and Warfield, G., *Phys. Rev. Lett.,* 8, 309, 1962.
96. Trukhan, E. M., *Prib. Tekhn. Eksp.,* 10(4), 198, 1965.
97. Rona, M., A measurement of the Hall effect in metal free pc (thesis), Univ. Microfilms, Ann Arbor, Mich., Order No. 67-5747; *Diss. Abstr. B,* 27(11), 3944, 1967.
98. Kubarev, S. I. and Mikhailov, I. D., *Teor. Eksp. Khim., Akad. Nauk Ukr. S.S.R.* 1(4), 488, 1965.
99. Hamann, C. and Starke, M., *Phys. Status Solidi,* 4(3), 509, 1964.
100. Starke, M. and Hamann, C., *Z. Anorg. Allg. Chem.,* 354(1—2), 1, 1967.
101. Atkinson, L., Day, P., and Price, M. G., *Phys. Status Solidi A,* 2(3) K157, 1970.
102. Hamann, C., *Phys. Status Solidi A,* 10(2), 509, 1972.
103. Kotani, M. and Akamatsu, H., *Discuss. Faraday Soc.,* 51, 94, 1971.
104. Rotenberg, Z. A. and Levina, S. D., *Elektrokhimiya,* 5(10), 1141, 1969.
105. Levina, S. D., Korob, L. A., and Kalenchuk, G. E., *Elektrokhimiya,* 6(9), 1334, 1970.
106. Moskalev, P. N. and Kirin, I. S., *Opt. Spektrosk.,* 29(2), 414, 1970.
107. Kirin, I. S. and Moskalev, P. N., Russian Patent 396,611, August 29, 1973.
108. Moskalev, P. N. and Kirina, N. I., *Zh. Prikl. Khim. (Leningrad),* 48(2), 371, 1975.
109. Fukuda, M., Iwaki, T., and Kobayashi, Y., Japanese Kokai 75 21, 225, March 6, 1975.
110. Gerischer, H., Petting, B., and Luebke, M., *Proc. Symp. Electrocatal. Soc.,* Princeton, N.J., 1974, 162.
111. Sharp, M., *Anal. Chim. Acta,* 76(1), 165, 1975.
112. Komissarov, G. G., Shumov, Yu. S., and Morozova, O. L., *Biofizika,* 15(6), 1120, 1970.
113. Alferov, G. A., Sevast'yanov, V. I., Ilatovskii, V. A., Shumov, Yu. S., and Komissarov, G. G., *Dokl. Akad. Nauk S.S.S.R.,* 207 (3), 628, 1972.
114. Ilatovskii, V. A. and Komissarov, G. G., *Zh. Fiz. Khim.,* 49(5), 1353, 1975.
115. Evstigneev, V. B., Savkina, I. G., and Gavrilova, V. A., *Biofizika,* 7, 298, 1962.
116. Evstigneev, V. B. and Savkina, I. G., *Biofizika,* 8(2), 181, 1963.
117. Schreiber, B. and Savy, M., *C. R. Hebd. Seances Acad. Sci. Ser. C,* 282 (17), 787, 1976.
118. Villar, J. G., *C. R. Acad. Sci. Ser. D,* 275 (4), 595, 1972.
119. Ghosh, A. K., Morel, D. L., Feng, T., Shaw, R. F., and Rowe, C. A., Jr., *J. Appl. Phys.,* 45(1), 230, 1974.
120. Hunger, H. F. and Ellison, J. E., *J. Electrochem. Soc.,* 122(10), 1288, 1975.
121. Putseiko, E. K., *Dokl. Akad. Nauk S.S.S.R.,* 132, 1299, 1960.
122. Liang, C. Y., Scalco, E. G., and Oneal, G., Jr., *J. Chem. Phys.,* 37, 459, 1962.
123. Heilmeier, G. H. and Warfield, G., *J. Chem. Phys.,* 38, 897, 1963.
124. Liang, C. Y. and Scalco, E. G., *J. Electrochem. Soc.,* 110(7), 779, 1963.
125. Kolomiets, B. T., Lyubin, V. M., Mostovskii, A. A., and Fedorova, E. I., *Elektrofotogr. Magnitografiya Vilnyus,* 36, 1965.
126. Benderskii, V. A. and Usov, N. N., *Dokl. Akad. Nauk S.S.S.R.,* 167(4), 848, 1966.
127. Zgierski, M. Z., *Phys. Status Solidi B,* 59(2), 589, 1973.
128. Glinchuk, K. D., Denisova, A. D., Litovchenko, N. M., and Vorobkalo, F. M., *Ukr. Fiz. Kh.,* 11(7), 745, 1966.
129. Hamann, C., *Wiss. Z. Tech. Hochsch. Karl-Marx-Stadt,* 10(2), 265, 1968.
130. Usov, N. N. and Benderskii, V. A., *Fiz. Tekh. Poluprov,* 2(5), 699, 1968.
131. Benderskii, V. A., Fedorov, M. I., and Usov, N. N., *Dokl. Akad. Nauk S.S.S.R.* 183(5), 1117, 1968.
132. Aoyagi, Y., Masuda, K., and Namba, S., *Sci. Pap. Inst. Phys. Chem. Res. Tokyo,* 63(3), 71, 1969.
133. Usov, N. N. and Benderskii, V. A., *Fiz. Tekh. Poluprov.,* 3(6), 943, 1969.
134. Usov, N. N. and Benderskii, V. A., *Zh. Strukt. Khim.,* 11(2), 281, 1970.
135. Day, P. and Price, M. G., *J. Chem. Soc. A,* 2, 236, 1969.
136. Day, P., Schregg, G., and Williams, R. J. P., *J. Chem. Phys.,* 38(11), 2778, 1963.
137. Day, P. and Williams, R. J. P., *J. Chem. Phys.,* 37, 567, 1962.
138. Day, P., Scregg, G., and Williams, R. J. P., *Nature (London),* 197, 589, 1963.
139. Usov, N. N. and Benderskii, V. A., *Phys. Status Solidi 1B,* 37(2), 535, 1970.
140. Brikenshtein, V. Kh. and Benderskii, V. A., *Dokl. Akad. Nauk S.S.S.R.,* 191(1), 122, 1970.
141. Kurik, M. V. and Vertsimakha, Ya. I., *Fiz. Tekh. Poluprov.,* 5(10), 1985, 1971.
142. Vertsimakha, Ya. I. and Kurik, M. V., *Mikroelektronika,* 1(3), 275, 1972.

143. Shumov, Yu. S., Sevast'yanov, V. I., and Komissarov, G. G., *Biofizika,* 18(1), 48, 1973.
144. Shumov, Yu. S. and Komissarov, G. G., *Biofizika,* 19(5), 830, 1974.
145. Sidorov, A. N. and Maslov, V. G., *Tr. Mosk. Obshchest. Ispyt. Prir.,* 49, 51, 1973.
146. Shorin, V. A. Fedorov, M. I., Borodkin, V. F., and Al'yanov, M. I., *Dokl. Nauch. Tekhn. Konf. Ivanov. Khim. Tekhnol. In-ta,* 87, 1973.
147. Fedorov, M. I., Shaposhnikov, G. P., Shorin, V. A., Borodkin, V. F., and Al'yanov, M. I., *Izv. Vyssh. Ucheb. Zaved. Fiz.,* 19(3), 157, 1976.
148. Shorin, V. A., Meshkova, G. N., Vartanyan, A. T., Pribytkova, N. N., Al'yanov, M. I., and Borodkin, V. F., *Izv. Vyssh. Ucheb. Zaved. Khim. Khim. Tekhnol.,* 16(12), 1904, 1973.
149. Alferov, G. A., *Tr. Mosk. Fiz. Tekhn. In-ta. Ser. Obshch. Mol. Fiz.,* 7, 64, 1975.
150. Yoshino, K., Kaneto, K., Tatsuno, K., and Inuishi, Y., *J. Phys. Soc. Jpn.,* 35(1), 120, 1973.
151. Yoshino, K., Kaneto, K., and Inuishi, Y., Energy Charge Transfer in Organic Semiconductors, *Proc. of the U.S.-Japan Seminar on Energy and Charge Transfer in Organic Semiconductors, Osaka,* 37, 1974.
152. Ghosh, A. K., Morel, D. L., Feng, T., Shaw, R. F., and Rowe, C. A., Jr., *J. Appl. Phys.,* 45(1), 230, 1974.
153. Hamann, C., *Phys. Status Solidi A,* 10(1) 83, 1972.
154. Groner, C. F. and Barney, R. D., *Res. Discl.,* February, 142, 19, 1976.
155. Chistyakov, E. A., Markova, I. Ya., Lutsenko, E. L., Shaulov, Yu. Kh., Rozenshtein, L. D., and Popov, Yu. A., *Dokl. Akad. Nauk S.S.S.R.,* 195(2), 413, 1970.
156. Komp, R. J. and Fitzsimmons, T. J., *Photochem. Photobiol.,* 8(5), 419, 1968.
157. Meier, H. and Albrecht, W., *Ber. Bunsenges. Physik. Chem.,* 69(2), 160, 1965.
158. George, A., *J. Appl. Phys.,* 44(11), 5148, 1973.
159. Hauffe, K., Ionescu, N. I., Meyer-Laack, A., and Petrikat, K., *Curr. Probl. Electrophotogr., 3rd Eur. Colloq.,* Walter De Gruyter, Hawthorn, N.Y., 1972, 271.
160. Hartmann, G. C., *Annu. Rep., 40th Conf. Elec. Insul. Dielec. Phenomena,* Natl. Acad. Sci., Washington, D.C., 1972, 24.
161. Schechtman, B. H., Photoemission and Optical Studies of Organic Solids: Pcs and Porphyrins, thesis, avail. Univ. Microfilms, Ann Arbor, Mich., Order No. 69-14,014, *Diss. Abstr. Int. B,* 30(3), 1295, 1969.
162. Hartmann, G. C. and Schmidlin, F. W., *J. Appl. Phys.,* 46(1), 266, 1975.
163. Kitzing, R. and Zographos, G., German Offenlegungsschrift 2,353,291, May, 9, 1974.
164. Hauffe, K. and Volz, H., German Offenlegungsschrift 2,356,687, June 20, 1974.
165. Groner, C. F. and Barney, R. D., *Res. Discl.,* 142, 19, 1976.
166. Iwasa, M., German Offenlegungsschrift 2,419,873, November 7, 1974.
167. Vilesov, F. I., Zagrubskii, A. A., and Garbusov, D. Z., *Fiz. Tverd. Tela,* 5(7), 2000, 1963.
168. Pong, W. and Smith, J. A., *J. Appl. Phys.,* 44(1), 174, 1973.
169. Harrison, S. E., *J. Chem. Phys.,* 51(1), 465, 1969.
170. Ilatovskii, V. A. and Komissarov, G. G., *Zh. Fiz. Khim.,* 49(5), 1352, 1975.
171. Sevastyanov, V. I., Alferov, G. A., Asanov, A. N., and Komissarov, G. G., *Biofizika,* 20(6), 1004, 1975.
172. Komissarov, G. G., Shumov, Yu. S., and Antonovich, V. A., *Biofizika,* 13(6), 981, 1968.
173. Alferov, G. A. and Sevast'yanov, V. I., *Zh. Fiz. Khim.,* 50(1), 214, 1976.
174. Villar, J.-G., *C. R. Acad. Sci. Paris,* 275(4), 595, 1972.
175. Putseiko, E. K., *Dokl. Akad. Nauk. S.S.S.R.,* 150(2), 343, 1963.
176. Putseiko, E. K., *Elem. Fotoprotsessy Mol. Akad. Nauk S.S.S.R.,* 371, 1966.

Chapter 17

PHTHALOCYANINE AND PHTHALOCYANINE-TYPE COMPOUNDS

I. SUPER PHTHALOCYANINES

"The reaction of *o*-dicyanobenzene with anhydrous uranyl chloride *does not* yield a cyclic, four subunit pc complex.[1] (2-iminoisoindoline) complex — a 'super pc'."*

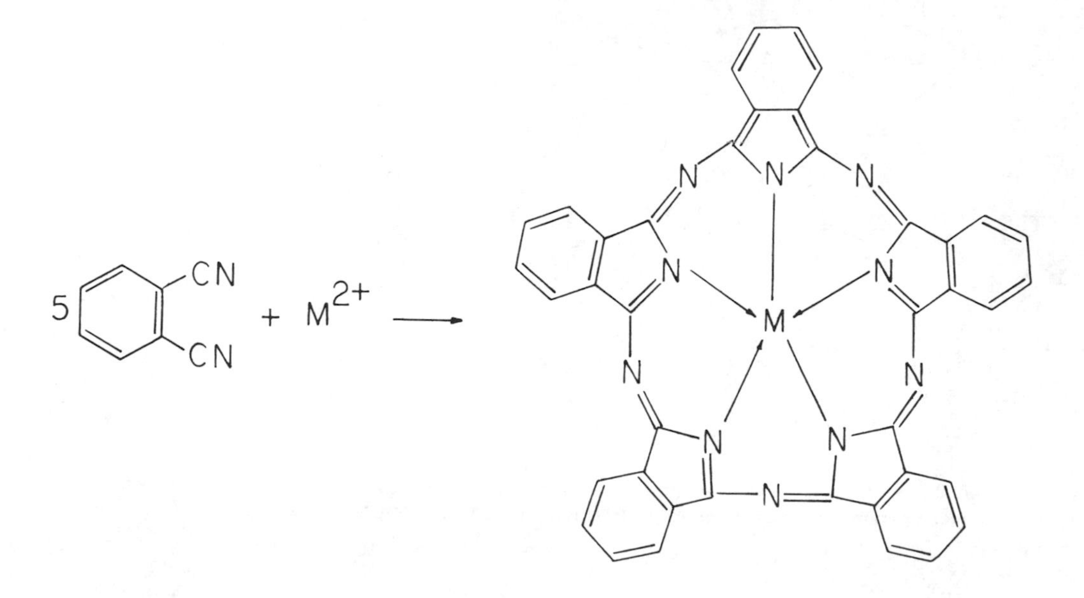

"Dioxocyclopentakis (2-iminoisoindoline) uranium (VI), UO_2 $(N_2C_8H_4)_5$, crystallizes in the monoclinic space group, P $2_{1/c}$, with a = 8.210 Å, b = 21.667 Å, c = 18.462 Å, β = 103.16°, and Z = 4 (ϱ_{calcd} = 1.891, ϱ_{obsd} = 1.882 g/cm³). . . . The coordination geometry of the uranium atom approximates an idealized compressed pentagonal bipyramid. The two axial ligands are oxygen atoms . . . The equatorial coordination is by five nitrogen atoms of the 20 atom 'inner' ring of the 50 atom (excluding hydrogens) macrocycle . . ."* The inner nitrogen atoms and their compression are shown in Figure 1.

"The pronounced tendency of the uranyl ion to form seven coordinate pentagonal bipyramidal complexes with U-N bond lengths of about 2.55 Å suggests that this group is ideally suited for constructing five membered macrocyclic porphyrin analogs via template reactions. Similar studies of other template reaction systems are underway, as are crystallographic studies of the 'free' super pc ligand. Substitution of the uranyl group by other metals may, in some cases, provide a series of interesting new compounds for chemical, spectroscopic, and crystallographic investigation. These studies are in progress. Studies employing metal atom clusters as templates have also begun."[1]* In addition to the formation of super pcs, it is possible to condense them to corresponding pcs according to the reaction[2]

* From Day, V. W., Marks, T., and Wachter, W. A., *J. Am. Chem. Soc.*, 97(16), 4519, 1975. With permission.

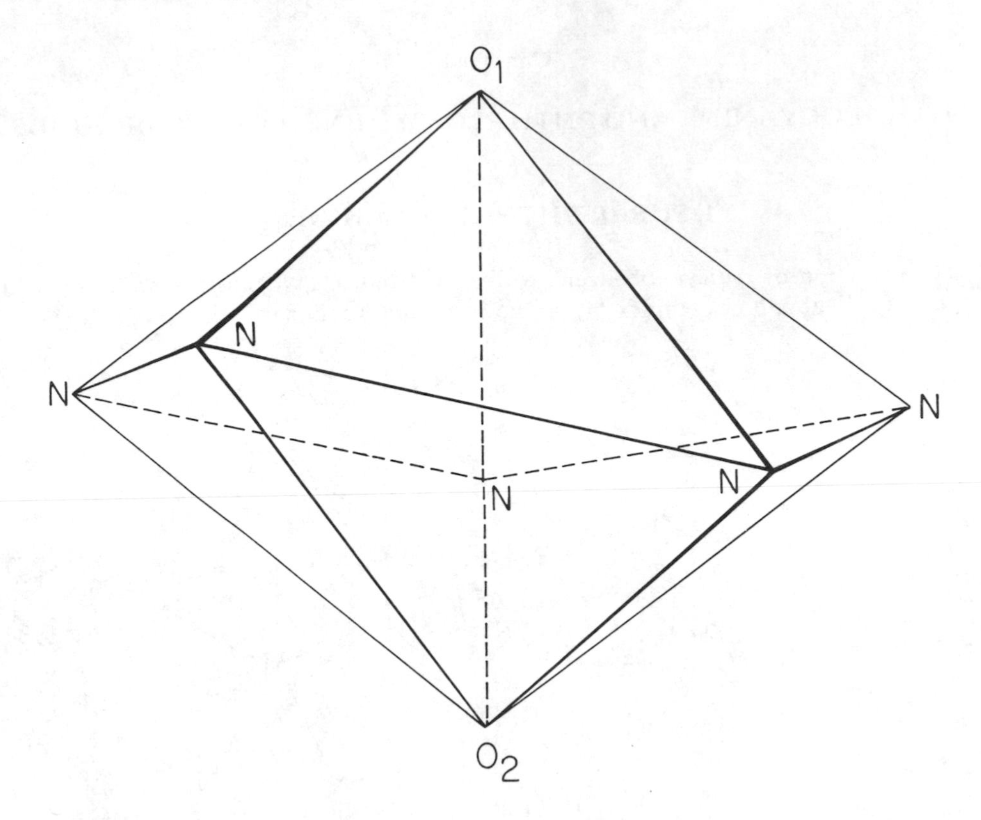

FIGURE 1. Perspective view (adapted from an ORTEP drawing) of the axially compressed pentagonal bipyramidal coordination polyhedron as observed in the $UO_2(N_2C_8H_4)_5$ molecule.

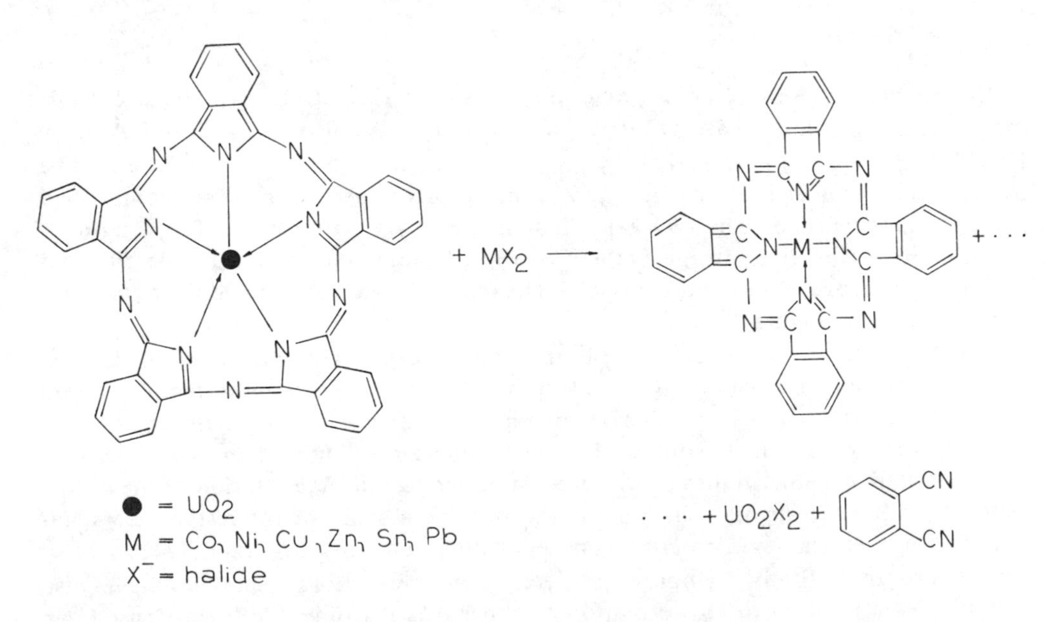

This reaction was announced in 1975 with the statement, "We now report a surprising ligand contraction process which is unprecedented in macrocycle coordination chemistry." "Electronic factors within the ligand do not appear to be important [for contraction] . . . the π energy of the H_2 super pc is not significantly different from that of H_2 pc plus phthalonitrile."[2]

II. PHTHALOCYANINE-TYPE COMPOUNDS

If pc-type compounds may be said to possess a minimum of four nitrogen atoms connected to other atoms (usually carbon) in a conjugated system, and within the core formed by the nitrogen atoms there is lodged two hydrogen atoms, or a metal atom with or without axial ligands, such as oxygen atoms, then there are at least several more novel pc-type compounds that have been synthesized.

Cobalt tribenzocyclohexadienotetrazaporphine is prepared from *cis*-1,2,3,6-tetrahydrophthalonitrile by treatment with cobalt acetate, urea, and ammonium molybdate.

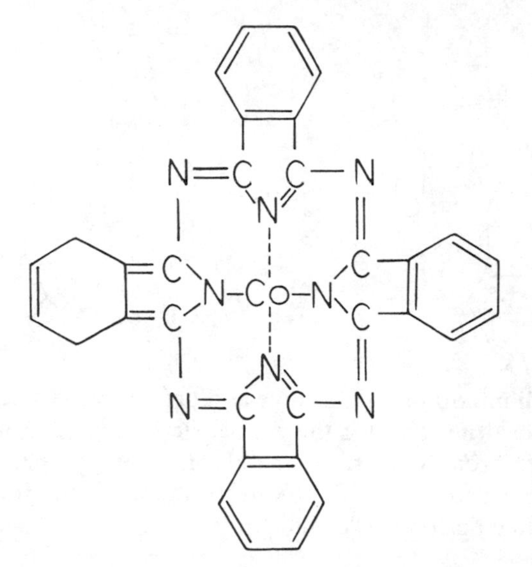

"The new compound resembles Co pc in various properties, but shows a characteristic intense absorption at 450 mμ and exhibits a strong resistance to both hydrogenation and dehydrogenation."[3]

The self-condensation of *o*-aminobenzaldehyde in the presence of Ni^{2+} and Cu^{2+} produces a tetrabenzo [b,f,j,n] [1,5,9,13] tetraazacyclohexadecine.[4] The four nitrogen atoms are planar with the central metal atom.

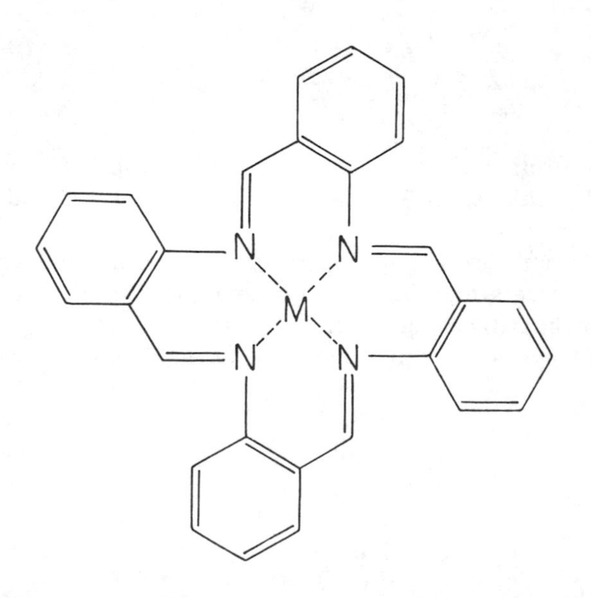

This pc-type macrocycle is "characterized by elemental analysis, ir spectra, magnetic moments, and visible reflectance spectra."[4]

The following pc-type compounds, tetrathio pcs, containing sulfur atoms.[5-7]

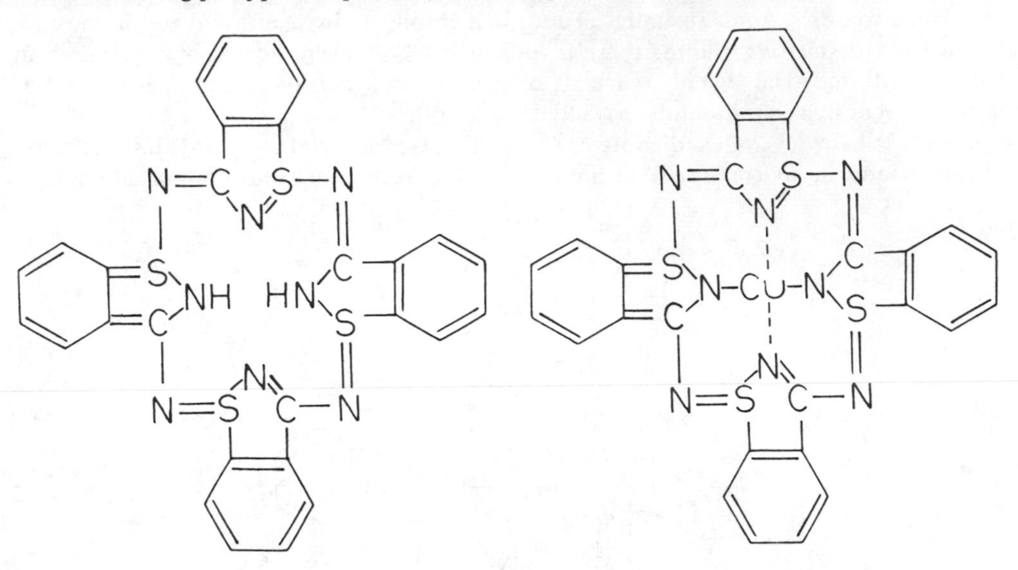

are prepared from diiminothioisoindolenine and urea in butyl alcohol for the H_2 derivative, sodium hydrosulfite, and for the copper derivative, cupric acetate. "The thio compounds are more strongly colored than the corresponding pcs."

Carbon atoms may also be replaced by sulfur atoms in the following pc-type compound, at the indicated positions 1, 2, and 3.[8]

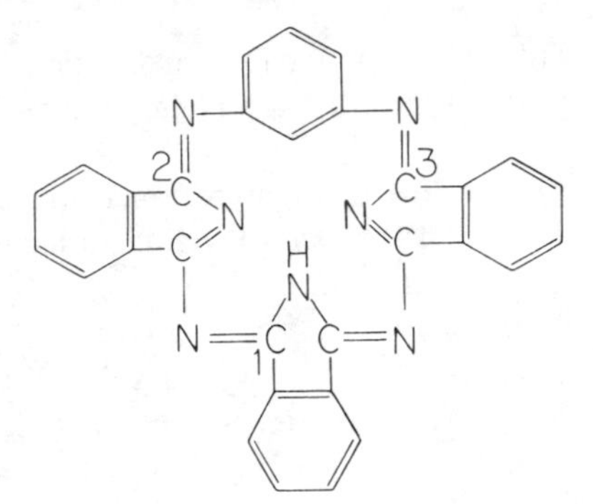

"The principal absorption bands in this macrocycle, and in it with S atoms in 1; 2 and 3; and 1, 2 and 3 positions, are 360 and 510; 360 and 500; 345 and 480; 335 and 430 nm."

"Tetravalent Ge complexes of the macrocyclic, tetradentate ligand hemiporphyrazine (hp H_2) have been synthesized . . . Treatment of the free ligand with $GeCl_4$ in quinoline gives Ge (hp) Cl_2 in good yield."[9]

Copper tetra-3,4-pyridineporphyrazine is synthesized from cinchomeronic acid,

N

COOH

COOH

by reacting a mixture of cinchomeronic acid, ammonium molybdate, urea, 1,2,4-$C_6H_3Cl_3$, and $CuCl_2$. It is "useful as a vat dye dyeing cotton cloth in bluish violet shades from an alkaline hydrosulfite bath."[10]

III. PHTHALOCYANINE ANALOGS

Pc analogs differ from pc-type compounds in that the structure of the pc molecule is retained. In the analog structure, the pc model is retained, to which are added additional carbon-carbon, carbon-nitrogen, or other bonds. A facile or favored route toward their preparation is by the use of dicyanophenyl derivatives.

For example,

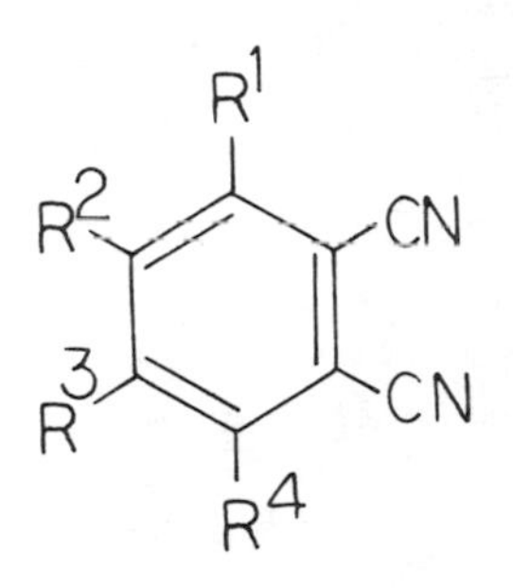

is synthesized as a point of departure for the preparation of pc analogs. In this instance,[11] $R^1 = R^4$ = phenyl, propyl, ethyl, or methyl, and $R^2 = R^3$ = phenyl.

2,3-Naphthalocyanines are prepared from 2,3-$C_{10}H_6(CN)_2$.[12]

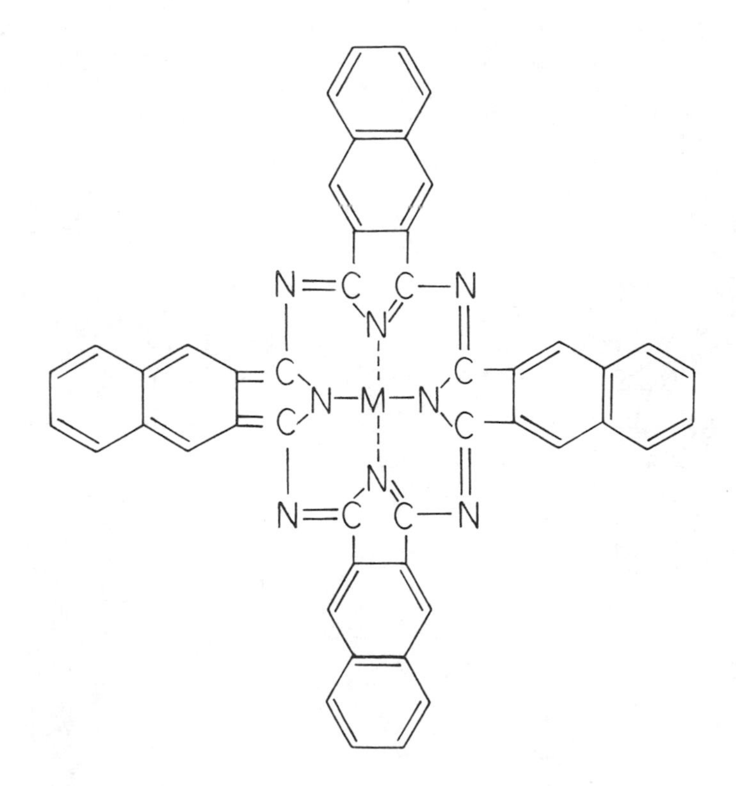

Octa-4,5-phenyl pcs and tetra-2,3-triphenylenoporphyrazines

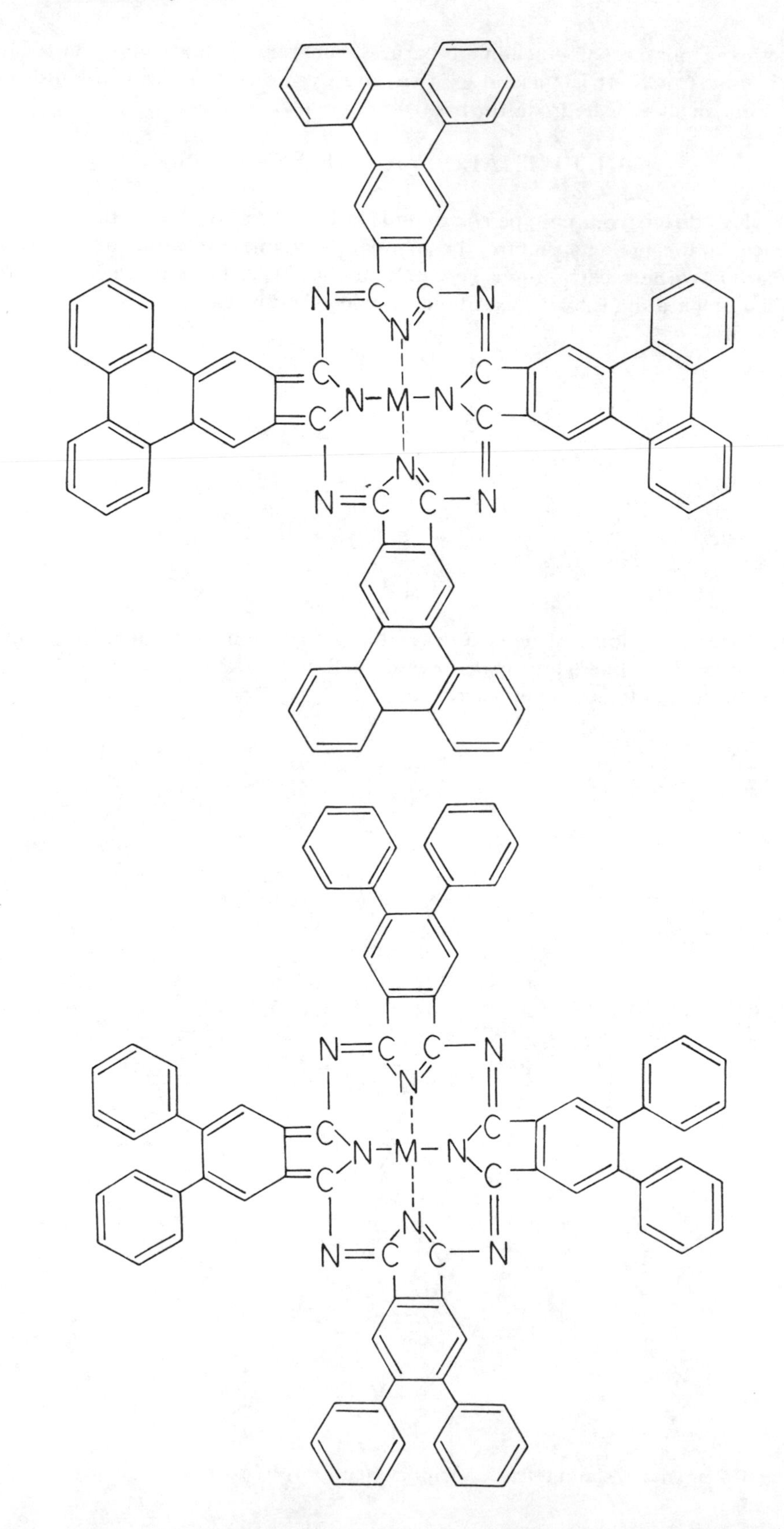
N=C C—N
N
C N-M-N C
C C
N
N=C C—N
N=C C—N
N
C N-M-N C
C C
N
N=C C—N

are prepared from 1,2,4,5-diphenyl $C_6H_2(CN)_2$ and 2,3-dicyanotriphenylene, respectively.[13]

An anthracene analog of pc is prepared by reacting 2,3-dicyano-9,10-diphenylanthracene with urea and the chloride of Cu, V, or Al at 250 to 300°C. Its electronic spectra shows "bands in the 900 nm region, showing a bathochromic shift (100 nm) owing to widening of the conjugated bond system of the ligand relative to pc."[14]

Another pc analog is made by condensing 2,3-quinoxaline-dicarbonitrile in trichlorobenzene and phenylmethanol to give dark green tetra-2,3-quinoxalinoporphyrazine.[15]

REFERENCES

1. Day, V. W., Marks, T., and Wachter, W. A., *J. Am. Chem. Soc.*, 97(16), 4519, 1975.
2. Marks, T. J. and Stojakovic, D. R., *J. Chem. Soc. Chem. Commun.*, 1, 28, 1975.
3. Kaneko, Y. and Fukada, N., *Sci. Rep. Niigata Univ. Ser. C*, 2, 1, 1970.
4. Melson, G. A. and Busch, D. H., *J. Am. Chem. Soc.*, 86 (22), 4834, 1964.
5. Klyuev, V. N. and Snegireva, F. P., *Izv. Vyssh. Ucheb. Zaved. Khim. Khim. Tekhnol.*, 14(2), 258, 1971.
6. Klyuev, V. N. and Snegireva, F. P., (to Ivanovo Chemical Technological Institute), Russian Patent 196, 856, May 31, 1967.
7. Klyuev, V. N. and Snegireva, F. P., Russian Patent 293,024, January 15, 1971.
8. Klyuev, V. N., Berezin, B. D., and Snegireva, F. P., *Izv. Vyssh. Ucheb. Zaved. Khim. Khim. Tekhnol.*, 13(2), 209, 1970.
9. Esposito, J. N., Rafaeloff, R., Starshak, A. J., Sutton, L. E., and Kenney, M. E., in *Proc. 8th Int. Conf. Coord. Chem., Vienna,* Elsevier, Amsterdam, Netherlands, 1964, 160.
10. Yokote, M., Shibamiya, F., and Tokairin, S., *Kogyo Kagaku Zasshi*, 67(1), 166, 1964.
11. Mikhalenko, S. A. and Luk'yanets, E. A., *Zh. Org. Khim.*, 6(1), 171, 1970.
12. Mikhalenko, S. A. and Luk'yanets, E. A., *Zh. Obshch. Khim.*, 39(11), 2554, 1969.
13. Mikhalenko, S. A., Yagodina, L. A., and Luk'yanets, E. A., *Zh. Obshch. Khim.*, 46(7), 1598, 1976.
14. Kopranenkov, V. N. and Luk'yanets, E. A., *Zh. Obshch. Khim.*, 41(10), 2341, 1971.
15. Gal'pern, M. G. and Luk'yanets, E. A., *Zh. Obshch. Khim.*, 39(11), 2536, 1969.

Chapter 18

MAGNESIUM PHTHALOCYANINE

I. INTRODUCTION

The evocative subject of the science and chemistry of Mg pc relates to the similarity between Mg pc and chlorophyll a. In both molecules, a central Mg atom binds covalently to four pyrrole rings, as shown in Figure 1.

The fact that all food energy comes from the sun and is trapped by green plants is a way of saying that chlorophyll a, as well as other chlorophylls, is as indispensable as the sun, carbon dioxide, and water. This great catalyst must be present over the surface of the earth in the amount of billions of tons at any one time. That it can act as the catalyst or intelligent principle for the combination of carbon dioxide and water to create energy rich carbon-carbon bonds that provide us with our sustenance, is one of the wonders of nature.

The elucidation by Linstead and colleagues of the structure and the synthesis of the pcs and of Mg pc took place over a 5-year span. However, this procedure of discovery for chlorophyll required over half a century, resulting in the total synthesis of chlorophyll by Woodward and co-workers in 1960. As part of this immense effort, Professor R. B. Woodward, in his address "The Total Synthesis of Chlorophyll", presented before the Robert A. Welch Foundation Conferences on Chemical Research at Houston, Tex., November 7 to 9, 1960, said, "in a series of elegant investigations completed only during the last few years, Linstead and his associates at Imperial College were able to solve the stereochemical problem [of chlorophyll] and to provide definitive confirmatory detail in respect to the number and disposition of saturated carbon atoms within the nuclear framework. A half century of structural study had culminated in the complete formula for chlorophyll a. . .

"Our active interest in the total synthesis of chlorophyll was initiated four years ago, in 1956. The first questions we asked were very general ones. The structural investigations had been carried out almost entirely during the twilight of the classical period of organic chemistry. Only the very simplest basic elements of theory played any role in the whole vast study. Neither was succor nor control sought in chemical principle, nor was any attempt made to place the often striking observations in any generalized framework. Would the conclusions from such a study stand scrutiny from the viewpoint of the present day? Was the structure proposed for chlorophyll correct? When we embarked upon the examination of these questions, we entered a chemical fairyland, replete with remarkable transformations which provide unusual opportunities for the testing and further development of principle, and we cannot but urge others to follow us in penetrating what must have seemed to many the monolithic wall of a finished body of chemistry."

Toward the conclusion of his discourse on the total synthesis of chlorophyll, Professor Woodward said, "as all projects of such magnitude must be, this one was planned in a fairly elaborate way at the outset. The measure in which our initial plans were realized is very gratifying, but it is at least equally so that major elements of discovery, and increase in understanding through observation and experiment, were involved in our progress. We learned and established much about this important class of compounds which could not have been known, or at best could only have been dimly foreshadowed, before our work was carried out. This fascinating aspect of work in chlorophyll chemistry has by no means been exhausted — indeed, we feel that our studies have opened up many more avenues than they have traversed, and we do not

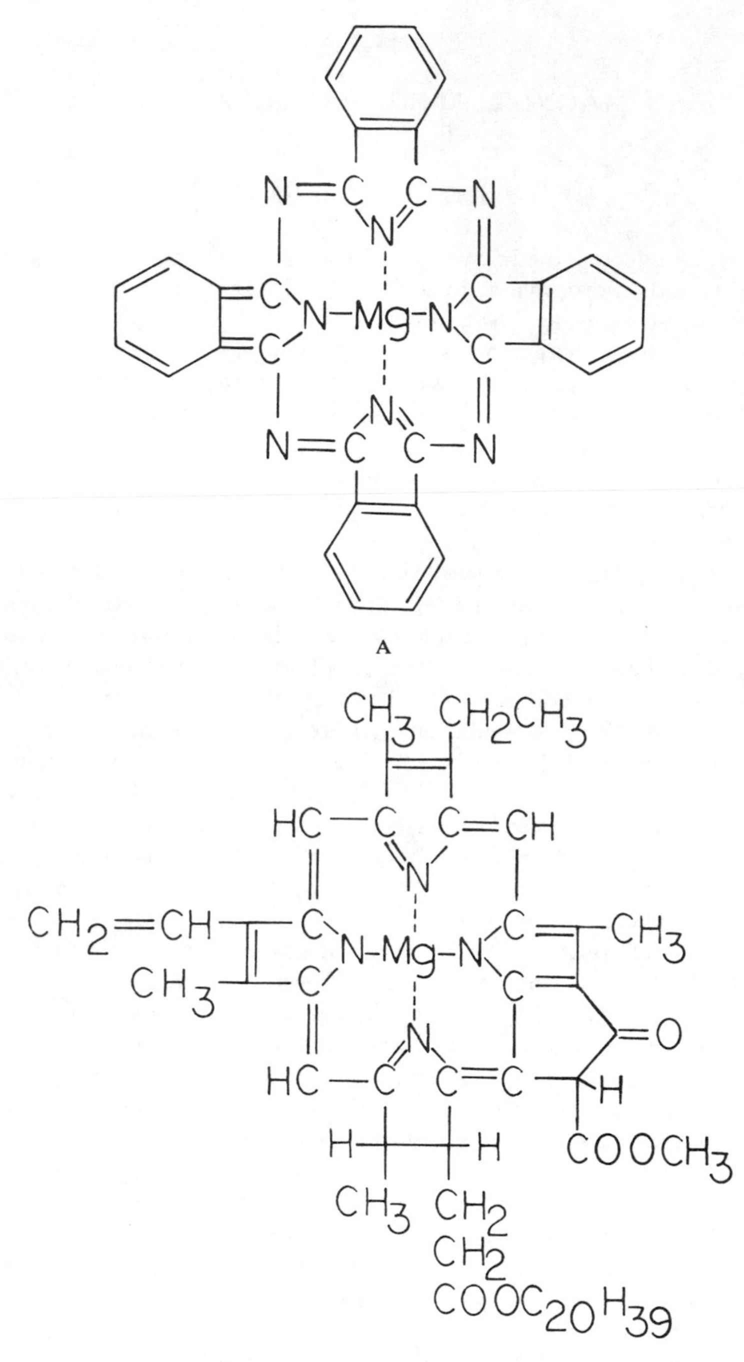

FIGURE 1. In both molecules, a central magnesium atom binds covalently to four pyrrole rings.

hesitate to hazard the opinion that the area is one from which much increase in chemical knowledge and understanding is to be had in the future."*

* From Woodward, R. B., The Total Synthesis of Chlorophyll presented at the Robert A. Welch Foundation Conferences on Chemical Research, Houston, Tex., November 7 to 9, 1960. With permission.

In the Priestley Medal Address "Green Factories", given by Professor Melvin Calvin at the American Chemical Society National Meeting in Anaheim, Calif., in March 1978,[1] Professor Calvin made the following observations. "Sunshine and its conversion into chemical energy and material through the pathways of carbon reduction and quantum conversion in the green plant is our ultimate energy source . . . What we as scientists need to do is to take the basic knowledge acquired in the lab and devise synthetic means of achieving the same kinds of energy processes as the green plant does all by itself, with only input from sunlight. This effort may be considered as having two distinct parts: first, the development of new kinds of 'petroleum plantation' agriculture, with a view toward using the actual substances of the plant for sources of fuel and material, and second, the creation of entirely new synthetic methods for producing in the lab (and ultimately on a much broader scale) the quantum conversion steps of natural photosynthesis with the resulting generation of hydrogen, one of the main energy sources of the future . . . there grows a plant called the 'gopher plant' *Euphorbia lathyris,* whose leaves produce a milky latex — latex, of course, being a hydrocarbon emulsion. The juxtaposition of the idea of using hydrocarbon-producing plants (with the latex emulsion) as a sourse of fuel or materials and the availability of at least one type of such a plant gave rise to an activity in our lab that is continuing to this day . . . our estimate of the hydrocarbon content of *E. lathyris* has been done in terms of organic extractables from the dried plant. This gives a figure of 8 to 10% of the dry weight of the whole plant for the hydrocarbon materials, which are mixtures of polyisoprene, sterols, glycerides, diterpenes, and others. The caloric value, estimated from a carbon-hydrogen-nitrogen analysis, is about 17,000 Btu per pound.

"It thus appears that the first and immediate possibility for the development of an economically useful solar energy capturing device we know, is the green plant, by selecting and modifying it to produce the materials we would like to have, namely, hydrocarbons of suitable molecular weight and structure."*

A university publication tells us "Photosynthesis . . . is absolutely free, if we can find a way to harness it.[2] It has been in use for millions of years by green plants, which convert light and water to energy at absolutely no cost. The main way scientists have obtained hydrogen until this point has been to use electrolysis to separate hydrogen from oxygen in water. This method requires electricity . . . In recent years, scientists have found it possible to redirect photosynthesis in green plants so that they produce oxygen and hydrogen, instead of oxygen along with the sugars, proteins and fats that nourish them.

"For photosynthesis to be changed in this way, the presence of an enzyme, hydrogenase, is necessary. Hydrogenase converts a chemical compound formed in plants into hydrogen. Dr. Krasna has spent twenty years studying the action of this enzyme, and since systems of producing hydrogen that depend on hydrogenase alone have proven unstable and of short duration, he is also studying variations, including synthetic substitutes for hydrogenase.

". . . It has been estimated that a single plant 'factory,' a body of water only a few hundred miles square filled with plants and bacteria, could fill the world's energy needs. The National Science Foundation has estimated that a covered water surface the size of Lake Erie could provide enough energy for the whole United States."**

In a news article entitled " 'Synthetic Leaf' Mimics Plants' Light Conversion", appearing in *Chemical & Engineering News,* February 16, 1976,[3] we read "Chlorophyll device can convert solar energy into electricity, may offer to study in vivo phenomena in green plants. Recent progress in photosynthesis research at Argonne National Lab-

* From Calvin, M., Preistley Medal Address, "Green Factories", *Chem. Eng. News,* March 20, 30, 1978. With permission.

** Reprinted from *Columbia* — the magazine of Columbia University, Spring 1978. With permission.

oratory has provided new insight into the role of chlorophyll in light energy conversion. It also has provided the impetus for developing new devices for solar energy conversion. The first device to issue from this work is a 'synthetic' leaf, a type of photoelectric cell employing chlorophyll as the primary element.

"Chlorophyll is a compound with unusual electron donor and acceptor properties. Most monofunctional nucleophiles tend to deaggregate chlorophyll dimers to form monomer solvates. However, bifunctional ligands, such as dioxane, can cross link chlorophyll molecules by coordination at the central magnesium atoms. A bifunctional ligand of particular interest is water. Water can coordinate to a magnesium atom with the oxygen atom in the water, and it has two hydrogen atoms available for bonding to another chlorophyll molecule. The interaction of water with a chlorophyll oligomer results in the formation of adducts having the general structure −Chl/HOH/Chl/HOH/Chl — which can reach colloidal size.

"Chlorophyll-water adducts are of great interest because they exhibit photoreactivity."[3]*

In "The Greening of the Galaxy: A Conversation with Freeman Dyson",[4] the author says that with respect to the spreading of life, "What will be the main issues that we must confront as a species? Dyson believes that we will be facing more of a biological than a physical question. He challenges the belief that 'we go out into space as just men and that's it. I would say that it is life that is going out into space . . . It will not be just men and machines, but life of all kinds, especially plants . . . *which are the basis of all existence,* and animals. So we shall be a catalyst for phase change in the universe which is not really so concerned with us, but with life as a whole . . . And life will establish itself all over the place."

In the 1960s, in a greenhouse in Ritter Park in Huntington, W. Va., the authors of this monograph experimented with the injection of Mg pc suspensions into the stalks of young plants, to observe whether the Mg pc would migrate upwards and towards the leaves and produce an effect along with the plant's chlorophyll, on the photoactivity of the plants in terms of plant growth, in terms of rapidity and full grown size. The injections, however, did not result in plant growth — upon sectioning the stalks, it was found the Mg pc had not migrated. A suggestion is to repeat the experiments using water-soluble forms of Mg pc. If the presence of Mg pc in a plant can affect or improve its photosynthetic efficiency, a new tool will be provided in the campaign for the use of green life as one of the best collectors of solar energy available to the service of man.

Magnesium pc and chlorophyll are often studied together. For example, there are such titles to their research as "Reversible Photooxidation of Mg Pc in Relation to Chlorophyll Photochemistry",[5] "Mg-25 and Nitrogen-14 Nuclear Quadrupole Resonances in Chlorophyll a and Mg Pc",[6] "Photogalvanic Effect in the Solid Chlorophyll and Mg Pc Films under Pulse Illumination",[7] "Flash Photoexcitation of Adsorbates of Hematoporphyrins, Chlorophyll, Leaf Pigments, and Mg Pc",[8] and "Mg Addend Bond in Complexes of Mg Pc and Chlorophyll".[9]

In the following discussion of research related to Mg pc and chlorophyll, institutions that have been active include: Bakh Institute of Biochemistry, Moscow; Filial Institute of Chemical Physics, Chernogolovka; Institute of Biological Physics, Moscow; Institute of Chemical Physics, Moscow; Institute of Photosynthesis, Pushcino-on-Oka; Institute of Physics, Minsk; Leningrad Gos. Univ. im. Zhdanova, Leningrad; Leningrad State University, Leningrad; V. V. Lomonosov Mosk. Gos. Univ., Moscow; Moscow State University, Moscow; Physics Energy Institute, Riga; and S. I. Vavilov State Optical Institute, Leningrad.

* From *Chem. Eng. News,* February 16, 32, 1976. With permission.

II. PHOTOREDUCTION AND PHOTOOXIDATION

By the method of ESR the formation of free radicals during the photoreduction of chlorophyll a, chlorophyll a and b, phenophytin a and b, hematoporphyrin, and Mg pc by ascorbic acid is detected.[10]

Illumination with an incandescent 300-V lamp through a combination of light filters, KC-14, and another retaining the IR, shows a rapid pH reduction reversible in the dark, in air free chlorophyll ethanol solutions containing hydroquinone.[11] A similar phenomenon is also observed in pheophytin solutions a and b, bacterioviridin, protochlorophyll, and Mg pc. Photooxidation of chlorophyll and its analogs in *p*-benzoquinone results in a higher pH reversible by darkness.

Mg pc is capable of reversible photooxidation with *p*-benzoquinone in ethanol.[5] This reaction is accompanied by the formation of a labile electrode active oxidized form. When the solution is illuminated with red light, the redox potential of the solution is reversibly shifted to the positive side. The main absorption maxima (400 to 600 nm) reversibly decrease and a new maximum (510 nm) appears. A slightly acid medium, pH 4, is favorable for the photooxidation.

The photoreduction of Mg pc in pyridine with phenylhydrazine is studied over a wide range of temperatures.[12] Below −55°C, the probability of proton addition decreases.

The ESR of the anion of Mg pc is observed after irradiation of Mg pc in piperidine with red light in a vacuum.[13] Addition of ascorbic acid to the piperdine solution causes reduction of Mg pc to the dihydro derivative.

The photochemical oxidation of chlorophyll, bacteriochlorophyll, Mg etioporphyrin, and Mg pc by such quinones as fluoranil and *p*-benzoquinone is studied in degassed ethanol solutions at temperatures from 20 to −196°C using EPR.[14] The reaction appears to follow a mechanism suggested by Stiehl and co-workers[72] for chlorophyll and its magnesium containing analogs. Pheophytin gives no such interaction, showing that the metal is required for the reaction with quinone.

Products of the photoreduction of pc, porphyrin, and of their Mg, Cd, Pb, Fe, Co, and Zn complexes obtained in DMF solutions at −60 and 20°C, in the presence of hydrazine, are studied spectroscopically in the 400 to 800-nm region. The results are compared with those in pyridine solutions.[15]

The products of the reduction of Mg pc are identified by EPR, UV, and visible spectra.[16] The THF solution of Mg pc is reduced by sodium or pyridine anions and the photodissociation, oxidation, protonization, and thermal stability of the compounds formed are studied. Photodissociation of Mg pc anions is reversible in frozen solutions.

The interconversion quantum yield of the triplet state of the photosensitizers chlorophyll a, tetraphenyl porphin, Zn tetraphenylporphin, diethyl ether of mesoporphyrin, Cu mesoporphyrin, Mg pc, rhodamine 6G, and 9,10-dimethylanthracene is obtained.[17] α- and β-pinene are oxidized in the presence of Mg pc as photosensitizer (Ne lamp, 30°C, 2-propanol as solvent) to produce 2(10)-pinen-3-yl hydroperoxide and 2-pinen-10-yl hydroperoxide, respectively.[18]

A dependence of photooxidation quantum yield of porphyrins with oxygen on their concentrations is shown.[19] The sensitized photooxidation of some porphyrins by other porphyrins in the presence of oxygen is studied in relation to concentration, intensity, and wavelength of absorbed light, quantum yield, and interconversion and lifetime of molecules in the triplet state.

The ESR spectral data of the triplet excited state of plant pigments including porphyrins, their metallized and metal-free forms, and Mg pc are tabulated.[20] In polymethyl methacrylate matrix the triplet signal persists to 335 K.

A comparison is made between monomer and dimer aggregate forms of Mg pc in its photochemical activity during reaction with oxygen.[21] Luminescence spectra are also obtained.

III. FLASH PHOTOLYSIS

Absorption spectra of the labile, short-lived products of the flash photolysis of Mg pc in pyridine, dioxane, $HCON(CH_3)_2$, and ethanol are observed.[22] Phenol, $(phenyl)_2NH$, $(phenyl)_3N$, and benzidine are added as reducing agents. The triplet state of Mg pc is observed in pyridine, but when phenol is added, a new band appears.

The decay of the triplet state of porphin derivatives (metal-free porphyrins, Zn, Cd, and Sn porphyrins, chlorophylls, pheophytins, Zn and Mg pcs) is studied by flash photolysis[23] in two viscous solvents, castor oil and propylene glycol, at 200 to 300 K, and in rigid glasses at 77 K. In general, the results indicate a decay mechanism of purely unimolecular deactivation, accompanied in some cases by temperature-dependent bimolecular quenching due to impurities in either solvent or solute.

Spectra of flash induced intermediates from hematoporphyrins, chlorophyll, leaf pigments, and Mg pc are determined by high-speed spectral scanning during the flash.[8] Chlorophyll, methylchlorophyllide, pheophytin, and macromolecular films of Mg pc triplets are reversibly formed in over 50% yield, with a greater yield at −180°C than at 20°C.

An electron paramagnetic resonance method for photoexcited triplet states with millisecond lifetimes is developed using modulated light excitation and phase-sensitive amplification.[24] It is successfully applied to chlorophylls, porphyrins, and Mg pc in glassy solutions at 77 K.

IV. PHOTO EFFECTS

Films or platelets are formed by evaporating solutions of polyvinyl butyral, polystyrene, styrene, or polyvinyl carbazole in pyridine or dioxane containing one of the following dyes: Mg pc, pc, methylene blue, a mixture of chlorophyllides a and b, chlorophyll, pheophytin, or pinacyanol.[25] The samples are irradiated with 270 mμ radiation and the photoelectric sensitization of the polymers is determined. There is no photosensitization of the polymer films or platelets by these dyes, only photoelectric effects due to the dyes.

Mg pc containing $< 0.1\%$ of organic impurities is used as a photoelement in the form of (2 to 5) $\times 10^{-5}$-cm-thick films with silver and aluminum as the ohmic and barrier layers, respectively.[26] Specimens are evacuated at 80°C and 10^{-5} torr and their electrical properties are measured. Saturation photocurrent in the diode operation occurs at 1 to 3 V according to irradiation intensity, and with increasing temperature up to 70°C, photocurrent increases by 30%, while photosensitivity decreases. On irradiation with pulsated light, photoresponse synchronizes with irradiation frequency within 3×10^{-8} sec.

The spectral dependence of quantum yield is experimentally investigated in 0.15- to 0.30-μm-thin films of Mg pc.[27] The photoemission threshold is 4.85 ± 0.05 eV. Mg pc is an *n*-type semiconductor which is attributed to the abundance of the donor levels lying at 0.60 eV below the conduction band. Heating in oxygen introduces the acceptor levels at 0.65 ± 0.10 eV above the valence band, causing *p*-type conduction.

A study is made of the effect of oxygen and the nonluminous conductivity, σ_T, and of the steady state and impulse photoconductivity, $\Delta\sigma$ and q, respectively, for α and β pcs and α and β Mg pcs.[28] When oxygen is added to α pc, α Mg pc, or β Mg pc, σ_T,

$\Delta\sigma$, and q first decrease, then increase. The sign of the major photocurrent changes from negative to positive during the increase.

The sign change and increase in the photocurrent are explained by the appearance of a new, more effective mechanism for generating the photocurrent carriers, i.e., by decomposition of singlet molecules excitations into impurity centers containing oxygen.

Determination of the threshold for photoionization of solids and liquids by the Millikan condenser technique can be made more accurate by using two different wavelengths of light, both shorter than the threshold wavelength.[29] Measurements made on graphite and Mg pc agree with values obtained by other techniques.

The interaction of a pigment of the chlorophyll type such as Mg pc with an electron acceptor under heterogeneous conditions is studied by the measurement of photopotentials of Mg pc in acid and alkaline media containing KCl electrolyte.[30] The effect of electron acceptance appears with lowering of the pH to the acid side and its magnitude increases with increasing dye concentration.

The absorption and photoelectric sensitivity spectra of amorphous and microcrystalline layers of Mg pc, either deposited *in vacuo* (10^{-5} mm Hg) on mica and quartz or precipitated from concentrated acetone and ether solutions, activated by electron accepting additives, are studied.[31] Layers of Mg pc-deposited *in vacuo* on mica or quartz at 20°C are amorphous. In their absorption spectrum maxima at 680 and 690 nm appear. In the evaporation method, the temperature of the mica and quartz and the pressure influence the state of aggregation. After the deposition of Mg pc on mica and quartz (100 to 150°C) on which previously a gold or platinum layer has been deposited, a new absorption maximum at 820 to 840 mμ appears. The same maximum appears when layers of Mg pc are treated with liquid acetone, ether, or ethanol, or when Mg pc is precipitated from concentrated acetone solution. This effect is attributed to the presence of a crystalline form, possibly β, with included molecules of the solvent. The thermal activation of Mg pc in water vapor *in vacuo* or the treatment by solvents causes an increase of photoelectric effect and the formation of a new photoelectric effect with the maximum at 840 mμ. Adsorption of quinone from dilute ethanol on amorphous Mg pc layers raises the photoelectric sensitivity 100 to 1000 times; at the same time the new photoelectric maximum at 840 mμ appears. This is attributed to the formation of superficial impurity centers on which electrons are fastened.

Experiments are made to determine the effectiveness of the interaction of electron acceptors with electron excited states of tetrapyrrole pigments — analogs of photosynthetic chlorophyll pigment — in degassed solutions of dioxane by a parallel study of electron acceptor deactivation of the singlet excited level, by using a fluorescence quenching method, and deactivation of the triplet level, by using a pulsed, photoexcitation method.[32,33] Chloranil, *p*-benzoquinone, *p*-dinitrobenzene, nitrobenzene, benzene, and nitromethane are used as electron acceptors. Pigments used include Mg pc, Zn tetraphenylporphyrin, and methylchlorophyllide a. Compounds with greater electron affinity have greater deactivating action in relation to both the triplet and the singlet excited states of the pigments.

The mechanism of the photoxidation of thiourea by oxygen in the presence of Mg pc is studied.[34] The photosensitized oxidation of thiourea proceeds with the participation of sensitizer molecules. Singlet excited oxygen molecules in the $^1\Delta_g$ state are formed in the reaction and they serve as oxidizers of thiourea. Photoelectrochemical effects in Mg pc films are measured.[35]

Films of chlorophyll and of Mg pc applied to a platinum electrode, immersed in an electrolyte solution, and exposed to light give rise to a positive electropotential.[36] Activation of Mg pc films is achieved by immersion for several hours in a *p*-benzoquinone

(10^{-4} mol/ℓ in ethylene) solution. The value of their photopotential increases after activation, with a maximum in the 825-nm region.

If a binary solution of Mg pc (I) and $UO_2(NO_3)_2 \cdot 6H_2O$ (II) is illuminated in the absorption band of II, a I·II complex precipitates from pyridine solution.[37] If the solution is illuminated in the absorption band of I, a change occurs, characterized by a decrease in the fluorescence of I. This change reverses in several minutes in the dark, and may result from a photoelectron transfer from I to II. If the solution in ethanol is illuminated in the band of II at −183°C, the absorption bands of I disappear reversibly, simultaneously with the appearance of a new band at 530 mμ. This may arise from a photoinduced hydrogen transfer from the solvent to I, since the change does not occur in pyridine solution.

The interaction of Mg pc with a high molecular weight biological environment is studied.[38] RNA and DNA samples are stained with different concentrations of Mg pc solutions. A stained sample is placed in a sandwich cell between translucent electrodes. Linear phototransmission of the system is determined in alternating monochromatic light of 150 Hz in air and in vacuum in the presence of water vapor and aqueous solutions of electron donors and acceptors. Absorption by diffuse reflection of the pigmented and unpigmented RNA and DNA layers is measured.

V. LUMINESCENCE

"The long lived afterglow caused by light is widely used for analysis of photophysical properties of molecules and crystals, as well as for investigation of photochemical and photobiological processes, and especially photosynthesis. However, in the latter cases the use of afterglow is complicated by lack of sufficient information on its molecular mechanisms. . . .[39]

"There are several types of afterglow of pigments which differ with respect to complexity and nature of intermediate stages. Phosphorescence is a result of a direct transition of light excited pigment molecules from the triplet to ground state; it has been described for chlorophylls a and b, protochlorophyll, pheophytins, and porphyrins.

"Delayed fluorescence is due to an intramolecular thermo-induced triplet-singlet transition or to bimolecular annihilation of the pigment triplet molecules in collision. Both types of delayed fluorescence have been observed for chlorophylls a and b, protochlorophyll, pheophytins, and porphyrins.

Photochemiluminescence occurs in the course of intermolecular photochemical redox reactions of pigments. Three principal ways of observing it are described: (a) immediately after illumination; (b) after addition of reagents into a preilluminated solution; (c) in the process of heating of the illuminated samples, photothermochemiluminescence.

"Photochemiluminescence of the first two types was observed in photoreactions of chlorophyll with isoamylamine, formaldehyde, oxygen, *p*-benzoquinone, as well as in oxidation of the primary photoreduced forms of pigments or in the reduction of the labile product of chlorophyll and bacteriochlorophyll photooxidation.

"Photochemiluminescence is the least studied type of afterglow . . . Therefore, a systematic investigation [is carried out] on one of the most simple cases of pigment photochemiluminescence — afterglow in photooxidation of pigments by oxygen . . ."* The parameters of photochemiluminescence of chlorophyll analogs and dyes are given in Table 1.[39]

* From Krasnovskii, A. A., Jr. and Litvin, F. F., *Photochem. Photobiol.*, 20(2), 133, 1974. With permission.

Table 1
PARAMETERS OF PHOTOCHEMILUMINESCENCE OF CHLOROPHYLL ANALOGS AND DYES

		Thermochemiluminescence		Maxima of emission spectra		
Pigment and wavelength of exciting light	Solvent	Temp during ilumination (°C)	Positions of peaks (°C)	Chemiluminescence at −20—20°C	Fluorescence (20°C)	Phosphorescence (−196°C)
Chlorophyll b	Acetone	−90	−40, −10, +5	630, 660, 720	660, 720	900, 970[a]
⩾600 nm	Ethanol	−70	−20, 0	665	665, 720	920, 980
		−130	−50, +20	685, 730		[37]
Protochlorophyll	Acetone	−90	−45, −15, +5	630, 650, 680	630, 680	—
⩾580 nm	Ethanol	−70	−40, +20	650, 680	640, 680	770, 860
		−130	−50, +10	680, 720		[38]
Bacteriochloro-	Acetone	−90	+5	650	(685[b]), 780	—
phyll ⩾730 nm	Ethanol	−70	−50, +20	665, 750		
Bacterioviridin	Acetone	−90	−50, −35, −15,	510, 580, 620,	670, 725	—
⩾620 nm			+5	670, 720		
	Ethanol	−70	−50, +5, +10	400—800	670, 730	—
		−130	−5	670		
Pheophytin α	Acetone	−90	−5	600—800	675, 730	—
⩾600	Ethanol	−70	−5, +20	600—800		940, 990
		−130	+10	600—800		[37]
Mg-phthalocyanine ⩾600 nm	Acetone	−90	−5	400—800	680	1100
Eosine, ⩾450 nm	Acetone	−90	−60, −40, −15, 0, +10, +20	500, 570	570	700 [9]
		−105	the same and −95	500, 570		
Methylene blue ⩾620 nm	Acetone	−90	−50, −20, −10	565, 700	700	—

[a] Unpublished data obtained in collaboration with Mr. N. N. Lebedev.
[b] The fluorescence maximum of impurity due to the bacteriochlorophyll dark oxidation.

Data are provided relating to the absorption and fluorescence spectra of pc, Mg pc, and protoporphyrin at 77K in *n*-octane solutions.[40]

Pheophytin a, chlorophyll a, Mg pc, and protoporphyrin IX are dissolved in a polar organic solvent and an increasing amount of water or glycerin is added to detect a change in the light absorption spectrum and fluorescence.[41] The changes are suggestive of molecular association.

The luminescence produced by chlorophylls a and b, pheophytin a, protochlorophyll, bacterioviridin, bacteriochlorophyll, and Mg pc is studied in solutions with several organic solvents after illumination in air or after evacuation.[42]

The addition of $FeCl_3 \cdot 6H_2O$ or HCl in dimethyl ketone to a dimethyl ketone solution of Mg pc leads to the displacement of Mg by H in a reaction analogous to the pheophytinization of chlorophyll.[43] This reaction is accompanied by the formation of an unstable intermediate complex which has characteristic bands in the absorption and luminescence spectra.

An investigation is made of the luminescence spectrum of Mg pc adsorbed at 77 K on silica gel and on alumina-silica gel.[44] The spectra have a single band with a maximum at 702 mμ. A similar band is observed for the frozen system Mg pc + $UO_2(NO_3)_2 \cdot 6H_2O$.

Ethanolic solution of chlorophyll a, methyl chlorophyllide, and Mg pc, a methanolic solution of chlorophyllin, and a dioxane solution of pc are studied with laser excitation.[45] The solutions are excited in the short wavelength absorption band and fluorescence results from deactivation from the S_2 level to the S_1 level.

Mg pc and chlorophyll form aggregates with poly(4-vinyl-pyridine).[46] The binding of chlorophyll to the polymer is followed by observing small changes in the position and width of the main red band of the absorption spectrum. "In dense aggregates of chlorophyll on polymer, fluorescence is largely quenched; on further addition of polymer, the fluorescence is restored and the degree of polarization of fluorescence excited by polarized light is increased. The dependence of quenching and of depolarization of fluorescence on polymer concentration suggest that there is energy transfer between pigment molecules attached to the same polymer molecule."[46]*

A spectroscopic study is made of the interaction of Mg pc with photoactive riboflavin.[47] Mg pc (10^{-5} *M*) and riboflavin (10^{-3} *M*) in ethanol is exposed to monochromatic light in the region of Mg pc absorption (436 mμ) at 77 K for 1.5 hr. Besides the main Mg pc luminescence and absorption bands at 678 and 670 mμ, respectively, an additional luminescence band at 686 mμ and the corresponding absorption band at 682 mμ are observed. The intensity of the new bands increases with the time of illumination and reaches a constant value before 1.5 hr. These bands are interpreted as arising from a protonated form of Mg pc resulting from the interaction of Mg pc with photoactivated riboflavin.

Intense light pulses excite new absorption bands of Mg pc with a maximum at 470 mμ.[48] The rate of extinction of the triplet state absorption is measured after flashing by means of a capillary discharge tube (500 J, 10 kV, 5 sec). The deactivation rate of the triplet state and the extinction rate of the fluorescence are measured in the presence of benzoquinone, trinitrobenzene, *m*-dinitrobenzene, and nitrobenzene.

Fluorescence quenching of protonated heterocycle mixtures including Mg pc containing aromatic hydrocarbons indicates that charge transfer complex formation occurs and that quenching results from electron transfer from the aromatic hydrocarbon to the heterocyclic cation.[49]

The suggested reaction of the transfer of hydrogen from a hydrogen donor molecule, HD, to the Mg pc-O complex adsorbed on MgO is

$$(pcMgO_2)_{ads} + HD \rightarrow (pcMgOOH)_{ads}$$

which is accompanied by the change in the spectrum 659 → 666 mμ is confirmed experimentally by measuring and analyzing the fluorescence and absorption spectra of Mg pc adsorbed on MgO at 20 to 375°C and various oxygen pressures, and in the presence of various hydrogen donor molecules (ethanol, diethyl ether, and dimethyl ketone.)[50]

For chlorophyll a, Mg pc, tetraphenylporphin, and mesoporphyrin, the duration of fluorescence, τ, the yield of fluorescence, ϱ, and the yield of intercombination conversion, γ, are measured at 2:5:5 isopropanol-petroleum ether-diethyl ether solutions at −196°C.[51]

Fluorescence spectra are reported for the oxygen complex of Mg pc adsorbed on Mg pc under various partial pressures of diethyl ether.[52]

The fluorescence spectrum of a Mg pc and oxygen complex adsorbed on MgO, exposed to ethanol or water vapor, changes immediately, the maximum at 659 mμ disappears and a new maximum at 666 mμ appears, but hydrogen and diethyl ether cause only a small change in the fluorescence spectrum of the complex.[53] This change indicates the initial stage of the reaction between the complex and attaching hydrogen (diethyl ether) molecule.

* From Seely, G. R., *J. Phys. Chem.*, 71(7), 2091, 1967. With permission.

Changes in the absorption and emission spectra, excitation, and luminescence polarization of Mg pc in the adsorbed state on transition from an aggregated to a monomeric state are studied.[54]

Absorption spectra and fluorescence spectra excited by the He-Ne laser line at 633 mμ are reported for H_2 pc, Mg pc, and Zn pc in the vapor. Cu pc shows no vapor emission.[55] The yields of fluorescence are temperature dependent but relatively independent of the presence of the foreign gases N_2 and CF_2Cl_2. There is no evidence for resonance fluorescence.

Fluorescence spectra, quantum yields, natural radiative lifetimes, and absorption oscillator strengths are reported for a number of porphyrins in benzene solution.[56] Emission yields for Mg pc and Zn pc in 1-chloronaphthalene solution are also cited. The pcs are considerably more fluorescent than the porphyrins, mainly because, perhaps, of their shorter radiative lifetimes. The heavy atom effect on fluorescence is strong with yields in the order $H_2 \sim Mg > Zn \gg Cd$.

The formation of donor-acceptor complexes of O_2, S_2, ethanol, diethyl ketone, diethyl ether, and other molecules with Mg pc molecules adsorbed on MgO crystals is studied by means of fluorescent spectra in a vacuum.[57] "The Mg pc molecule is very actively adsorbed onto the surface of the small crystal face of MgO and under these conditions it fluoresces with a bright red color. In the monolayer adsorbed state it exhibits extremely interesting general features, which allow important conclusions to be drawn about some properties of this molecule as an intercomplex compound, due to the magnesium atom . . . Complexes are formed with donor molecules having an unshared pair of electrons and an ionization potential of not more than 14 eV. The Mg addend complex bond is very strong. The complexes are very active with regard to hydrogen containing molecules, the reaction involving hydrogen transfer to the complex donor component. The phenomena described also hold for the chlorophyll molecule."[57]*

VI. SPECTRA

The IR spectra of a series of porphyrin derivatives are measured and compared in KBr or in solution, including pyrrole, bilirubin, protoporphyrin, mesoporphyrin, Mg mesoporphyrin, tetraphenylporphin, Mg tetraphenylporphin, tetraazaporphin, and Mg pc.[58]

Absorption spectra are obtained of the products of the first stage of reaction of Mg pc and complex forming hydrogen containing molecules adsorbed on magnesium oxide.[59]

Changes in the absorption and emission spectra, excitation, and luminescence polarization of Mg Pc in the adsorbed state on transition into the monomeric state are studied.[54]

The complex refractivity index $m + n - i\chi$, where n is the refractivity index and χ the absorption coefficient is determined for the solid layers of Cu, Zn, and Mg pcs, chlorophyll, methylchlorophyll, tetraphenylchlorophyll, and pheophytin at 360 to 800 nm.[60] From these values, the coefficient k_λ of the losses of the electromagnetic radiation is calculated for the colloidal water solutions of these substances.

Absorption spectra of Mg pc in tetrahydrofuran containing negative pyridine ions in vacuo and in air are determined.[61] The pyridine ions are produced by contact with a sodium mirror.

The magnetooptical rotation spectra of a number of porphyrins, pcs, and their metal

* From Gachkovskii, V. F., *Adv. Mol. Relaxation Processes*, 5(1—3), 107, 1973. With permission.

derivatives are determined in an effort to correlate magnetooptical rotation spectra with absorption spectra.[62] Of the compounds studied, Zn pc and Mg pc have the largest magnetic rotation: -8×10^5 and -8.8×10^5, respectively. No evidence is obtained for relations between the shape or the magnitude of the observed magnetic rotations and the ground state para or diamagnetism of the molecules.

VII. PARAMAGNETISM

A high frequency, 465 kc/sec, electron paramagnetic resonance study is made, *in vacuo,* in a glass ampoule, of chlorophyll a and b, chlorophyll a, ethylchlorophyllide, pheophytin, protoporphyrin, hematoporphyrin, pc, and Mg pc.[63] In the dark, all give a symmetrical singlet signal with a G value close to one of a free electron. Light-induced signals are 10 to 25% stronger.

Formation and decomposition of paramagnetic complexes at elevated temperatures during the reaction of oxygen with β Mg pc crystals are studied using EPR methods. The concentration of the paramagnetic complexes increases exponentially with temperature.[64]

The nature of paramagnetic centers in Mg pc crystals is studied.[65] Recrystallized and sublimed Mg pc gives a weak EPR signal, g = 2.0024, about 10^{16} unpaired electrons per gram. Amorphous Mg pc prepared from it by sublimation in air or exposed to oxygen after sublimation gives a much stronger signal, 10^{18} to 10^{19} electrons per gram, while amorphous Mg pc obtained by the sublimation of crystals *in vacuo* does not show any increase in the signal magnitude as compared to crystal samples. Under the effect of oxygen, the compound Mg pc^+ O^-_2 presumably forms in Mg pc (the electrical conductivity of Mg pc is increased by oxygen) and the formation of local paramagnetic centers of impurity type take place.

VIII. STRUCTURE AND MAGNESIUM ADDEND BOND

The crystal and molecular structure of the monohydrated dipyridinated Mg pc complex $MgC_{32}H_{16}N_8 \cdot H_2O \cdot 2C_5H_5N$ are determined by X-ray diffraction.[66] The crystals are monoclinic, space group $P2_1/n$, with cell parameters a = 17.098 ± 0.003, b = 16.951 ± 0.003, c = 12.449 ± 0.003 Å, and β = 105.88 ± 0.003°. The structure is solved by a combination of statistical and Fourier methods. The pc ring deviates from a plane and the magnesium atom is 0.496 ± 0.004 Å out of the plane of the inner nitrogen atoms toward the water molecule. The pc molecules are close together in pairs.

The complex formation *in vacuo* of Mg pc adsorbed on MgO with nucleophilic molecules having ionization potentials of 8.3 to 15.6 eV is investigated.[9] The complexes are formed with molecules whose ionization potential is < 14 eV, proving the great electron affinity of the Mg pc molecule, where the Mg atom is close to Mg^{++}, causing a stable covalent donor-acceptor bond to be formed in the ground state between Mg pc and the addend.

^{25}Mg and ^{14}N zero field nuclear quadrupole resonance is observed in chlorophyll a and Mg pc by proton nuclear magnetic double resonance.[6] In Mg pc, the observed resonances correspond to a quadrupole coupling constant of 3.79 MHz for ^{25}Mg and 1.81 MHz for the ^{14}N ligands. In chlorophyll, the ^{25}Mg coupling constant is 3.73 MHz.

IX. NEGATIVE IONS

The visible spectrum from 4000 to 10,000 Å of the negative ion of Mg pc is recorded at 10^{-5} torr pressure in a THF solvent.[67] The anion is prepared by reaction with sodium.

Fast regeneration of the parent compound takes place when the negative ion reacts with air.

The reactions between sodium and a number of metal complexes of porphins, tetraazaporphins, tetrabenzoporphin, and pcs including Zn pc and Mg pc are discussed.[68]

A novel method to prepare thin solid layers of Mg pc negative ions is described.[69] Electronic and vibrational absorption spectra of these ions are measured for varying degrees of ionization and compared with the spectra of the neutral molecules.

The magnetic circular dichroism spectra and the absorption spectra of the first four negative ions of Mg pc are reported.[70] The magnetic CD measurements may be the first to be made on radicals and radical ions.

X. ELECTRICAL PROPERTIES

Current-voltage dependencies characteristic of uniform trapping are observed in Mg pc sandwich cells with silver electrodes.[71] Annealing the samples promotes them from an α to β phase involving an increase in resistivity of four orders of magnitude. The resulting parameters of the system are a Fermi level of 0.373 eV above the valence band, a trap concentration of $2.09 \times 10^{17}/cm^3eV$, and a mobility-density-of-states product of 4.22×10^{11} $(cmVsec)^{-1}$.

The dependence of pulsed photogalvanic potential in solid chlorophyll a and Mg pc films on the intensity of illumination, temperature, presence of oxygen in the electrolyte, and polarization of the electrode with the film is investigated.[7] In Mg pc films the photogalvanic potential is generated with a potential rise time $\leqslant 1$ μsec, and a decay time from 1 to 10 msec depending on the sample. The pulsed photopotential in the chlorophyll a films is characterized by a potential rise time of about 5×10^{-4} sec and decay times of tens or hundreds milliseconds. The photopotential value in Mg pc films is linearly dependent on intensity of illumination which indicates a one photon mechanism of photopotential generation. Acceptor molecules adsorbed on the surface and pores of the layer play an important role in photogalvanic potential generation. The bound energy of molecular oxygen on the Mg pc surface is estimated to be 0.27 eV. Also, the primary photochemical stage of electron transfer can probably take place not only at the pigment-electrolyte interface but also at the pigment-metallic electrode boundary.

REFERENCES

1. Calvin, M., Priestley medal address, "Green factories", *Chem. Eng. News,* March 20, p. 30, 1978.
2. Noring, S., Solving the energy crisis: will green plants replace oil?, [discusses the work of Dr. A. I. Krasna], *Columbia,* p. 11, Spring 1978.
3. Synthetic Leaf [discusses the work of Drs. J. J. Katz, T. R. Janson, J. R. Norris, W. Oettmeier, L. L. Shipman, M. C. Thurnauer, M. Wasielewski, and M. Bowman], *Chem. Eng. News,* February 16, p. 32, 1976.
4. McVay, C., The greening of the galaxy: a conversation with Freeman Dyson, *The Princeton Eng.,* p. 10 and 22, February 1980.
5. Evstigneev, V. B. and Gavrilova, V. A., *Biofizika,* 14(1), 43, 1969.
6. Lumpkin, O., *J. Chem. Phys.,* 62(8), 3281, 1975.
7. Evstigneev, V. B., Shkuropatov, A. Ya., and Stolovitskii, Yu. M., *Stud. Biophys.,* 49(1), 27, 1975.
8. Terenin, A. N., Dmitrievskii, O. D., and Glebovskii, D. N., *Biofizika,* 9(1), 25, 1964.
9. Gachkovskii, V. F., *Zh. Fiz. Khim.,* 47(2), 368, 1973.
10. Bubnov, N. N., Krasnovskii, A. A., Umrikhina, A. V., Tsepalov, V. F., and Shlyapintokh, V. Ya., *Biofizika,* 5, 121, 1960.

11. Evstigneev, V. B. and Gavrilova, V. A., *Mol. Biol.*, 2(6), 869, 1968.
12. Stolovitskii, Yu. M. and Evstigneev, V. B., *Mol. Biol.*, 3(2), 176, 1969.
13. Bobrovskii, A. P. and Kholmogorov, V. E., *Zh. Fiz. Khim.*, 47(7), 1740, 1973.
14. Bobrovskii, A. P. and Kholmogorov, V. E., *Dokl. Akad. Nauk S.S.S.R.*, 208(6), 1472, 1973.
15. Maslov, V. G. and Sidorov, A. N., *Teor. Eksp. Khim.*, 7(6), 832, 1971.
16. Sidorov, A. N. and Kholmogorov, V. E., *Teor. Eksp. Khim.*, 7(3), 332, 1971.
17. Petsol'd, O. M., Byteva, I. M., and Gurinovich, G. P., *Opt. Spektrosk.*, 34(3), 599, 1973.
18. Kropf, H. and Kasper, B., *Justus Liebig Ann. Chem.*, 12, 2232, 1975.
19. Byteva, I. M., Gurinovich, G. P., and Petsol'd, O. M., *Biofizika*, 20(1), 51, 1975.
20. Gribova, Z. P., Kayushin, L. P., and Sibel'dina, L. A., *Dokl. Akad. Nauk S.S.S.R.*, 181(5), 1266, 1968.
21. Rapoport, V. L. and Zhadin, N. N., *Dokl. Akad. Nauk S.S.S.R.*, 212(5), 1155, 1973.
22. Shakhverdov, P. A. and Terenin, A. N., *Dokl. Akad. Nauk S.S.S.R.*, 150(6), 1311, 1963.
23. McCartin, P. J., *Trans. Faraday Soc.*, 60(502), 1694, 1964.
24. Lhoste, J. M. and Grivet, J. Ph., *Adv. Radiat. Res., Phys. Chem.*, 1, 327, 1973.
25. Myl'nikov, V. S., *Zh. Fiz. Khim.*, 42(9), 2168, 1968.
26. Fedorov, M. I. and Benderskii, V. A., *Fiz. Tekh. Poluprov.*, 4(7), 1403, 1970.
27. Belkind, A. I. and Aleksandrov, A. B., *Latv. PSR Zinat. Akad. Vestis Fiz. Teh. Zinat. Ser.*, 2, 37, 1971.
28. Usov, N. N. and Benderskii, V. A., *Zh. Strukt. Khim.*, 11(2), 281, 1970.
29. Pope, M., *J. Chem. Phys.*, 37, 1001, 1962.
30. Evstigneev, V. B. and Savkina, I. G., *Dokl. Akad. Nauk S.S.S.R.*, 163(5), 1270, 1965.
31. Putseiko, E. K., *Dokl. Akad. Nauk S.S.S.R.*, 148, 1125, 1963.
32. Shakhverdov, P. A., *Tepl. Dvizhenie Mol. Mezhmol. Vzaimodeistvie Zhidk. Rastvorakh*, p. 214, 1969.
33. Shakhverdov, P. A. and Terenin, A. N., *Dokl. Akad. Nauk S.S.S.R.*, 150(6), 1311, 1963.
34. Gurinovich, G. P., Petsol'd, O. M., and Byteva, I. M., *Biofizika*, 19(2), 249, 1974.
35. Shkuropatov, A. Ya. and Vankevich, M. M., *Biol. Nauch. Tekhn. Prog.*, p. 45, 1974.
36. Stolovitskii, Yu. M., Shkuropatov, A. Ya., Evstigneev, V. B., and Shiohkov, V. V., *Biofizika*, 19(5), 820, 1974.
37. Kobyshev, G. I., Lyalin, G. N., and Terenin, A. N., *Dokl. Akad. Nauk S.S.S.R.*, 153(4), 865, 1963.
38. Putseiko, E. K. and Chalyi, A. K., *Biofizika*, 19(1), 57, 1974.
39. Krasnovskii, A. A., Jr. and Litvin, F. F., *Photochem. Photobiol.*, 20(2), 133, 1974.
40. Litvin, F. F. and Personov, R. I., *Fiz. Probl. Spektroskop. Akad. Nauk S.S.S.R. Mater. 13-go*, 1, 229, 1960.
41. Gurinovich, G. P. and Strelkova, T. I., *Biofizika*, 8, 172, 1963.
42. Litvin, F. F. and Krasnovskii, A. A., *Dokl. Akad. Nauk S.S.S.R.*, 173(2), 451, 1967.
43. Akimov, I. A. and Korsunovskii, G. A., *Opt. Spektrosk.*, 8, 427, 1960.
44. Kobyshev, G. I., Lyalin, G. N., and Terenin, A. N., *Opt. Spektrosk.*, 15(6), 837, 1963.
45. Kobyshev, G. I. and Terenin, A. N., *Proc. Int. Conf. Lumin.* Vol. 1, Szigeti, G., Ed., Akad. Kiado, Budapest, Hungary, 1968, 520.
46. Seely, G. R., *J. Phys. Chem.*, 71(7), 2091, 1967.
47. Terenin, A. N., Lyalin, G. N., and Sirota, V. G., *Mol. Biol.*, 2(2), 131, 1968.
48. Shakverdov, P. A. and Terenin, A. N., *Dokl. Akad. Nauk S.S.S.R.*, 160(5), 1141, 1965.
49. Krasheninnikov, A. A., *2nd Mater. Resp. Konf. Molodykh Uch. Fiz.*, Akad. Nauk Beloruss. S.S.R. Inst. Fiz., Minsk, U.S.S.R., 1972, 32.
50. Gachkovskii, V. F., *Dokl. Akad. Nauk S.S.S.R.*, 193(3), 618, 1970.
51. Dzhagarov, B. M. and Gurinovich, G. P., *Opt. Spektrosk.*, 30(3), 425, 1971.
52. Gachkovskii, V. F., *Dokl. Akad. Nauk S.S.S.R.*, 186(2), 471, 1969.
53. Gachkovskii, V. F., *Zh. Fiz. Khim.*, 44(3), 664, 1970.
54. Kompaniets, V. V., Rapoport, V. L., and Zhadin, N. N., *Opt. Spektrosk.*, 33(4), 646, 1972.
55. Eastwood, D., Edwards, L., Gouterman, M., and Steinfeld, J., *J. Mol. Spectrosc.*, 20, 381, 1966.
56. Seybold, P. G. and Gouterman, M., *J. Mol. Spectrosc.*, 31(1), 1, 1969.
57. Gachkovskii, V. F., *Adv. Mol. Relaxation Processes*, 5(1—3), 107, 1973.
58. Gurinovich, I. F. and Gurinovich, G. P., *Opt. Spektrosk. Akad. Nauk S.S.S.R. Otd. Fiz. Mat. Nauk, Sb. Statei*, 2, 196, 1963.
59. Gachkovskii, V. F., *Dokl. Akad. Nauk S.S.S.R.*, 190(6), 1370, 1970.
60. Pribytkova, N. N. and Savost'yanova, M. V., *Opt. Spektrosk.*, 26(5), 765, 1969.
61. Sidorov, A. N. and Kholmogorov, V. E., *Dokl. Adak. Nauk S.S.S.R.*, 170(5), 1202, 1966.
62. Shashoua, V. E., *J. Am. Chem. Soc.*, 87(18), 4044, 1965.
63. Umrikhina, A. V., Golubev, I. N., Kayushin, L. P., and Krasnovskii, A. A., *Biofizika*, 9(4), 423, 1964.
64. Sharoyan, E. G., Tikhomirova, N. N., and Blyumenfel'd, L. A., *Zh. Strukt. Khim.*, 6(6), 843, 1965.

65. Sharoyan, E. G., Tikhomirova, N. N., and Blyumenfel'd, L. A., *Zh. Strukt. Khim.*, 5(5), 697, 1964.
66. Templeton, D. H., Fischer, M. S., Zalkin, A., and Calvin, M., *J. Am. Chem. Soc.*, 93(11), 2622, 1971.
67. Shablya, A. V. and Terenin, A. N., *Fiz. Probl. Spektroskop. Akad. Nauk S.S.S.R. Mater. 13-go 1*, 203, 1960.
68. Dodd, J. W. and Hush, N. S., *J. Chem. Soc.*, November, p. 4607, 1964.
69. Sidorov, A. N., *Dokl. Akad. Nauk S.S.S.R.*, 215(6), 1349, 1974.
70. Linder, R. E., Rowlands, J. R., and Hush, N. S., *Mol. Phys.*, 21(3), 417, 1971.
71. Morel, D. L. and Berger, H., *J. Appl. Phys.*, 46(2), 863, 1975.
72. Stiehl, H. H. and Witt, H. T., *Z. Naturforsch. B*, 24(12), 1588, 1969.

Chapter 19

PHTHALOCYANINES IN BIOLOGY

I. STAINING

Metachromasia, the property of certain dyes to color different tissue components in various colors, thus serving as a means to distinguish them, may be exhibited by pc dyes[1] according to a study entitled "Metachromasia of Alcian Blue, Astrablau and other Cationic Pc Dyes". "The relevance of Alcian blue aggregation to the speed of staining, its use in critical electrolyte concentration techniques, and in microspectrophotometry, is considered."[1]

The effect of calcium ions on the sorption of pc dye by membranes of individual nerve fibers is studied by dissecting single nerve fibers from the walking legs of green crabs and staining them with Heliogen Blue K.[2] The fibers are then washed in various hypotonic and/or isotonic solutions. No evidence is found for a direct interaction between Heliogen Blue K and Ca^{++} ion.

Glycerolized muscle fibers are stained with direct light-fast turquoise M (C. I. Direct Blue 86).[3] The isotherms of its absorption resemble the Langmuir adsorption isotherms at very low or very high dye concentrations. The dye is fully eluted from the fibers at pH 12 and the changes are observed in the ultrastructure of the muscle fibers treated with dyes at high concentrations.

Pb pc diazotate has a high affinity for animal lysosomes and plant vacuoles.[4] "When Pb pc is used as a coupling salt in ripe fruits it is found in the Golgi apparatus and a variety of different sized vacuoles . . . In the insect midgut Pb pc is found in lysosomes, Golgi apparatus, endoplasmic reticulum and the nuclear membrane."[4]

The use of tri(4-amino) Pb pc diazotate for high resolution studies, both optical and electron microscopy, of lysosomal acid phosphatases, esterases, and β-glucuronidase in plant and animal cells is assessed.[5]

II. BINDING WITH PROTEINS

The reaction between Fe and Co tetrasulfonated pcs with globin results in the formation of green complexes[6] at a molar ratio of 1:1 as observed from spectrophotometric titration data. "The results presented . . . although incomplete, show that metal tetrasulfonate pc-globin complexes may serve as the models of hemoprotein."[6]

The interaction between phenol-melamine copolymers and pc tetrasulfonic acid is studied as a model for the immunological binding between protein chains and ionic groups.[7]

REFERENCES

1. **Scott, J. E.**, *Histochemie,* 21(3), 277, 1970.
2. **Rozental, D. L. and Levin, S. V.**, *Tsitologiya,* 18(9), 1090, 1976.
3. **Shapiro, E. A. and Komissarchik, Ya. Yu.**, *Biofiz. Osn. Regul. Protsessa Myshechnogo Sokrashcheniya, Mater. Vses. Simp.* Akad. Nauk S.S.S.R., Pushchino-on-Oka, U.S.S.R., 1971, 230.
4. **Beadle, D. J., Dawson, A. L., James, D. J., Fisher, S. W., and Livingston, D. C.**, *Electron Microsc. Cytochem., Proc. 2nd Int. Symp.,* North-Holland, Amsterdam, Netherlands, 1974, 85.
5. **Livingston, D. C., Coombs, M. M., Franks, L. M., Maggi, V., and Gahan, P. B.**, *Histochemie,* 18(1), 48, 1969.
6. **Przywarska-Boniecka, H., Trynda, L., and Antonini, E.**, *Eur. J. Biochem.,* 52(3), 567, 1975.
7. **Rackow, B. and Suessenbach, D.**, *Biokybernetik, Mater. 1st Int. Symp, 1967,* Vol. 1, Karl-Marx University, Leipzig, East Germany, 1968, 246.

INDEX

A

B

C

D

G

H

I

K

L

M

N

O

P

Q

R

S

T

U

V

W

X

Y

Z